MEDIEVAL COMBAT IN COLOUR

MEDIEVAL COMBAT IN COLOUR

Hans Talhoffer's Illustrated Manual of Swordfighting and Close-Quarter Combat from 1467

Edited and introduced by DIERK HAGEDORN
With a foreword by MARK RECTOR

GREENHILL BOOKS

Published in 2018 by Greenhill Books,
c/o Pen & Sword Books Ltd,
47 Church Street, Barnsley, S. Yorkshire, S70 2AS
www.greenhillbooks.com

ISBN 978-1-78438-285-8

Library of Congress Cataloging-in-Publication data available

Typeset in Arno by Dierk Hagedorn

Images used by kind permission of
Bayerische Staatsbibliothek München, Cod. icon. 394a

Printed and bound in India by Replika Press Pvt. Ltd.

CONTENTS

FOREWORD

This is a brand new edition of Hans Talhoffer's *Fechtbuch*, or 'Fight Book', from the year 1467, one of the most lavishly drawn fencing manuals of the fifteenth century. It comes with an exciting new English translation and introductory material by Dierk Hagedorn.

Fencing scholar Gustav Hergsell discovered this manuscript at the end of the nineteenth century in the library of Ernst II, Duke of Saxe-Coburg-Gotha, and translated the original text into modern German, publishing his version in Prague in 1887.

Hergsell describes Talhoffer's manuscript in his introduction:

> The original drawings were made with pen and ink on parchment sheets, and boldly coloured. The cross is drawn in red upon the shields, caps, chests and backs of the shield-fighters. In some of the drawings, blood spurts from wounds, and the shields are coloured yellow. The drawings appear on both sides of each leaf.
>
> The captions [...] are inscribed in a bold hand. They alone set out the particular fights, as this codex contains neither a title nor any other text.
>
> Next to the final illustration, showing a fight on horseback with crossbow and lance, Talhoffer himself is depicted, sword in hand, with the inscription, 'This book was written by Hans Talhoffer, who posed for this portrait.'[1]

As Hergsell notes, this book contains no explanatory text beyond the captions to the illustrations, and these are often oblique and open to multiple interpretations. The illustrations themselves are not always arranged to form coherent sequences. Personal combat training was a highly competitive field in medieval Europe and fencing masters zealously guarded the secrets of their craft. While a certain amount of martial knowledge and skill is assumed of the reader, Talhoffer's manuscript may not have been crafted as a fencing treatise or a 'how-to' book. It may have been composed as a declaration of his ability as a master of arms, a professional manual for the practical application of killing techniques with the prescribed weapons, or an expensive *objet d'art* for his wealthy and noble patron.

Two plates in Talhoffer's *Fechtbuch* are not included: 127 (an almost blank leaf, with barely distinguishable outlines showing a shield fight), and 222 (a blank leaf).

The individual sections in Talhoffer's book may be divided into two sorts: those dealing with the 'judicial duel' and those concerned with

1 HERGSELL (1887). Translations from Hergsell are by Mark Rector.

personal combat. The judicial duel was an officially sanctioned fight to resolve a legal dispute, and was common in Western Europe from the tenth through the fifteenth century. Talhoffer illustrates noblemen fighting a trial by combat in full armour with sword and spear. For commoners, the judicial duel involved unique and occasionally bizarre weapon combinations and ritual costumes: the combatants were stitched into cowled leather suits, greased with pig fat, and armed with a selection of wooden maces, swords, and spiked and hooked shields. These trials by combat were duels to the death.

One of the most curious episodes Talhoffer illustrates is a judicial duel between a man and a woman.

The man stands in a circular pit up to his waist, armed with a club in his right hand, with which he assays to strike at the woman. He is severely prohibited, upon pain of forfeit, from stepping out of the pit. He may, however, steady himself with a hand on the edge of the pit or on the ground.

The woman has a veil in her hand, in which a stone of several pounds is knotted. With it she attempts to strike at the man. As soon as the woman is able, she moves behind the man's back, endeavouring to drag his head to the edge of the pit and to strangle him. The woman strikes a blow with her veil, which the man parries with his club. But the veil wraps itself around the club and the woman seizes upon this advantage to wrench the club from the man's hand, disarming him, ending the combat and levying guilt unto the man.

However, if the man parries the blow with his free left arm, he is presented with the opportunity to grab the woman about the waist, and to drag her into the pit, ending the fight unfavourably for her.[2]

In this particular duel, if the man is dragged from the pit or the woman into it, the loser is to be taken off and immediately executed.

The majority of the manuscript deals with personal combat. Talhoffer includes lengthy illustrated passages on wrestling, unarmoured fighting with the long sword, messer (the 'long knife', a civilian weapon similar to the hunting hanger, falchion, or ancient saxe), pollaxe, dagger, sword and buckler, and a series of mounted techniques. Talhoffer's manual amply proves that a unique, indigenous, and systematic method of fencing and training with a wide array of weapons and techniques had developed in Germany during the Middle Ages, reaching its height in the fifteenth century. This is the German *Kunst des Fechtens*, the 'art of fighting'.

THE GERMAN ART OF FIGHTING

> Princes and Lords learn to survive with this art, in earnest and in play. But if you are fearful, then you should not learn to fence,

2 HERGSELL (1887).

> because a despondent heart will always be defeated, regardless of all skill.[3]

Germany in the Middle Ages was a conglomeration of small principalities ruled by dynastic houses, bishoprics, municipal leagues, towns, and villages, all struggling against each other for advantage and privilege. Warfare was a part of everyday life. Not only were there the full-scale wars of conquest and plunder sanctioned by the German Emperor, but petty 'private' wars also flourished at this time, fought between princeling and town, town and village, bishop and town, and every combination thereof. Knights and men-at-arms banded together into societies of self-interest and hired out their services. Renegade noblemen preyed upon travellers on the lonely trails through the dense German forests. Tradesmen and artisans in the larger towns formed fencing societies, which undertook the training and licensing of teachers.

The two most famous of these societies, the Marxbrüder (The Brotherhood of St. Mark) and the Federfechter, provided champions to the court of the Holy Roman Emperor.[4] In Talhoffer's native Swabia, Count Eberhardt of Württemberg maintained one of the best military forces in Germany, consisting of a small knightly cavalry and an infantry levy of commoners trained on regular exercises and frequently lent to neighbouring powers.[5] The inherently violent nature of medieval society meant that little distinction was made between the civilian and military application of the German martial arts.

And they were arts, worthy of much study and practice, since one's life could depend on their mastery. Medieval swords were amazingly light and well balanced, weighing between only two and four pounds. Combatants in the Middle Ages used footwork, actions on the sword, leverage, timing and distance to exploit and enhance the sword's inherent cutting and thrusting capabilities. These skills were supplemented with techniques for disarming, grappling, wrestling, kicking, and throwing the opponent.

TALHOFFER'S ARSENAL

LONG SWORD (PLATES 1–67, 74–8)

The long sword, war sword, hand-and-a-half sword, or bastard sword is the chief weapon of Talhoffer's arsenal, and the fundamental techniques for its use may be applied to every other weapon in his book. The long sword is between four and four-and-a-half feet in length with a sharply tapered double-edged blade of diamond profile, a simple cross hilt, a fig-shaped pommel, and a handle long enough to be gripped with both

3 Dresden, Sächsische Landesbibliothek, *Mscr. Dresd. C 487*, fol. 16v. Translation by Jörg Bellinghausen

4 Hergsell (1887).

5 Du Boulay (1983).

hands. This weapon is surprisingly light, capable of making blindingly fast attacks, and is suitable for both cuts and thrusts. As the plates illustrate, the entire sword is used, either by gripping the blade with the off-hand for strong half-sword attacks, by reversing the sword to strike with the pommel or by employing the hilt as a hammer to batter through the opponent's defences.

ARMOURED FIGHT WITH SPEAR AND LONG SWORD (PLATES 68–73)

The armoured fight with spear and sword is a judicial combat between noblemen in full plate armour. They begin the ritual by sitting on a platform in the lists (the barriers surrounding the duelling ground) and displaying their weapons, with their attendants beside them and their coffins before them. The spear is used as the primary weapon, with the sword held in reserve for the *coup de grâce.*

POLLAXE (PLATES 79–103)

The pollaxe is most commonly used in armoured fighting, and is specifically designed as an 'armour-breaker'. Talhoffer, however, shows it in an unarmoured fight. Talhoffer's weapon is a 'Lucerne Hammer': a long staff weapon with a notched hammer head surmounted and surrounded by sharp spikes. The shaft is used to trip the opponent by sweeping his legs, or to hook and drag the opponent to the ground, and the long, blade-like spike on top is used to make killing thrusts. Although the pollaxe is primarily a knightly weapon, the techniques for its use are equally applicable to the axe-bladed halberd, the weapon of choice for the select combatants held in reserve for counter-attack in the centre of the formidable pike formations of the Swiss Confederates and German *Landsknechte.*[6]

DUELLING SHIELD (PLATES 104–64)

The duelling shield is used in judicial combat. It is a slightly concave oblong, approximately six-and-a-half feet tall, with a rectangular coffered boss along the longitudinal axis allowing room for the hand to grip a central pole, which is sharply spiked at both ends. The shield is made of wood, possibly covered with cloth or leather, and bears the image of the cross. It is used both defensively and offensively to deflect blows and hammer into the opponent's shield to create openings, as well as to stab with the spikes. The combatants are barefoot and sewn into tight-fitting cowled leather suits with a red cross on their chests and backs and the back of their cowls. When fighting according to the Franconian law, the combatants use a large wooden mace with their shields. When fighting according to the Swabian law, they employ the long sword.

THROATED HOOKING SHIELD (PLATES 165–9)

The throated hooking shield is perhaps the most unusual weapon in Talhoffer's arsenal. It is a cello-shaped affair with four protruding hooks at

6 Oman (1924).

the top and bottom for catching, pulling, and stabbing the adversary. Like the simple duelling shield, it has a sharply-spiked central pole which may be used to impale the opponent.

DAGGER (PLATES 170–89)

The dagger is the ultimate weapon of last resort. Talhoffer shows the rondel variety with an extremely long, triangular blade and a grip with flanges at top and bottom to protect the hand. The dagger was the simplest and most commonly carried weapon of medieval Germany. Techniques are illustrated for using the dagger to shield against stabs, gripping the blade with the off-hand to block the opponent's dagger and grappling.

WRESTLING (PLATES 190–221)

Nearly a third of the armed combat plates show some kind of wrestling or grappling, and Talhoffer devotes thirty-two plates to wrestling alone. Wrestling is essential to the German art of fighting, since all martial art comes from it, and because the fight must become hand-to-hand once distance is closed within the effective range of the weapons, or if either of the combatants should be disarmed.

MESSER (PLATES 223–30)

The messer, or 'long knife', has a short, single-edged blade with a simple cross hilt. The blade is slightly curved and the last few inches of the clipped back or false edge are sharp. It is carried in one hand, with the other hand held behind the back, as in modern sabre fencing. The messer resembles the ancient saxe, falchion, the Italian *storta*, the Bohemian *dussak* (a wooden training weapon used in Germany into the seventeenth century), the Polish sabre, and the modern machete. Commoners in medieval Germany carried it as an everyday side-arm. The messer is capable of making shearing cuts fierce enough to remove a hand from the arm, as can be seen in plate 228. Albrecht Dürer devotes a large section of his *Fechtbuch* to the messer.[7]

SWORD AND BUCKLER (PLATES 251–70)

We get the word 'swashbuckler' from the British proclivity for the sword and buckler. This weapon combination was in use across Europe, beginning in the twelfth century. The oldest fencing manual in existence, the anonymous thirteenth-century German manuscript *I.33* in the collection of the British Museum, is entirely devoted to the sword and buckler. The buckler is a small, round shield with a central bowl-shaped boss allowing room for the hand and handle. It is used with the single-handed or 'arming' sword with a simple cross hilt and a large circular pommel. Talhoffer's buckler is rather ornate, with hook-like protrusions at the top and bottom for catching and deflecting the oncoming blade. At the end of this section, Talhoffer shows techniques for one man fighting against two.

7 Vienna, Albertina, *Hs. 26-232*

COMBAT BETWEEN A MAN AND A WOMAN (PLATES 242–50)

The fight between a man and a woman, as described earlier, is another judicial duel in which the man is placed up to his waist in a pit and armed with a large wooden mace. The woman is free to move around him on the ground, and is armed with a three- or four-pound stone knotted into a long veil.

MOUNTED COMBAT (PLATES 251–70)

The final series of plates shows methods of mounted combat with the long sword, long sword against the lance, lance against crossbow, and mounted hand-to-hand fighting.

FOR THE MODERN READER

Since this edition was first published in 2000, scholarship in the medieval German art of fighting and the practical study and application of its techniques by seasoned and talented martial artists have produced a wealth of research, books, fighting manuals, and modern fencing schools. For more detailed instruction on the use of the sword and messer, I would recommend reading the works of Christian Tobler. For the dagger, Guy Windsor. And for grappling and wrestling, Jessica Finley. Each of them is a modern master.

Talhoffer's book provides us with an unparalleled visual reference. Each drawing is a carefully composed snapshot isolated in time, revealing the elegant intricacies of the German art of fighting. We do not know for certain who drew the figures in Talhoffer's manual. Whether it was Talhoffer himself or an anonymous Swabian scribe, the artist took painstaking care to give each of the figures life and expression. We are treated to images of everyday dress, with doublet, codpiece, hose, a variety of headgear, and the extravagantly pointed shoes of the late Middle Ages; full suits of armour; ritual battle dress; and equestrian equipage. This book is a time machine, providing us with a glimpse of real people really fighting with all the skill, sophistication, and ruthlessness of the *Kunst des Fechtens*.

INTRODUCTION

This edition of Hans Talhoffer's book about medieval combat is completely new in most respects, although the story of the manuscript and its many previous editions goes back several centuries. The original codex is dated to 1467, and it is the last one in a series of manuscripts he authored. Its contents have been copied a number of times, either partially or completely: Manuscripts based on the original were created from the 16th until the beginning of the 19th century. The first printed edition finally appeared in 1887, published by Gustav Hergsell, a fencing master from Prague. At his own expense, he edited three of Talhoffer's manuscripts. He certainly targeted an eager audience, and his edition of the 1467 codex, embellished with lavish lithographs, was a fine achievement in its time. This edition was re-published with black-and-white images by VS Books in 1998, which was in turn the base for the first English edition, published by Greenhill Books in 2000. Since then, a partial edition with an accompanying attempt to reconstruct the techniques was issued in three parts, but the entire project was never completed. In consequence, the present edition is something very special: it is the first to present Hans Talhoffer's most famous manuscript to a broad audience in full colour and all its splendour.

HANS TALHOFFER'S MUNICH CODEX

Master Hans Talhoffer is one of the best-known medieval fencing masters of all, which is particularly due to the sheer number of surviving sources: five contemporary manuscripts have survived, plus at least ten copies.[1] On the other hand, the scarce biographical material – mainly based on the dating of his manuscripts between 1443 and 1467 – makes the man himself difficult to grasp.

Talhoffer's art of fencing does not stand in a vacuum; instead, many techniques – or 'pieces,' to use the original 15th-century term – can only be understood within the tradition in which Hans Talhoffer lived and acted. Fencing with the long sword makes some use of the nomenclature previously established by Johannes Liechtenauer. This is partly taken up

1 Original manuscripts: *78 A 15*, Berlin, Kupferstichkabinett der Stiftung Preußischer Kulturbesitz; *Ms. Chart. A558*, Gotha, Forschungsbibliothek Schloss Friedenstein; *Hs. XIX, 17-3*, Königseggwald, Castle of the Count; *Thott 290 2°*, Copenhagen, Det Kongelige Bibliotek; *Cod. icon. 394a*, Munich, Bayerische Staatsbibliothek
Copies: *Cod.I.6.2°.1*, Augsburg, Universitätsbibliothek; *Hz.014*, Coburg, Kunstsammlungen der Veste Coburg; *Philos. 61*, Göttingen, Niedersächsische Staats- und Universitätsbibliothek; *2° Ms. iurid. 29*, Kassel, Landesbibliothek; *Cod. icon. 394*, Munich, Bayerische Staatsbibliothek; *Cod. icon. 395*, Munich, Bayerische Staatsbibliothek; *26.236*, New York, Metropolitan Museum of Art; *KK 5342*, Vienna, Kunsthistorisches Museum; *Cod Vindob. Ser. Nov. 2978*, olim Vienna, Österreichische Nationalbibliothek, now in a private collection; *Cod. Guelf. 125.16 Extrav.*, Wolfenbüttel, Herzog August-Bibliothek.

verbatim, for example on fol. 11r, where it says: *Krüm vff behend*. Nevertheless, special designations appear, such as 'point', which for Liechtenauer stands exclusively for the tip of the sword, but usually stands for the thrust for Talhoffer.

Talhoffer also uses terms that have not appeared in any surviving source by other masters; they may be entirely his own invention – such as the *Hinterbinden* which appears repeatedly, or the *Strebe*. Since the captions are often very brief and do not always contain instructions for action, but instead only the name of a specific technique, statements about the sequence of techniques are difficult and must sometimes remain hypothetical.

The perspective in Talhoffer's texts changes constantly: Mostly the fighters are referred to in the third person singular or plural, but often the perspective changes to the first person singular. In addition, the reader is also addressed directly in the second person.

Furthermore, the reading direction is not universally from left to right. Often the picture on the right and the accompanying text precede the picture on the left; this is particularly clear on fol. 96v.

TRANSCRIPTION GUIDELINES

This edition of the manuscript *Cod. icon. 394a* of the Bayerische Staatsbibliothek in Munich follows the original text to the letter. Abbreviations are dissolved and marked in *italics*, a single missing part on fol. 13r, caused by a damaged margin, is [added]. Comments or additions in the translation are also set in [square brackets].

The vowels 'u' and 'o' are often overwritten by the scribe with a tilde-like character. In the case of 'u', the reason may have been to make it easier to distinguish it from 'n'. The tilde marks are ignored in the transcription, since there is no additional benefit in understanding the text.

A similar sign can be found in a more or less closed form, especially over the 'u', where it can take on an almost circular shape and stands for a long vowel. This is reproduced as 'ů'.

Occasionally, the 'o' is also overwritten with a trema for the umlaut 'ö'. Since the differences between the tilde-like sign and the umlaut are only minimal, it has been transcribed according to the context as either 'o' or 'ö'.

Another diacritical sign, a single dot or small dash above the 'y' also remains untranscribed. In many cases, particularly in the case of 'u' and 'o', it could not always be determined whether there was such a sign at all. When in doubt, I have not written a diacritical sign.

Furthermore, numerous dots between individual words, which serve to separate these words rather than serving as punctuation, remain untranscribed. This becomes particularly clear in the section of the fight with the shield according to Franconian law.

Some letters can be clearly identified as capitals, especially the beginnings of a section. These are also rendered as capital letters in the transcription. The letters 'S' and 'I' are often used in their capital form, regardless of whether the word is a verb, a pronoun or a noun which is rather uncommon in modern German, which only capitalises nouns. The letters 'v' and 'w', especially at the beginning of the line or in the first lines, are often provided with an initial curve. In case of doubt I have decided in favour of lower case.

Hergsell's old edition from 1887 numbers the manuscript according to plates that correspond to a pagination; the present new edition with transcription and translation also lists the folio with recto and verso pages.

NOTES ON THE TRANSLATION

For this new edition, I have created a completely new translation of Hans Talhoffer's text. For the first edition, Mark Rector chose an approach that explains the techniques presented, going beyond the pure text to make it easier for the modern reader to understand. A similar approach was taken by Gustav Hergsell, who obviously tried to find a language that offered an understandable text for those who were familiar with the fencing terms of his time in his edition more than 100 years ago. But translating Early New High German terms into Italian- and French-inspired expressions naturally results in blurred and sometimes questionable interpretations.

Nevertheless, my translation must also have a certain interpretive character; the captions are sometimes extremely short and sometimes refer to the dark and 'secret' verses of Johannes Liechtenauer, and it is in their nature to be obscure and elusive. Furthermore, many terms have a double meaning, which cannot always be rendered with a single modern term. Some technical expressions are simply not transferable, especially those that describe a technique with a very general vocabulary, such as *überfallen* – falling over (fol. 13v), which – in today's modern German – conjures up the connotation of a robbery, but must in Talhoffer's case be understood literally: one fencer falls over the sword and arms of the other.

In order to make the technical procedure more comprehensible, I have most frequently called the two fighters 'this one' and 'the other' – notwithstanding Talhoffer's original, where only 'he' is mentioned, regardless of whether it is the attacker or the defender – except of course in cases where the fighters are named in the first or second person.

I have translated some of Talhoffer's fencing terms into English, mostly in those cases when they describe a more or less commonplace action. However, the majority of technical terms have retained their German form. But since the orthography in Early New High German was not always uniform and several readings of the same word can be found within one text, I have chosen a slightly modernized version based on modern German.

THE GLOSSARY

The glossary lists the most important terms from Hans Talhoffer's Munich fight book. In the translation, the terms in question are underlined like a standard link on the internet. Each term is only marked the first time it appears in the text. Cross-references in the glossary itself are indicated by an → arrow. Many of the technical procedures can be found in detailed form in other fencing books, which is why only a very brief explanation may suffice here. In addition, not all techniques are explained in the glossary, especially those that are rarely or even only once explicitly mentioned, such as 'Zulegen' on fol. 5v.

THE IMAGES

While the manuscript is in a portrait format, the images are in a landscape format. This means that the original codex must be turned clockwise in order to see the images properly. In some cases, however, the book must be turned counter-clockwise to comfortably peruse the image. For the reader's convenience, we have rotated the images 90 degrees clockwise or counter-clockwise, depending on whether the fighters are facing the binding or the margin of the pages. Fol. 112v/plate 222 is empty in the original, and fol. 65r/plate 127 contains a very vague, hardly recognisable sketch of two fighters with duelling shields. This image, in contrast to the preceding and following ones, is oriented upside down. The drawing is very subtle as if executed with only very little or very light ink. It appears again, but this time in full colour and with accompanying captions a bit later on fol. 76v/plate 150. Neither page is reproduced in this edition.

The size of the images is smaller than in the original manuscript: all plates are reduced to roughly 50 per cent of the original.

ORGANISATION OF THE QUIRES

The diagram on the following page was created especially for this edition. I would like to thank Karin Eckstein from the Bayerische Staatsbibliothek, Institut für Bestandserhaltung und Restaurierung, for providing me with this valuable information; and particularly Juliane Trede for granting me permission to peruse several pages of the original codex in order to clarify some open questions.

Quire 1 ($\text{VI}+1^{13}$)

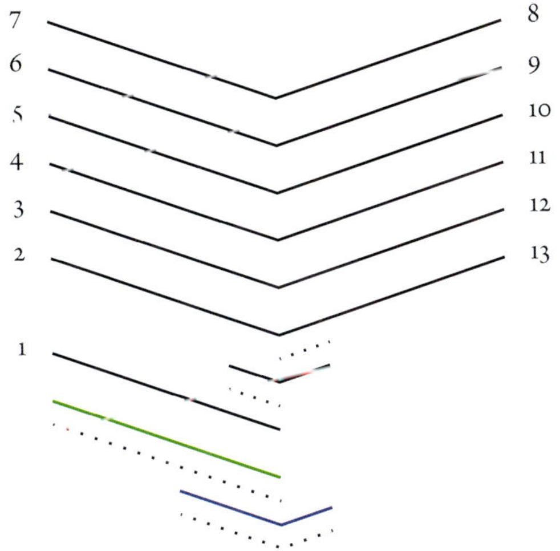

Quire 2 ($\text{VI}+1^{26}$)

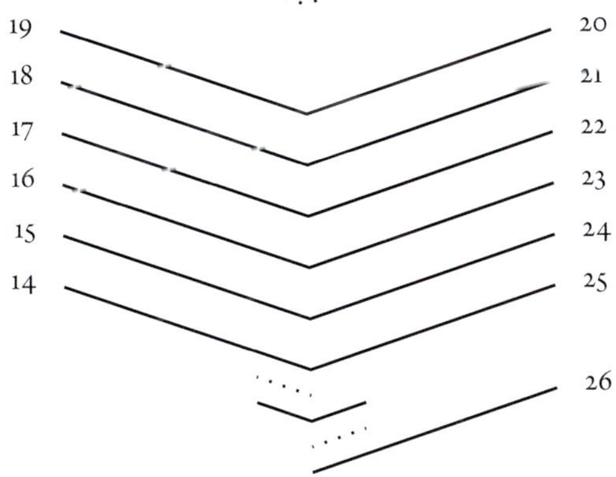

IV^{34}, III^{40}, IV^{48}, IV^{56}, IV^{64}, VI^{76}, VI^{88}, VI^{100}, VI^{112}, VI^{124},

Quire 13 (VI^{136})

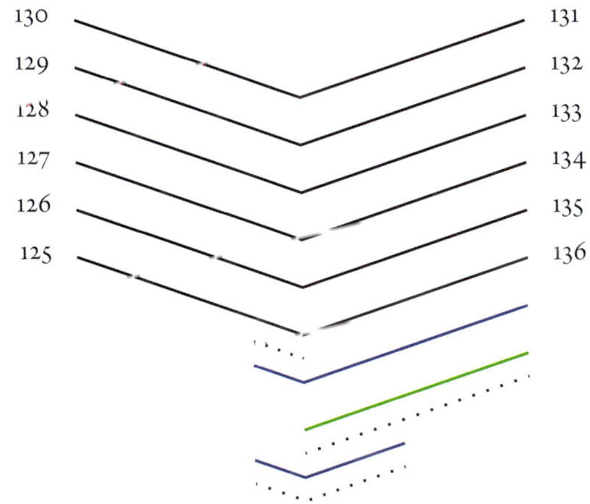

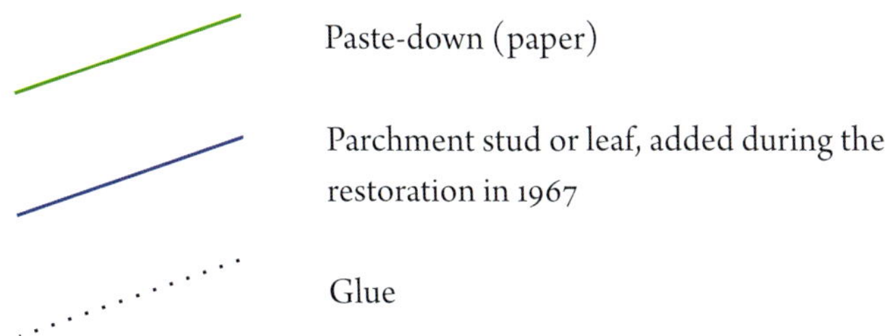

TALHOFFER GALORE

In this article, I want to touch on two major subjects: the first is to highlight the tradition line in which fencing master Hans Talhoffer lived and worked, while the second is to give a brief overview of the fencing treatises that are connected to his name. For this paper, I have re-used some passages from my essay 'German Fechtbücher from the Middle Ages to the Renaissance'.[1]

A BRIEF HISTORY OF BOOKS

The two doyens of research as far as the German sources are concerned, Martin Wierschin[2] and Hans-Peter Hils,[3] have both published extensive volumes about the German medieval fencing literature. In their works, they list and describe a considerable number of fencing manuscripts, called fight books (*Fechtbücher*, in German). Over the past decades, numerous fencing books have been rediscovered or found that were believed lost or were previously unknown. Martin Wierschin listed 47 manuscripts in 1965, and twenty years later, we find 55 treatises in Hans-Peter Hils' catalogue. Speaking of catalogues, the *Katalog der deutschsprachigen Handschriften des Mittelalters*[4] from 2009 offers 47 again – but these are only the illustrated ones. At the time of writing, 74 fight book manuscripts from the German speaking regions are known.

The earliest known fight book worldwide is German, known by its signature *I.33*,[5] which dates to the beginning of the 14th century. It is kept in the Royal Armouries in Leeds and is – despite its origin – written in Latin, the language of scholarship of the time, the language of the church. And indeed, in this book we meet a priest who teaches his student the art of fencing. We do not know the reason for this. Was fencing a popular pastime among the clergy? Was he a younger son of the nobility trained in the martial arts who had to seek his career as a clergyman? Was such teaching among monks the rule or the exception? The Cracow *Gladiatoria* manuscript[6] on fol. 54v, which speaks of the *chuttennische puckler*, the buckler of the monks wearing the habit (*Kutte* in German), provides only a brief reference.

The following two or three centuries see a surprising amount of handwritten fight books – in the German-speaking regions, that is. With their

1 Hagedorn (2016).
2 Wierschin (1965).
3 Hils (1985).
4 Leng et al. (2008).
5 Leeds, *I.33*.
6 Cracow, *Ms. Germ. Quart. 16*.

total of 74 known manuscripts from this period, the German-speaking regions of Europe were far more prolific than any other area.

There was a certain transitional period from the handwritten codices to printed volumes. The development was a slow but steady process. The year 1452, in which Johannes Gutenberg began to work on his Bible, can be regarded as the beginning of the art of printing – and coincidentally, this was the same year in which one of the most extensive German manuscripts was completed, the Peter von Danzig manuscript.[7] Nevertheless, it took a considerable amount of time until printing finally took over. The first printed German fight book appears almost three-quarters of a century later (Andre Paurenfeindt's book, published in 1516).[8] Only about a dozen fight books were printed in the 16th century, but still some 30 manuscripts were produced. But when we talk about manuscripts, we cannot ignore one overshadowing figure, a fencing master who appears either by name or by his teachings in 27 out of 74 manuscripts: one cannot write about German fight books without paying tribute to grandmaster Liechtenauer.

7 Rome, *44 A 8*.

8 PAURENFEINDT (1516).

HE WHO MUST BE NAMED

Much ink – real and digital – has been spent on the shadowy figure of the medieval fencing master Johannes Liechtenauer, yet still we know virtually nothing about the man. We can obtain only very scarce biographical data from an anonymous codex from Nuremberg[9] that mentions him. On fol. 13v, we learn that Master Liechtenauer did not in fact invent the art of swordsmanship, but he travelled far and wide in order to further his knowledge:

> And above all things shall you note and know that there is only one art of the sword which may have been invented and devised many hundred years ago. It is a foundation and core of the entire art of fencing, and Master Liechtenauer knew it entirely and perfectly and was wholly capable of it. However, he did not invent it himself, as is written above, instead he has travelled and visited many a country because of that true and sincere art, since he wanted to get to know it thoroughly.

Nevertheless, the codex does not provide us with certain dates from Master Liechtenauer's life. So far, no reliable biographical evidence whatsoever has been found. His origin remains mysterious, though his name itself

9 Nuremberg, *Cod. Hs. 3227a*.

may refer to his birthplace. Unfortunately, there are too many towns called Liechtenau or Lichtenau for this to narrow it down much.

Although the person himself remains obscure, Liechtenauer's influence was quite long-lasting. The manuscript from Nuremberg probably offers us the first encounter with the art according to Liechtenauer. A living tradition of roughly 250 years ensued.

Although Liechtenauer's influence was (and among modern practitioners still is) huge, his teachings are actually fairly limited in scope. He has only left us a couple of verses that rhyme more or less elegantly. These verses are obscure and elusive; and without proper explanation they remain almost opaque gibberish, and are thus next to useless for any practical application. Luckily, we gain much more insight into Liechtenauer's teachings through secondary sources. The master bequeathed his obscure poems to his pupils and thus to posterity – possibly only orally, possibly already in written form. After all, there may be a reason why they are called *zedel* in the old sources: '*Zedel*, Latin *schedula*, usually signifies an ephemeral way of notation on single leaves – in contrast to the more durable form of books.'[10] Without comments, these verses are hard or even impossible to understand. Liechtenauer produced them in this way, according to the writings of his successors and/or students, in order to prevent all the world from being able to understand these skills:

> Therefore he had them written down in hidden and secret words so that not everybody shall hear and understand them. He has done so because of the careless fencing masters who don't value their art. Thus he wants to avoid these masters either making his art public or spreading it.[11]

Fortunately for us, Liechtenauer's students didn't care about that very much, for in the very next sentence of these introductory words it says:

> And these very same hidden and secret words of the notes (*Zedel*) are written hereafter in the commentaries explained and interpreted so that everybody, who already is able to fence, can read and understand them.[12]

So, although Master Liechtenauer himself wished to hide his knowledge from a broader audience, his disobedient pupils did not share his methods – or rather, they did. Nevertheless, we must not forget that the books that contain Liechtenauer's knowledge are manuscripts – one-off items that were not widely available but rather aimed at a single client in order to serve as a memory aid.

Master Liechtenauer has spawned at least one follow-up generation of masters. These are known as the *Gesellschaft Liechtenauers* (Liechtenauer

10 MÜLLER (1992), p. 256.

11 Rome, *44 A 8*, fol. 9v.

12 Ibid., fol. 9v.

Society) as the name appears in Paulus Kal's manuscripts, as codified for instance in his manuscript from Munich:[13]

> Master Johannes Liechtenauer, Master Peter Wildigans from Glaz, Master Peter from Danzig, Master Hans Spindler from Znaim, Master Lamprecht from Prague, Master Hans Seidenfaden from Erfurt,
> Master Andre Liegnitzer, Master Jacob Liegnitzer (brothers),
> Master Sigmund Amring, Master Hartmann from Nuremberg,
> Master Martin Huntfelt, Master Hans Pegnitzer,
> Master Philip Perger, Master Virgili from Cracow,
> Master Dietrich, dagger fencer from Brunswick,
> Master Ott, the Jew, who was the wrestler of the lords of Austria, the noble and strong Stettner, who above all was the master of all pupils, and I, Master Paulus Kal, was his pupil, and may God have mercy upon him.

This is not necessarily a formal society of individuals that met on a regular basis; it can rather be seen as a list of masters that were in one way or the other associated with Liechtenauer, the grandmaster. Paulus Kal lists many masters we have references to in other manuscripts, such as Peter von Danzig, Andre Lignitzer and others; some on the list have left no trace, like Peter Wildigans; some may be identical to authors that have bequeathed material to us: Sigmund Amring for instance is certainly identical to Sigmund Emring[14] and Sigmund ain Ringeck;[15] but it is a curious circumstance that there is at least one master contemporary to Kal who is not mentioned: Hans Talhoffer, the fencing master from whom we have the most surviving fight books by far.

THE MAN AND HIS MANUSCRIPTS

Only scarce biographical material exists about Hans Talhoffer. Even his name appears in various forms: Talhoffer,[16] Dalhofer[17] or Talhofer.[18] Although some recent publications[19] deal with Master Talhoffer, the person behind the manuscripts remains mostly elusive. Most of the obtainable data can be gathered from online sources:[20] The first time we encounter Talhoffer in historical records is in 1433, when he served

13 Munich, *Cgm 1507*, fol. 2r.

14 Glasgow, *E.1939.65.341*.

15 Dresden, *Mscr. Dresd. C 487*.

16 K, p. 47, 49, 71, 121; M, fol. 136v; the abbreviations are explained in the next section.

17 B, fol. 63r.

18 K, p. 121; C, fol. 2r, 101v, 103v.

19 BURKART (2014), ISRAEL (2017).

20 wiktenauer.com/wiki/Hans_Talhoffer; talhoffer.wordpress.com/category/a-life-like-that-of-talhoffer/biography/

Johann II von Reisberg, archbishop of Salzburg, in a Vehmic court. A year later, during or after his service for the archbishop, he was accused of murdering Wilhelm von Villenbach. Talhoffer confessed to having been hired to abduct Hans von Villenbach, the deceased's brother, but testified that others committed the murder. The trial caused some upheaval, but Talhoffer remained in the archbishop's service.

We meet Talhoffer again in the records of the city of Zurich in 1454, when he supervised a public fencing event and had to intervene in a fight that got out of hand.[21]

Talhoffer continued to serve various patrons, such as Leutold von Königsegg, the brothers David and Buppelin vom Stain and Eberhard I, Duke of Württemberg. An extensive manuscript was produced for each of these noble employers, and it is a curious circumstance that both the Königsegg and the Stain version contain the identical illustrated duel – thus it is doubtful whether either duel actually took place or whether the pictures were created for advertorial reasons or as a sign of homage to the principal. While in Württemberg's service, his name is recorded for the last time when Eberhard paid Talhoffer ten guilders and some rye and oats.

Up to this day, there is speculation whether Talhoffer was a member of the Marxbrüder fencing guild (brotherhood of St. Mark) since a lion, their heraldic animal, appears in his coat of arms in one of his manuscripts.[22] Nevertheless, no conclusive evidence has been found to date.

21 Hils (1985), p. 176.

22 Copenhagen, *Thott 290 2°*, fol. 102r.

Most fight books are compendia with various sections, sometimes written by a single person, such as Hans Talhoffer, sometimes a compilation of works from various masters. These anthologies differ in extent, and the individual components are not identical from one volume to another. Some authors, such as Paulus Kal or Jörg Wilhalm, commissioned copies of their own works more or less unaltered; in contrast, Hans Talhoffer's books vary considerably. We have five archetypical manuscripts from his lifetime, plus one contemporary copy, plus some nine later copies. This makes a total of fifteen Talhoffer-related manuscripts.

ABBREVIATIONS

A: Augsburg, Universitätsbibliothek, *Cod.I.6.2°.1.*

B: Berlin, Kupferstichkabinett der Stiftung Preußischer Kulturbesitz, *78 A 15.*

Co: Coburg, Kunstsammlungen der Veste Coburg, *Hz.014.*

C: Copenhagen, Det Kongelige Bibliotek, *Thott 290 2°.*

G: Gotha, Forschungsbibliothek Schloss Friedenstein, *Ms. Chart. A558.*

Gö: Göttingen, Niedersächische Staats- und Universitätsbibliothek, *2° Cod. Ms. philos. 61.*

Ka: Kassel, Landesbibliothek, *2° Ms. iurid. 29.*

K: Königseggwald, Gräfliches Schloss, *Hs. XIX, 17-3.*

M: Munich, Bayerische Staatsbibliothek, *Cod. icon 394a.*

M2: Munich, Bayerische Staatsbibliothek, *Cod. icon 394.*

M3: Munich, Bayerische Staatsbibliothek, *Cod. icon 395.*

N: New York, Metropolitan Museum of Art, *26.236*.
P: Private collection, olim: Vienna, Österreichische Nationalbibliothek, *Cod. Vindob. Ser. Nov. 2978*.
V: Vienna, Kunsthistorisches Museum, *KK 5342*.
W: Wolfenbüttel, Herzog August-Bibliothek, *Cod. Guelf. 125.16 Extrav.*

Though Talhoffer did not personally write any of the manuscripts, he claims to have posed for at least two of them, namely C and M. He is also shown helping his patrons don their suits of armour in K and B. Eric Burkart,[23] however, doubts that any of these images can be assigned the status of a true portrait.

To explore thoroughly the contents of each of these manuscripts is beyond the scope of this paper. Much information, however, can be gathered from Hils (1983), Leng (2008) and Burkart (2014). From these sources, we can assume for sure that Talhoffer was a man of multiple interests: apart from fencing in various disciplines, he offers valuable information on why and how a judicial duel was to be held and fought; some of his books contain portions of war books with siege engines and war machines; and he also looked into onomatomantia (working out who will be victorious in a fight, based on his name and the time at which it is to happen).

In order to shed some light onto the web of the connections of the various manuscripts, I shall identify which are the archetypes, which is to say the original version of the text, and which are the copies. Archetypes are set in bold, corresponding copies are indented. Subsequently, however, I will look at the archetypes in more detail.

23 Burkart (2014).

ARCHETYPES AND COPIES

- **G** **Gotha**, 1443/48
  - Ka Kassel (partially), 17th c.
  - W Wolfenbüttel (partially), late 17th c.
    - Co Coburg (partially), 18th c.
  - N New York, 17th c.
  - Gö Göttingen, late 17th c.
  - M3 Munich, *ca.* 1820
- **K** **Königseggwald**, middle of 15th c.
  - V Vienna, end of 15th c.
  - A Augsburg, before 1561
- **B** **Berlin**, before 1459
- **C** **Copenhagen**, 1459
- **M** **Munich**, 1467
  - P Private collection, 16th c.
  - Ka Kassel (partially), 17th c.
  - W Wolfenbüttel (partially), late 17th c.
    - Co Coburg (partially), 18th c.
  - G Göttingen, late 17th c.
  - M2 Munich, ca. 1820

G is an anthology of many disciplines and a collection of various artists. At least four scribes and 11 artists were involved in the process of creating the codex.[24] The entire manuscript has a rather heterogenous appearance and may have been compiled from various sources and then bound into one single volume, a procedure which was common in the Middle Ages and the Renaissance.

- Unarmoured longsword – illustrated: fol. 2r–6r (blank pages: 2v, 3v, 5v, 6v)
- Johannes Hartlieb's onomatomantia: fol. 7r–17r (blank pages: 12v, 17v)
- Master Liechtenauer's teachings for longsword and mounted combat: fol. 18r–23r (blank pages: 21r, 23v)
- Duel with large shields – illustrated: fol. 24r–48r (blank pages: 42v, 43v, 44v, 45v, 46v, 47v; fol 24r with a pencil sketch; 41v traced-through sketch from overleaf)
- Harness fencing (Liechtenauer's verses): fol. 48v
- Harness fencing – illustrated: fol. 49r–72r (all verso pages are blank)
- Harness fencing – illustrated (outline sketches, no captions): fol. 73v–80v (all recto pages are blank; 81r–82 blank)
- Dagger – illustrated: fol. 83r–104v (fol. 105r–109r blank)
- Ott's wrestling: fol. 109v–114v (fol. 115r–116r blank)
- Wrestling – illustrated: fol. 116v–133v (fol. 134–140 blank)
- War book – illustrated: fol. 141r–148r
- Ammunition (17th c. [?] addition) – illustrated: fol. 148v

K has been in the possession of the Königsegg family from its inception to the present. It describes the training of squire Leutold von Königsegg and a stylised duel with an unknown adversary. The codex has spawned two copies, V and A. V is so similar to K that it is not unlikely that both manuscripts stem from the same workshop or even the same artist. The images and the handwriting, however, are a bit less elaborate in V. According to Leng, the watermarks indicate a slightly later date, *ca.* 1480–1500.[25] Altogether, K is the more complete version since V lacks several captions, particularly in the back of the codex. A is an artistically rougher copy, likely of K, not of V, since it contains those captions not included in V.

- Introduction and verses: pp. 1–2
- Stances for harness fencing: pp. 3r–17 (some pages are blank: 4, 6, 8)
- Leutold's duel in armour: pp. 18–48 (some pages are blank: 46, 48)
- Dagger: pp. 49–70
- Wrestling: pp. 71–88
- Spear: pp. 89–98
- Fighting on horseback (lance): pp. 99–100, 107
- Fighting on horseback (sword against lance): pp. 101–106
- Fighting on horseback (sword): pp. 109–112

24 According to LENG (2008).

25 LENG (2008), p. 58.

- Fighting on horseback (wrestling): pp. 113–119
- Jumping into the saddle: p. 120
- End page: p. 121

B offers an armoured combat scene in the beginning that is distinctly different from K. This is followed by the same armoured, unarmoured and mounted techniques we are familiar with from K, although here the sequence is torn apart (due to rebinding the manuscript). But while it is Leutold von Königsegg who is being taught by master Talhoffer in K, it is here either David or Buppelin vom Stain. Additionally, the various heraldic shields are different. The overall style of the illustrations is rather similar to K, although some of them, particularly in the dagger and wrestling sections, look much more refined. Some additional material concerning how to dress for a duel and the appropriate weapons is also provided, which is absent from its predecessor, and we are introduced to fencing with the *langes Messer* and the buckler as well as with the flail. A section with the large duelling shields, following Franconian and Swabian law, is also included. In summary, B is a heavily extended version of K.

- Harness fencing: fol. 1r–27v
- Fighting on horseback: fol. 28r–32r
- Wrestling: fol. 32v, 38r/v, 43r–44v, 46r, 48r/v, 51r–53v, 57r–59v, 61r
- Staff/spear/flail: fol. 33r/v, 45r/v, 55r–56v, 60r/v
- Dagger: fol. 36v, 39r–42v, 46v–47v, 49r–50v
- Dressing the fighter: fol. 37r/v
- Buckler: fol. 54r/v
- Weapons for the fight: fol. 62r/v
- Brothers vom Stain, Talhoffer: fol. 62r
- Long shield: fol. 62v–77v

C constitutes the next step in the evolutionary process. Talhoffer has – allegedly – not only posed in person for the illustrations, he expands the scope considerably: A lengthy introduction, a war book and sections about cryptography and the Hebrew alphabet are included, as well as chapters about the planets and anatomy. The fight book part has also been extended by fighting with the pollaxe and by duels between man and woman.

- Introduction and verses: fol. 1r–10v (6 and 7 are blank leaves)
- Illustrations about cryptography (?): fol. 11r/v
- Bellifortis (war book): fol. 12r–48v
- Wrestling: fol. 49r–60v, 138r–139r (139v is blank)
- Dagger: fol. 61r–71r
- Pollaxe: 71v–74v, 131r–137v
- Spear against sword and other weapons: fol. 75r–78v
- Sword against mace: fol. 79r/v
- Man against woman: fol. 80r–84r
- Harness fencing: fol. 84v–94r (from 89r onwards shown without armour)
- Fighting on horseback: fol. 94v–97r, 124r–130r
- Duelling shield and sword: fol. 97v–99r, 102v, 116v–117r

- Duelling shield and club: fol. 99v101r, 103r, 110v–111r, 116r
- Image and crest of Hans Talhoffer: fol. 101v–102r
- Colophon: fol. 103v
- Arms: fol. 104r–110r
- Duelling shield: fol. 111v–115v
- Sword/*langes Messer* and buckler: fol. 117v–123v
- Tournament scene: fol. 130v

The following leaves are inscribed upside down, starting at the last leaf; therefore the foliation is reversed:

- Cryptography/Hebrew alphabet: fol. 150v–149v (149r is blank)
- Planets: fol. 148v–142v
- Anatomy: 142v–140v (140r is blank)

M, finally, is stripped again of the non-fencing-related material. The chapters are much more coherent than in the previous books, and for the first time we find a whole chapter about true unarmoured combat with the sword. Before, it was mostly armoured combat depicted sans armour – for whatever reason, possibly convenience on behalf of the models. The artistic standard, certainly enhanced in C, reaches its peak in this volume with its slender and elegant late Gothic figures.

Apart from general bibliographical information such as the material (parchment) and the size (333 × 215 mm), Leng[26] relates that four artists were responsible for the images of the codex, with I and II being hard to distinguish – I: 2r–39v, 40v–52r; II: 40r, 86v–109r, 110v–112r; III: 52v–53r, 109v–110r, 113r–122r, 127r–136v; IV: 53v–86r, 122v–126v.

- Sword (unarmoured): fol. 2r–17v
- Halfsword (armoured techniques performed out of armour): fol. 18r–35r
- Harness fencing: fol. 35v–40v
- Pollaxe: fol. 41r–53r
- Duelling shield and club: fol. 53v–64v (65r is blank)
- Duelling shield and sword: fol. 65v–76v
- Duelling shield: fol. 77r–86r
- Dagger: fol. 86v–96r
- Wrestling: fol. 96v–112r (112v is blank)
- *Langes Messer:* fol. 113r–116v
- Sword and buckler: fol. 117r–121r
- One against two: fol. 121v–122r
- Man against woman: fol. 122v–126v
- Fighting on horseback (sword): fol. 127r–131v
- Fighting on horseback (wrestling): fol. 132r–133v
- Fighting on horseback (lance): fol. 134r/v
- Fighting on horseback (crossbow against lance): fol. 135r–136v

26 Leng (2008), pp. 54–6.

OWNER OR AUTHOR

For many scholars, it is a mystery why Paulus Kal omitted Talhoffer from Liechtenauer's society. Hans-Peter Hils suggested that this omission was inspired by envy on Kal's side.[27] I find this unlikely since Paulus Kal found himself in frequent employment by dukes and other noblemen, whereas Talhoffer seems to have been more or less a journeyman in fencing affairs. But what if Talhoffer had never been a member of this illustrious circle anyway?[28]

This suggestion is largely based on the fact that Liechtenauer's verses appear in G. But if Talhoffer merely acquired the book without being its author, this of course does not mean that he must have been a member of this society. It is possible that a work from Liechtenauer or his circle may have come into his hands as an outsider. Of course, this does not exclude that he found such an authority as Liechtenauer interesting, or that he possibly admired or even revered him. Perhaps he even took over, adapted or plagiarised parts of Liechtenauer's teaching. After all, a lot of Liechtenauer-related terminology has been preserved in Talhoffer's treatises. But as the anonymous first glossator on Liechtenauer's verses said, the revered master had already travelled to many a country in order to learn and perfect his skills. There may have been fencing terms before Liechtenauer – and parallel to him. After all, there is also a whole chapter in the early Nuremberg fight book about techniques of other masters, such as the priest Hanko Döbringer, Andre the Jew, Jost from the river Neiße and Niclas from Prussia. Talhoffer could therefore have learned the material that goes back to Liechtenauer on his own, in order to modify and design it according to his taste or his own understanding of art. After all, he had travelled widely, and his experience was considerable.

So we need to take Talhoffer's authorship of G with a considerable grain of salt. The most obvious circumstantial evidence is his name on fol. 1r. In this regard, Hans-Peter Hils says:

> The manuscript is Talhoffer's oldest known creation and presumably his first work. It was apparently at least temporarily in his possession, indicated by a painted text banner on fol. 1r (today only legible if the page is held against the light): 'Daz bůch ist maister hanssen talhoffersz'. The last three letters are destroyed by worm damage, but can be completed according to the (possessive) genitive 'hanssen' and the same formulation in another manuscript of the Master – Thott 290 2°, fol. 103v.

That, however, does not qualify whatsoever as a proof of authorship. Another fight book author was Paulus Hector Mair, treasurer of the city of Augsburg. Although he was indeed the patron of three luxuriously

27 HILS (1983), p. 112; HILS (1985), pp. 179–80.
28 As already suggested by MÜLLER (1992), p. 262.

illustrated fight books,[29] several additional manuscripts also bear his name – as owner, not as author.[30]

Apart from the name on the first page of G, there is one more important clue that Talhoffer at least owned the manuscript – leaving it an open question nevertheless whether authorship can be attributed to him: a coat of arms is displayed on fol. 28r which is similar although not identical to examples from K, B and C. Curiously enough, the jousting helmet of the coat of arms is adorned with a whip, which was rather crudely painted over with an anchor by another artist – possibly to look similar to the coat of arms of Paulus Kal (see for instance Bologna, *Ms. 1825*, fol. 5v; Gotha, *Ms. Chart. B1021*, fol. 1v) – for an unknown reason. After all, the codex differs considerably from Talhoffer's other manuscripts (K, B, C, M) in many ways, as Hans-Peter Hils discovered more than 30 years ago.[31] Nevertheless, being a compilation, the various fighting disciplines appear overarchingly in many independent sources. Hils provides us with a detailed concordance of the images in the manuscripts he examined (G, K, V, B, C, M).[32] His overview, however, is flawed insofar as some images show generic postures that indeed look similar but only to an extent that is to be expected in a discipline such as fighting with the large duelling shields in which certain positions are commonplace. Upon closer examination, the similarities in this sequence are only superficial or coincidental. This is also true for harness fencing, see for instance fol. 56r in G and fol. 5v in the *Gladiatoria* manuscript from Cracow.[33] Nevertheless, occasionally there is a striking resemblance among the Talhoffer-related manuscripts. This is particularly noteworthy with a technique called 'the scissors' which is shown almost identically in G (fol. 66r), C (fol. 92r) and M (fol. 31r), and in a mirrored version in B (fol. 27v) and K (p. 34). A similar but not identical version can be seen in *Gladiatoria* (fol. 22v), but here the kick to the backside is omitted. In summary, we cannot determine whether Talhoffer actually wrote G, but he certainly was the owner of the manuscript.

29 Dresden, Sächsische Landesbibliothek, *Mscr. Dresd. C 93/94*; Munich, Bayerische Staatsbibliothek, *Cod. icon. 393* (1 and 2); Vienna, Österreichische Nationalbibliothek, *Cod. 10825/10826*.

30 *Codex I.6.2°.1* – see abbreviation A; *Codex I.6.2°.2* – a compilation of Jörg Wilhalm's longsword treatise and Lienhart Sollinger's manuscript version of Andre Paurenfeindt's printed fight book from 1516; *Codex I.6.2°.3* – Jörg Wilhalm's fencing in harness and on horseback; *Codex I.6.2°.5* – a compilation of records of the Marxbrüder fencing guild, a glossed version of Johannes Liechtenauer's longsword verses and engravings of Maarten van Heemskerck's fencing and wrestling; *Codex I.6.4°.2* – the so-called *Codex Wallerstein* or *von Baumann's fight book*, a compilation of armed and unarmed fighting disciplines with detailed descriptions, and an older, uncaptioned part that is connected to the *Gladiatoria* group of fencing treatises; *Codex I.6.4°.5* – a draft version of *Codex I.6.2°.3*; *MS E.1939.65.354* – Gregor Erhart's fight book; *Schätze 82 Reichsstadt* – Anton Rast's fight book, although in this case Mair claims some editorial influence.

31 Hils (1983), p 102.

32 Hils (1985), appendix.

33 Cracow, *Ms. Germ. Quart. 16*.

THRUSTS OF DOUBT AND VOWEL STRIKES

It can be fun to make an attempt to reconstruct master Talhoffer's teachings. But mostly, it is a somewhat tedious effort, due to the utter lack of exhaustiveness of the sources. His short captions offer little insight into the required movements, and much guesswork ensues. If, however, one has begun to develop an understanding of medieval fencing arts by the means of manuscripts by more verbose masters, such as Liechtenauer for instance, one may indeed come to more or less plausible solutions. It is particularly advantageous if a certain technique like the *Krummhaw,* which is shown in Talhoffer's 1467 codex M, fol. 11r/v (reproduced in this volume on pp. 51/52), is explained in great detail by Liechtenauer's commentators. However, it is almost impossible to make sense of other techniques.

In order to pick two examples, let us briefly examine the *Zwyuelstich'* (thrust of doubt) and the *funff focal höw* (five vowel strikes).

The thrust of doubt is a unique thing: its name appears only once in M on fol. 40r. Although performed by a fighter sans armour, the context makes it very clear that it is a technique intended for armoured combat with the sword. The combatant holds his sword by the blade with both hands, the handle protruding over the right ear, the point directed at his adversary's breast or groin. The posture seems to be more in line what in other situations is called a *Mordschlag* – a 'murder strike' with the pommel of the sword. This particular image is very difficult to interpret, to say the least. No further text offers us any insight, nor can we gather any from other related or even unrelated sources. For the time being, this technique remains a complete mystery.

The other example refers to five vowel strikes. Unlike the previous example from M, this is a text passage from C and doesn't have an accompanying image. The text on fol. 2r says: 'Der höw sind fünff und haissent funff focal, Die lern recht und mercks fürwär.' – 'There are five strikes and they are called five vowels. Learn them correctly and truly remember them.' This is not an easy passage to grasp. On the one hand, the two lines don't rhyme, unlike the rest of the text in this section; on the other, interpreting *focal* as 'vowel' (*Vokal* in modern German) may be a bit daring at first glance. But then again, taking late medieval German orthography into consideration, maybe it's not: the letter 'v' was used either as a vowel or as a consonant – as it used to be in Latin. In German, the vowel 'v' was pronounced 'u', the consonant as either 'v' or 'f'. So we see that it's actually not too far a stretch to come to that conclusion. As far as the actual strikes with the sword are concerned, master Liechtenauer has the 'five hidden strikes' (that were re-phrased as 'master strikes' a century later): *zornhaw, krumphaw, twerhaw, schilhaw, schaitelhaw* (in order of appearance, quoted according to *Cod. 44 A 8*). Orthography was, as mentioned above, sometimes arbitrary, and particularly the last strike appears frequently as *scheittelhaw*, but if we stick with the example given, we notice that every strike contains one distinct vowel: z**O**rnhaw, kr**U**mphaw, tw**E**rhaw, sch**I**lhaw, sch**A**itelhaw.

Furthermore, if we bring the vowels in their correct order, we will find, that an astonishing sequence ensues: **A**, the schaitelhaw has hands and point of the sword in the highest position; **E**, the twerhaw still has the hands high, but the point can also attack low targets; **I**, the schilhaw has the hands slightly lower, with the point tilted downwards; **O**, the zornhaw has the hands in a middle position, point slightly upwards; **U**, the krumphaw finally has the hands in a middle position as well, with the point going downwards. So when we follow Liechtenauer's five hidden strikes in alphabetical order, we come from the highest to the lowest position.

I have no idea whether this is pure coincidence, whether Liechtenauer was already aware of this circumstance, whether Talhoffer made this up, or whether it is only me, as a practitioner, interpreting too much into the old verses. It is a fun thought, though, and a great way of providing modern students of the old art with a handy mnemonic. Interesting to see, by the way, that this bold idea of mine, spread on the internet a couple of years ago, has already been echoed in genuinely serious dissertations.[34]

In any case, Talhoffer's manuscripts still hold many a secret that is worthwhile to examine and explore. The more information we can gather about the fencing systems of the old master, the more likely we will come to solid and plausible solutions.

34 Leiske (2018), p. 148.

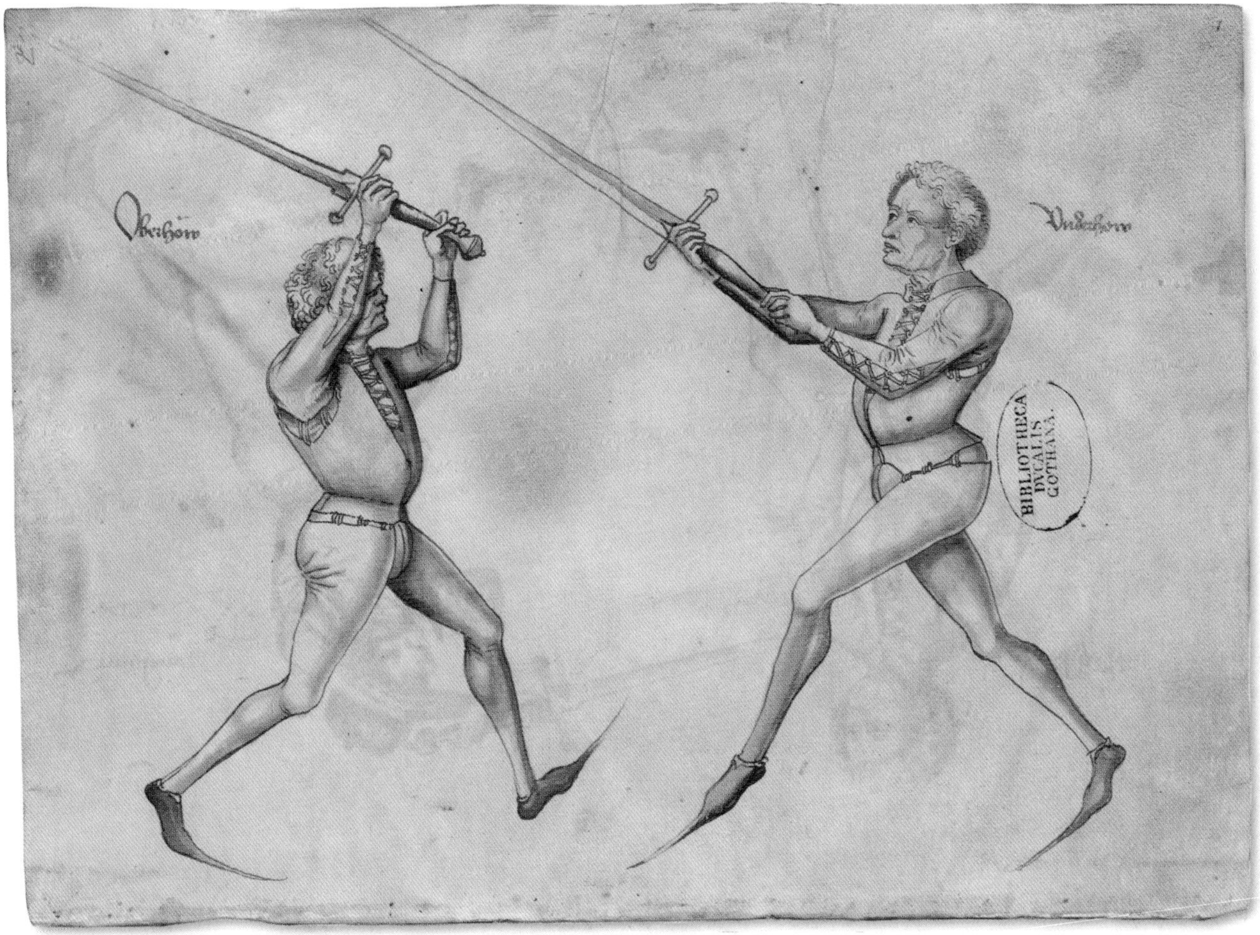

Oberhau.

~

Unterhau.

Sturzhau.

~

Wechselhau.

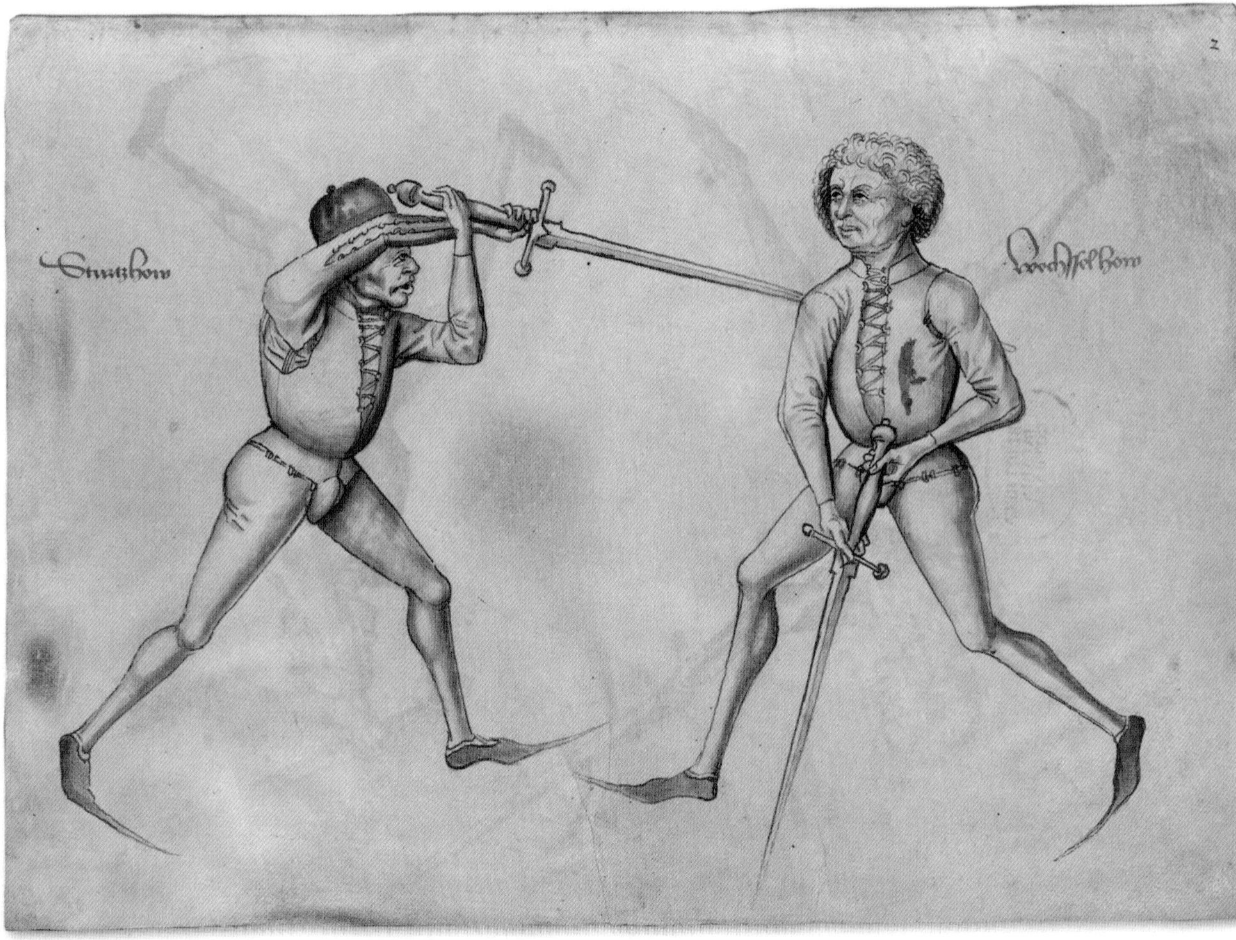

The *Zornort* threatens he ...

... who strikes an *Oberhau*.

The long *Zornort*.

The clasped thrust defends against this.

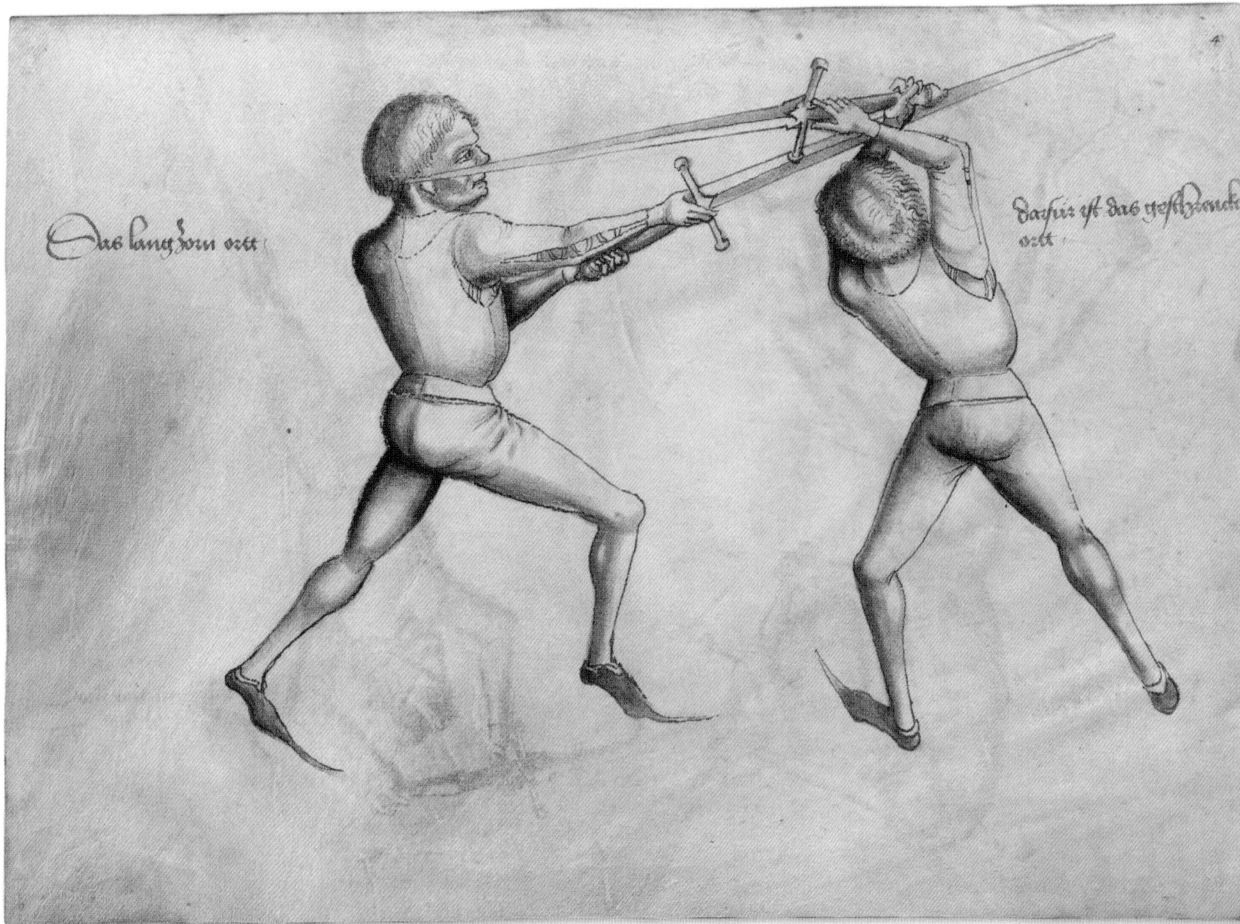

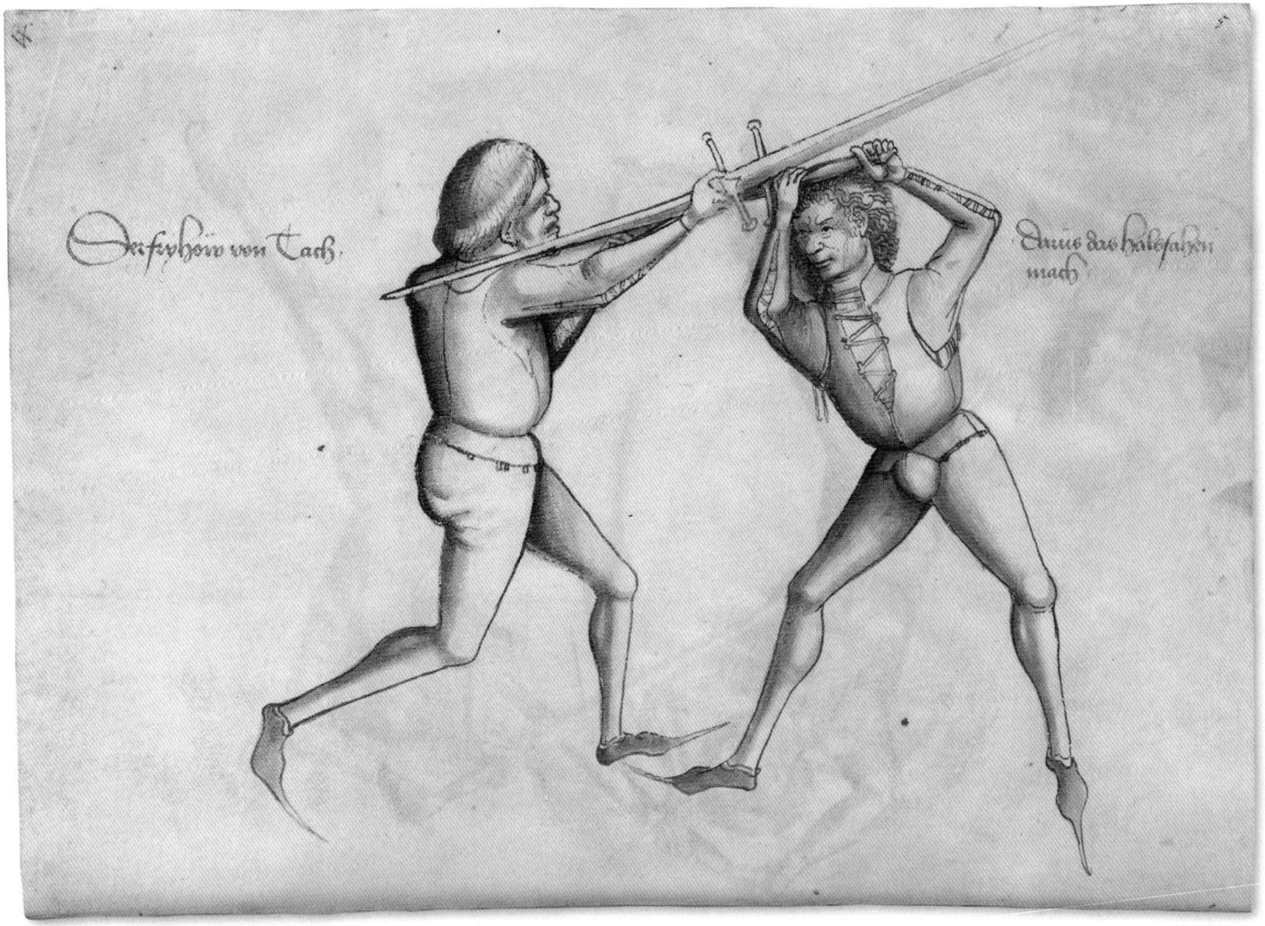

The free strike *vom Tag*.

From this, catch the neck.

A lower opening.

Another a lower opening.

Taking the sword.

Drawing out or anticipating.

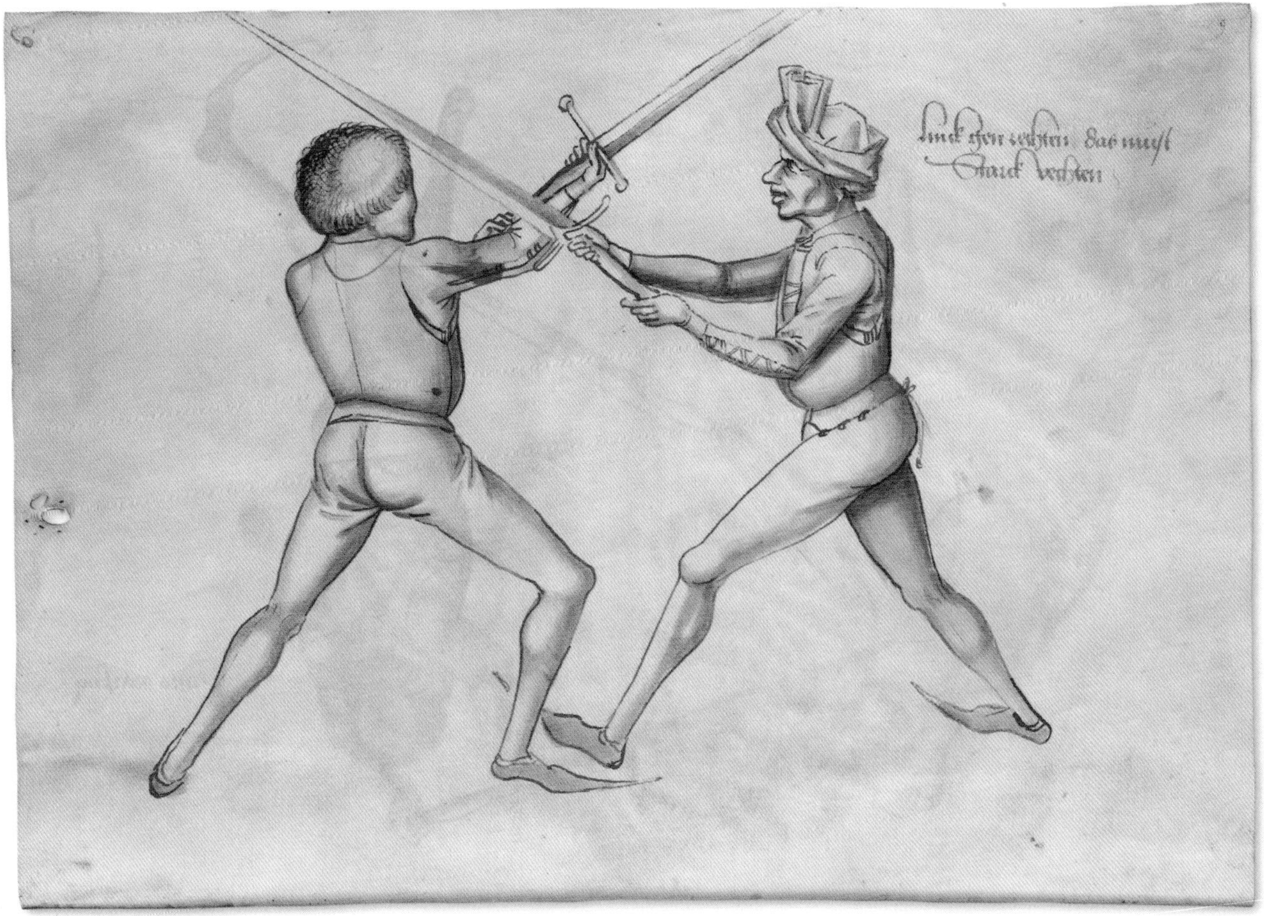

Left against right, here you must fence strongly.

A free thrust.

~

Whipping.

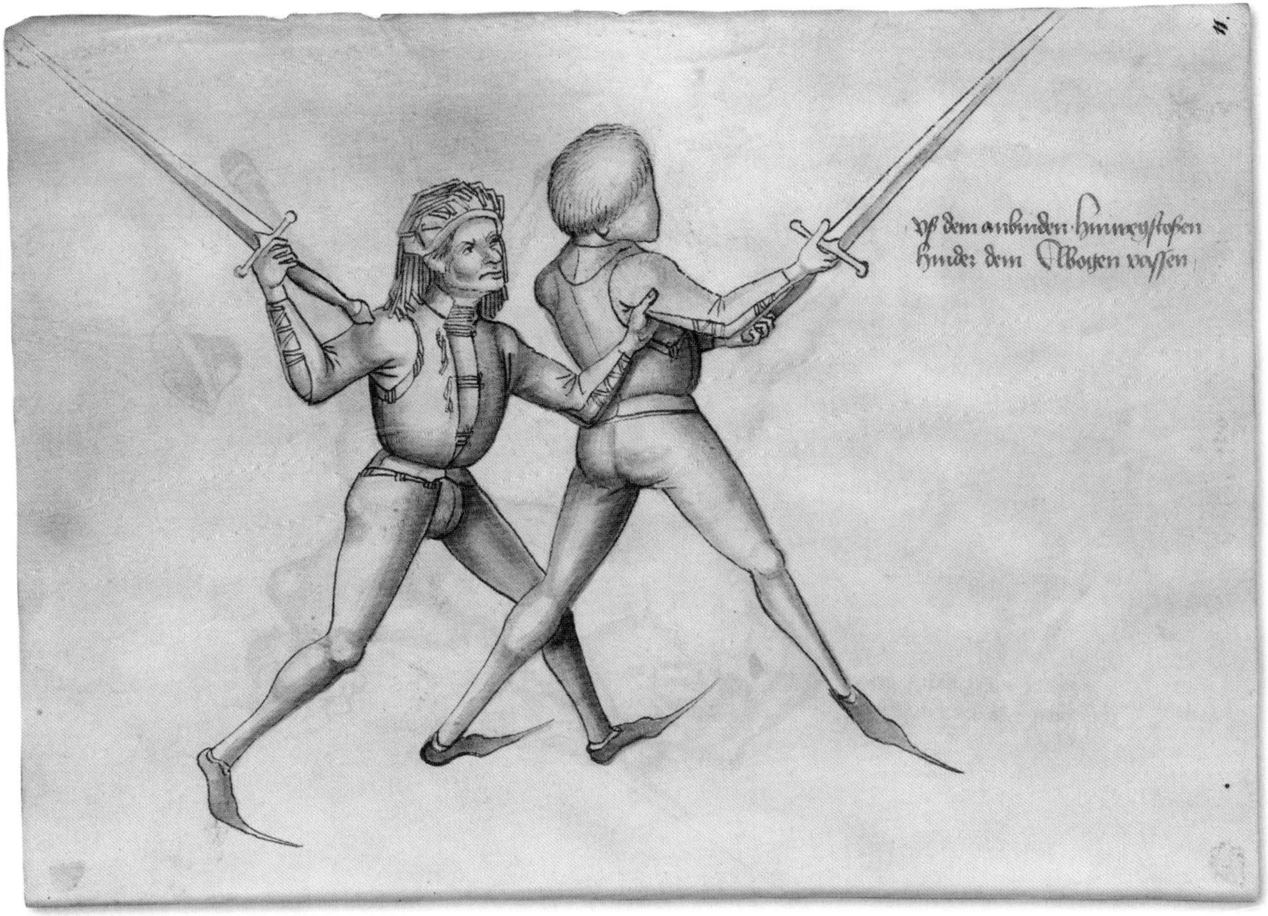

From binding: pushing away, grasping behind the elbow.

Running in with a free thrust and kicking into the belly.

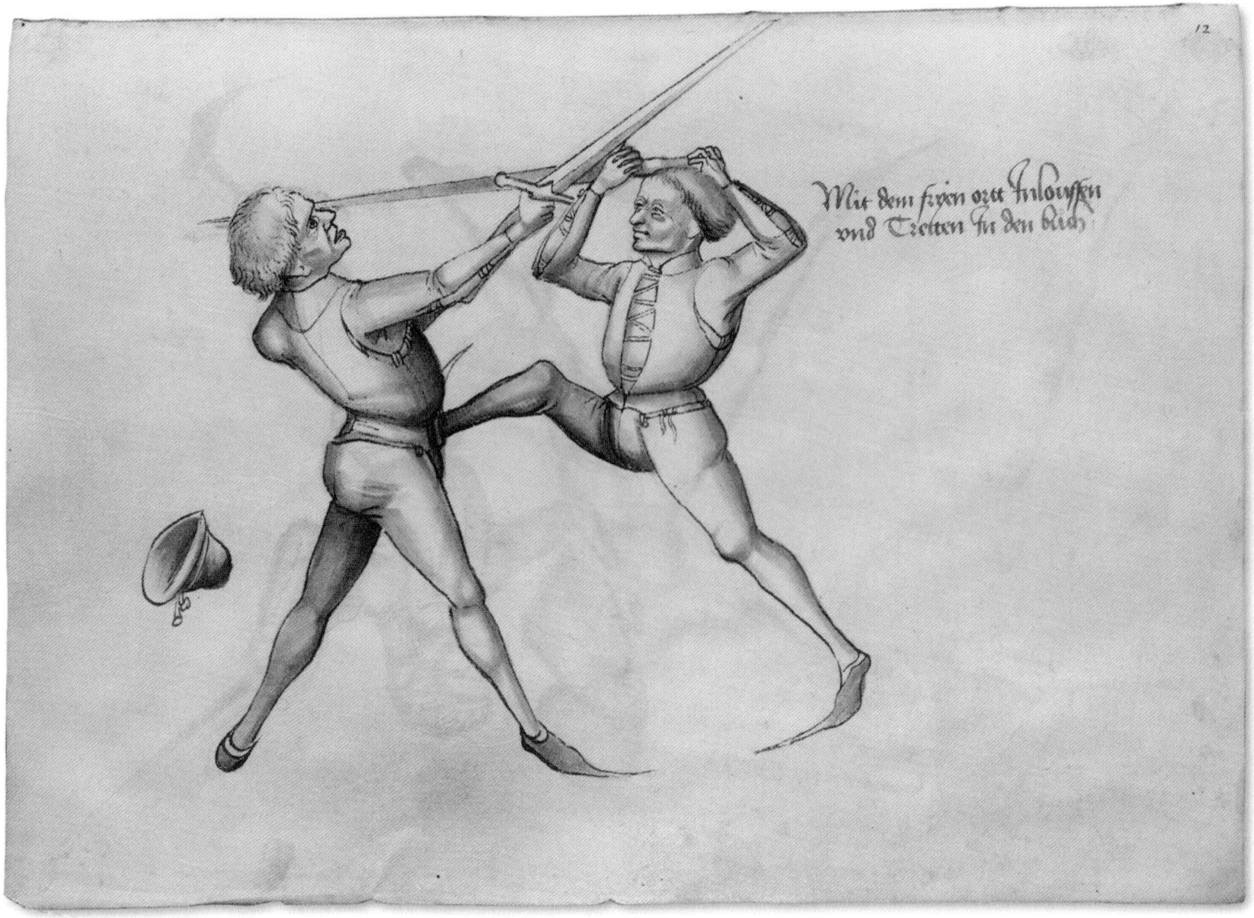

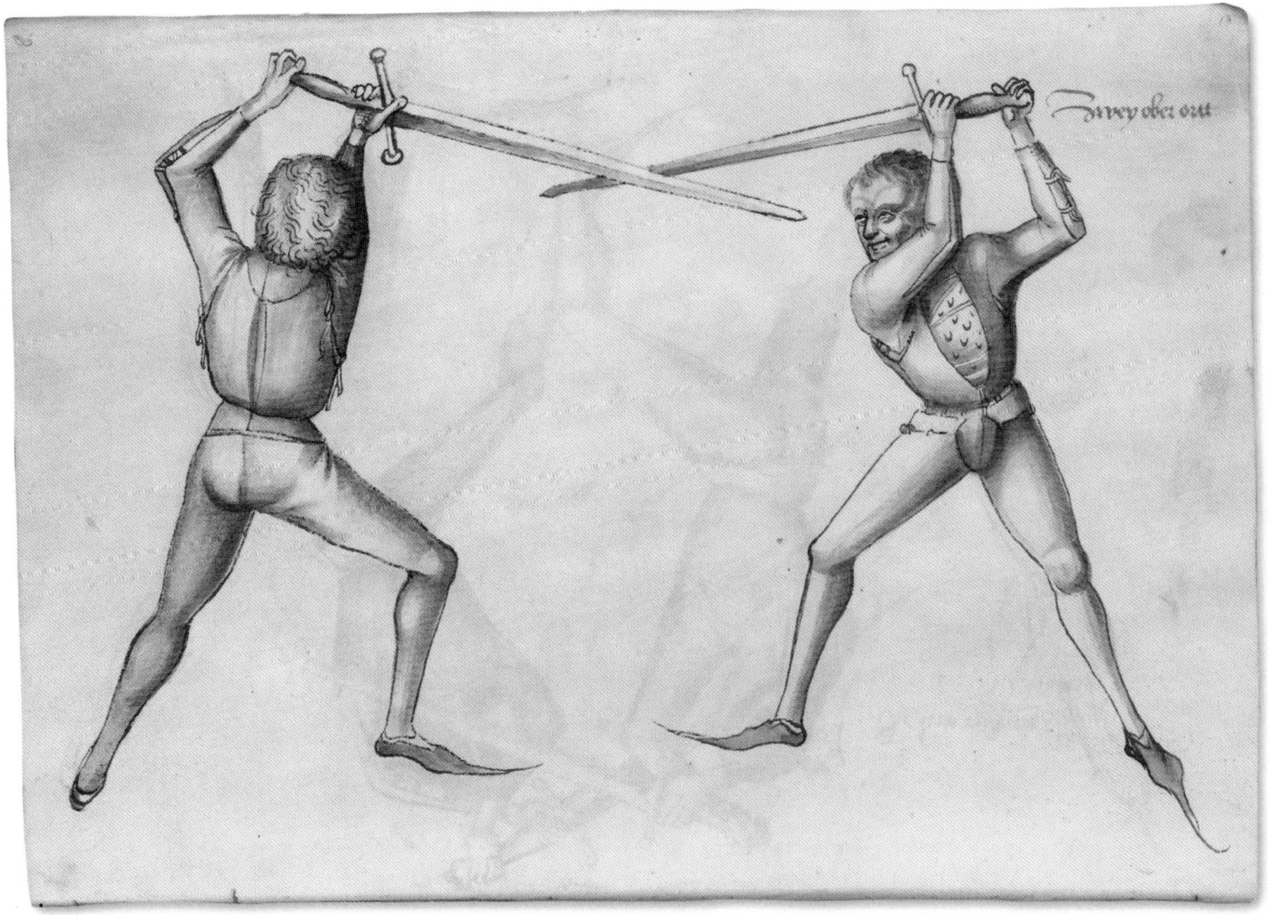

Two upper thrusts.

The two lower openings.

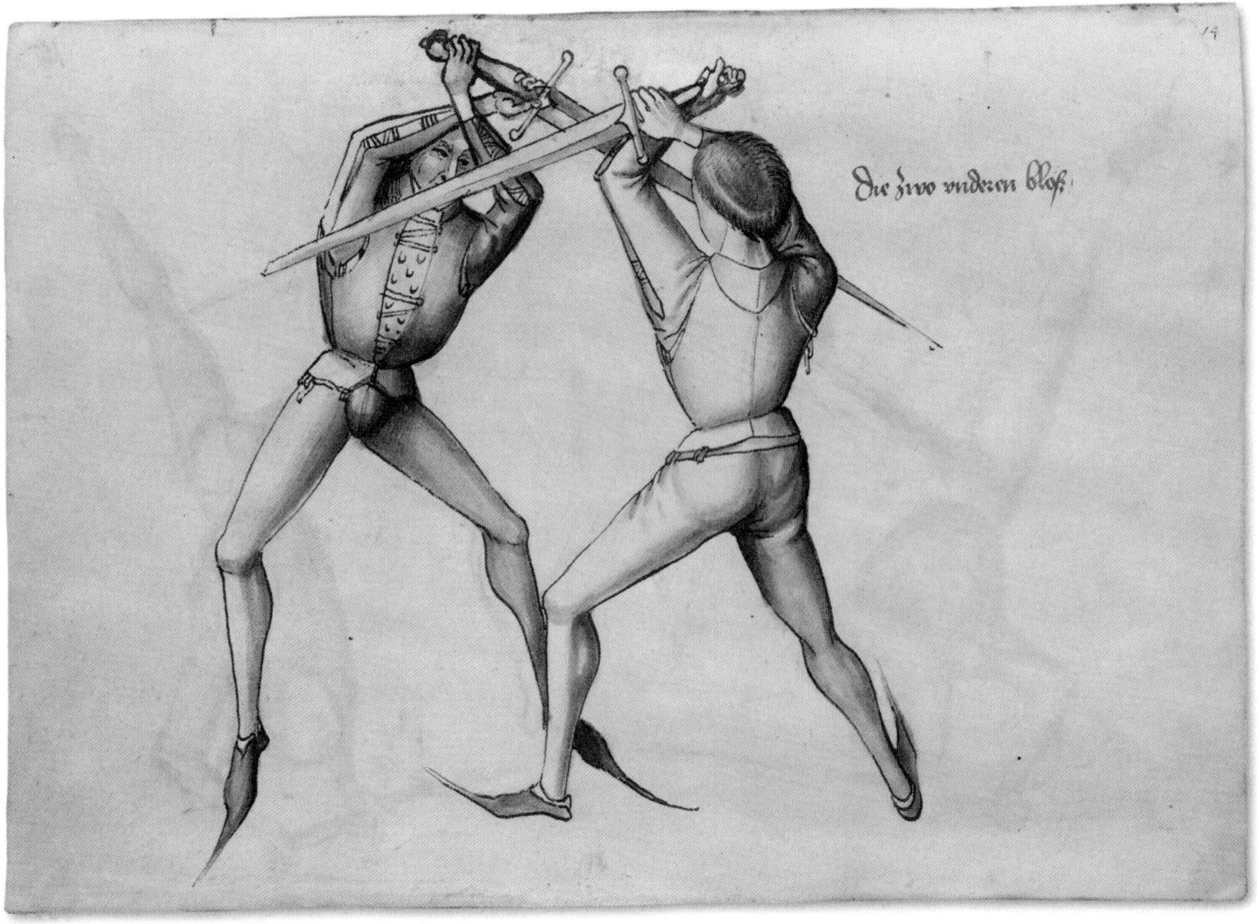

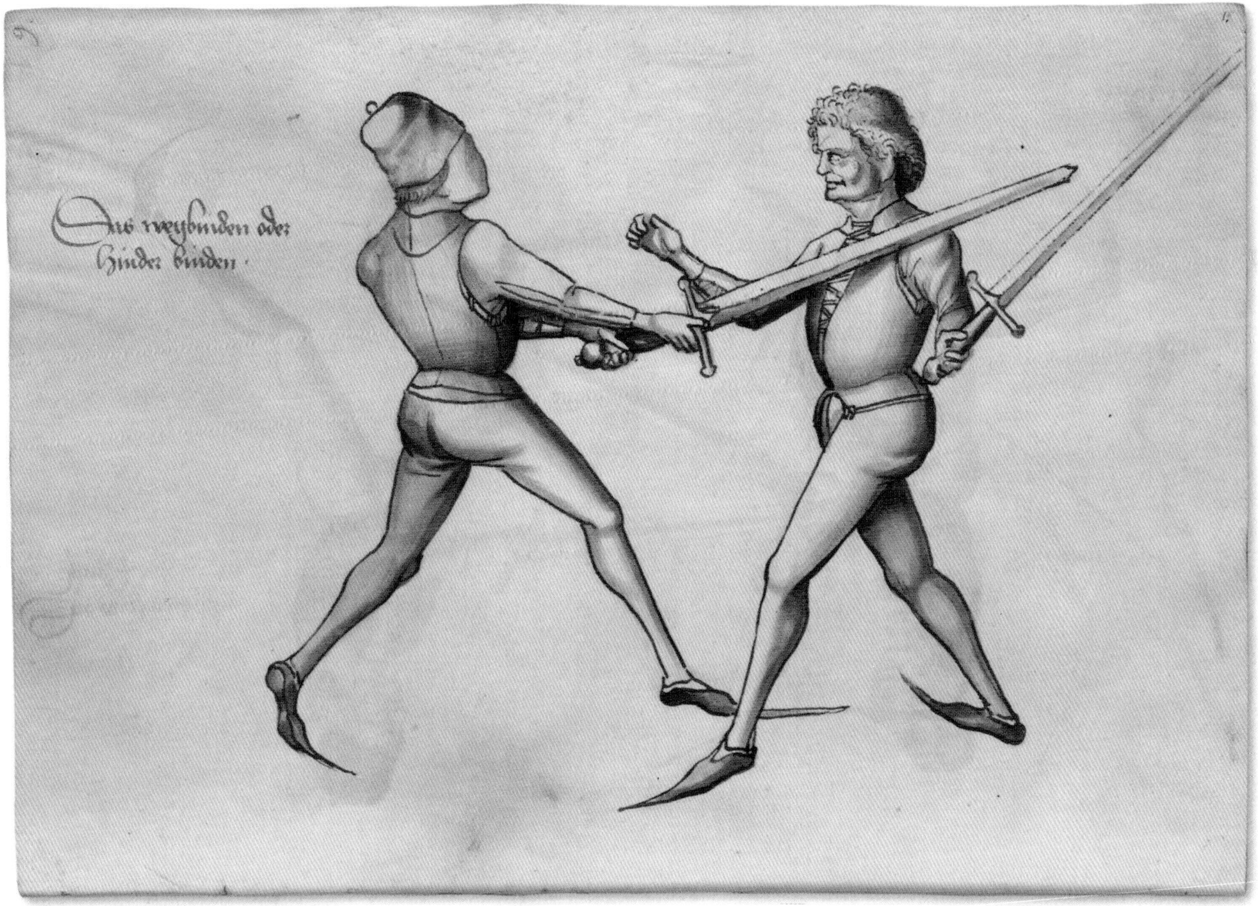

Binding away or binding behind.

Reaching over the iron gate.

The iron gate.

Feinting an *Oberhau* and striking into his arms.

This one executes the entangled thrust.

He strikes *vom Tag*.

Krummhau upwards nimbly.

~

Turn the point against your opponent.

Here, the *Krummhau* is accomplished.

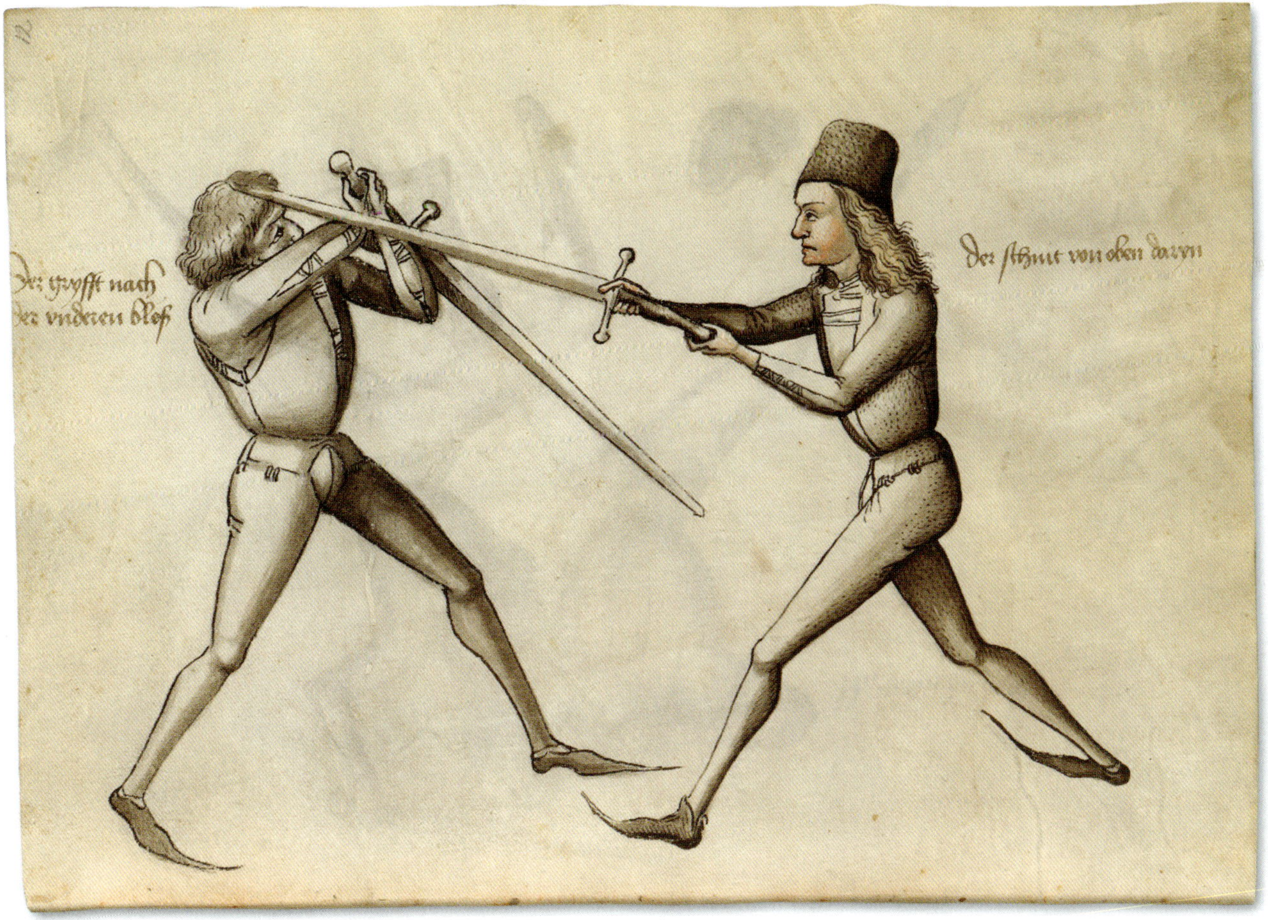

This one attacks the lower opening.

~

That one cuts into it from above.

Working in war.

The entangled weak.

This one stands in a guard.

Falling over.

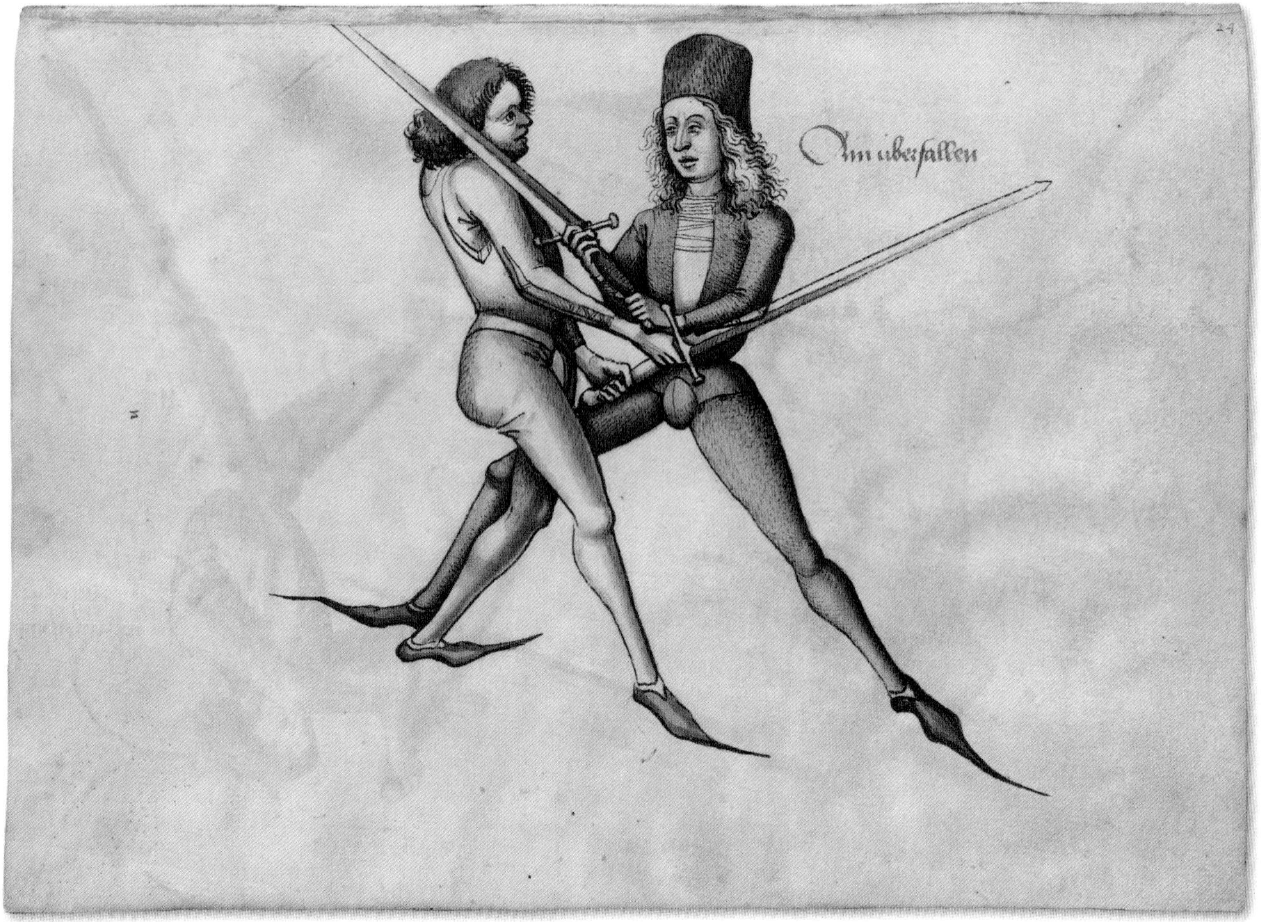

Both stand in a guard.

An armoured foray and taking the sword.

This one stands freely in the thrust.

~

That one strikes left over the thrust.

Falling into an armoured thrust from the *Oberhau*.

From the binding: grasping above and throwing.

From the binding: another forceful catch.

~

1467

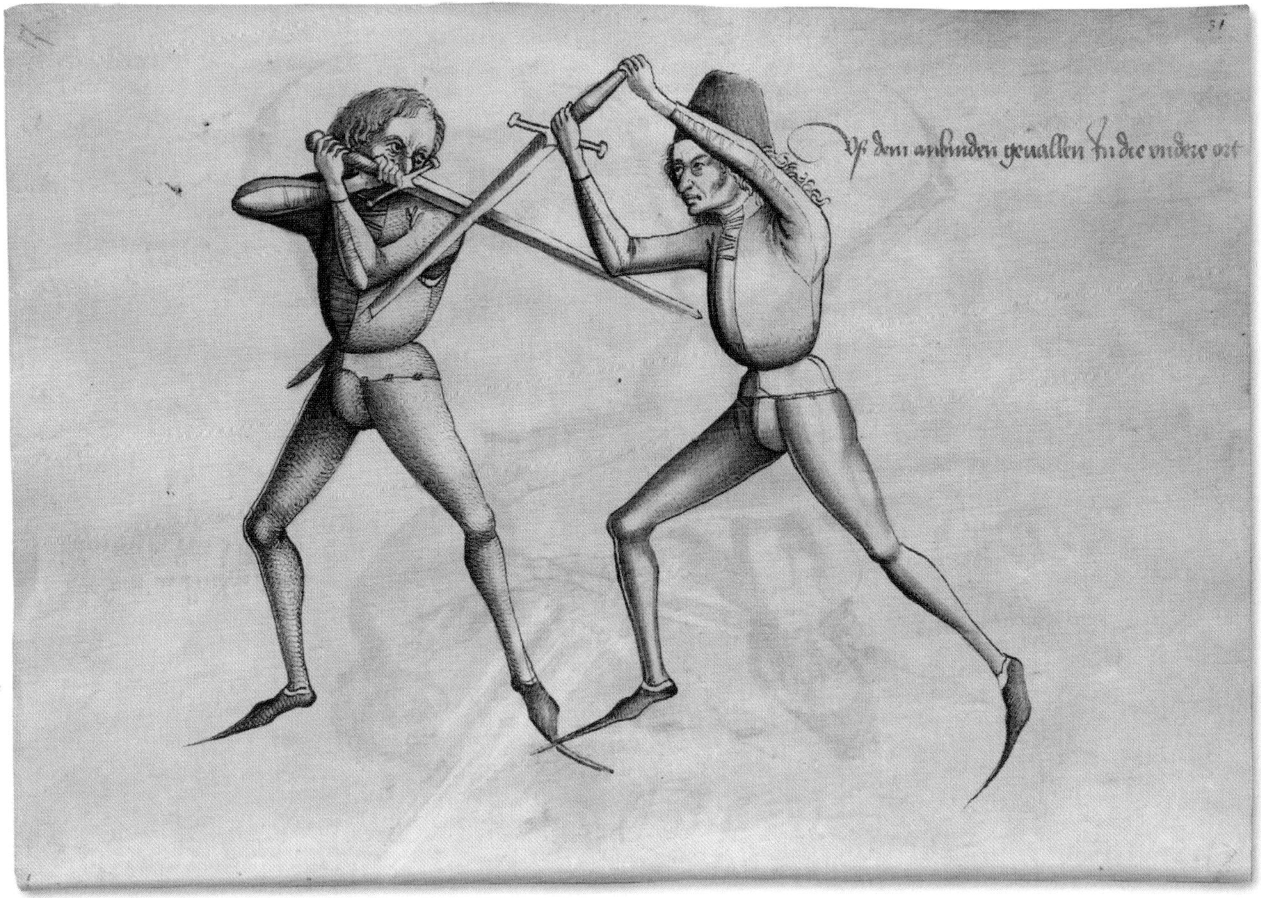

From the binding: falling into a lower thrust.

From the binding: pushing away at the elbow.

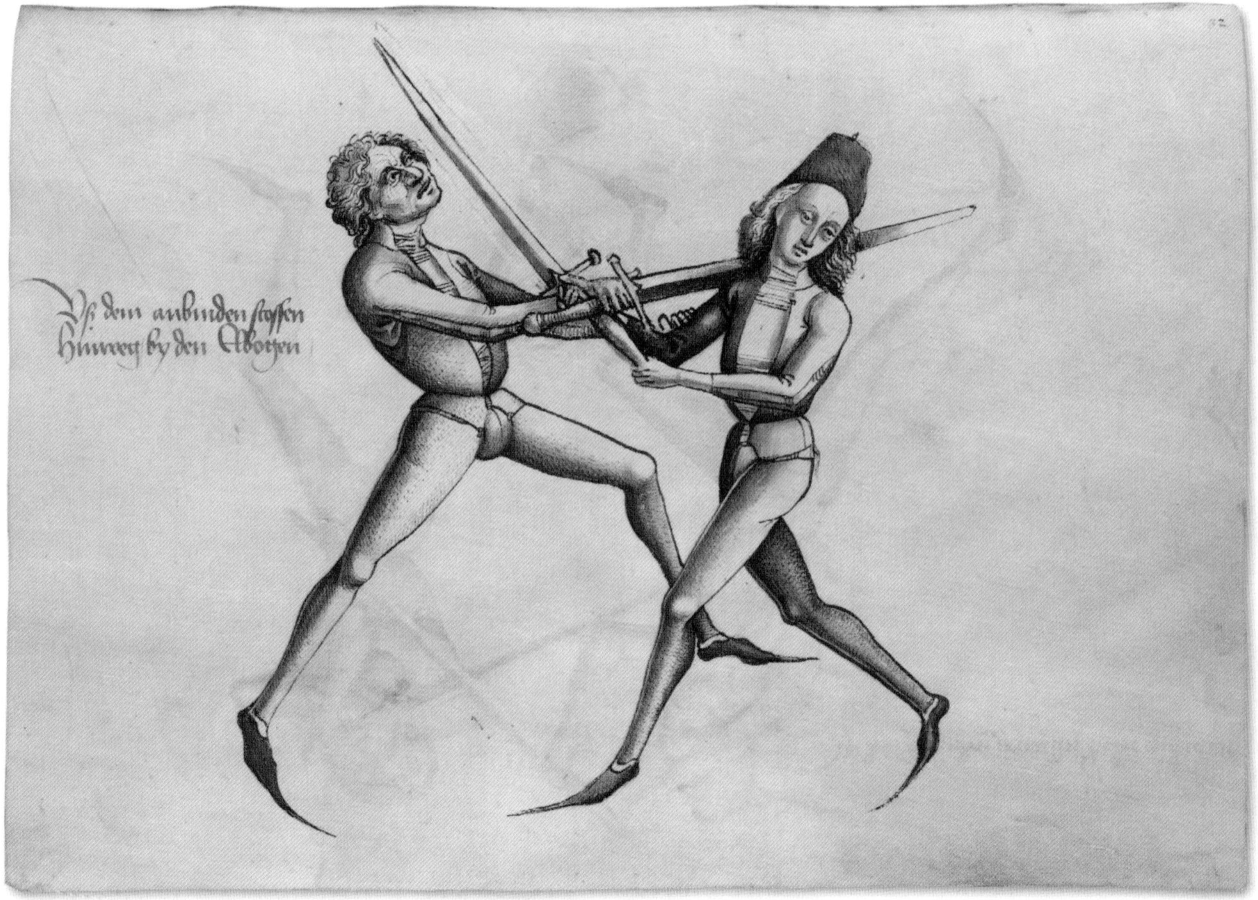

The *Mordschlag*.

A throw from the *Ortschlag*. [Apparently, this should have been the previous *Mordschlag*.]

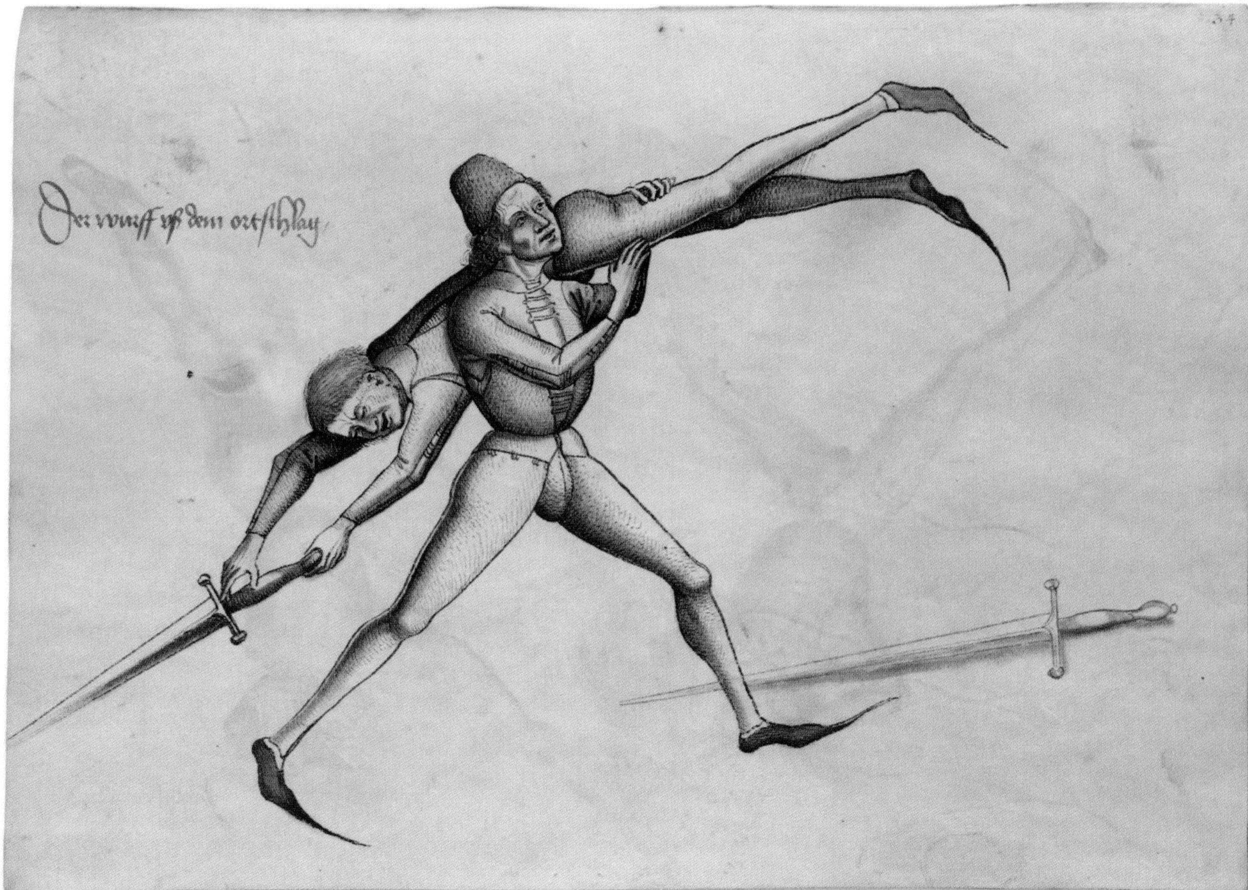

Falling into an armoured thrust from the free strike.

Falling into an armoured thrust from the *Schielhau*.

From the *Donnerschlag*: a throw.

From the *Donnerschlag*: a wrenching [at the sword].

After the wrenching:
the hit to the face.

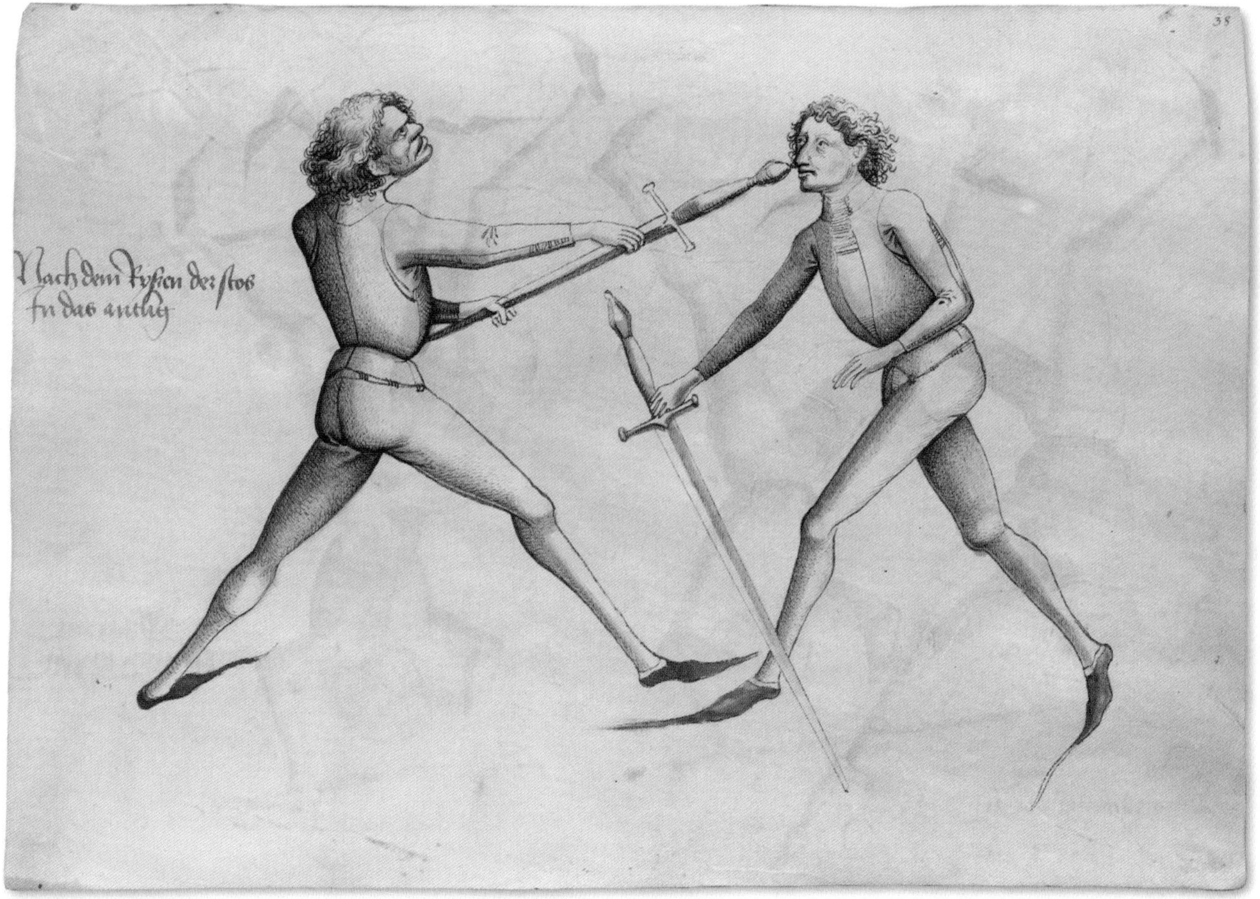

The *Brentschüren* …

… or the binding with the short sword.

From the binding with the short sword.

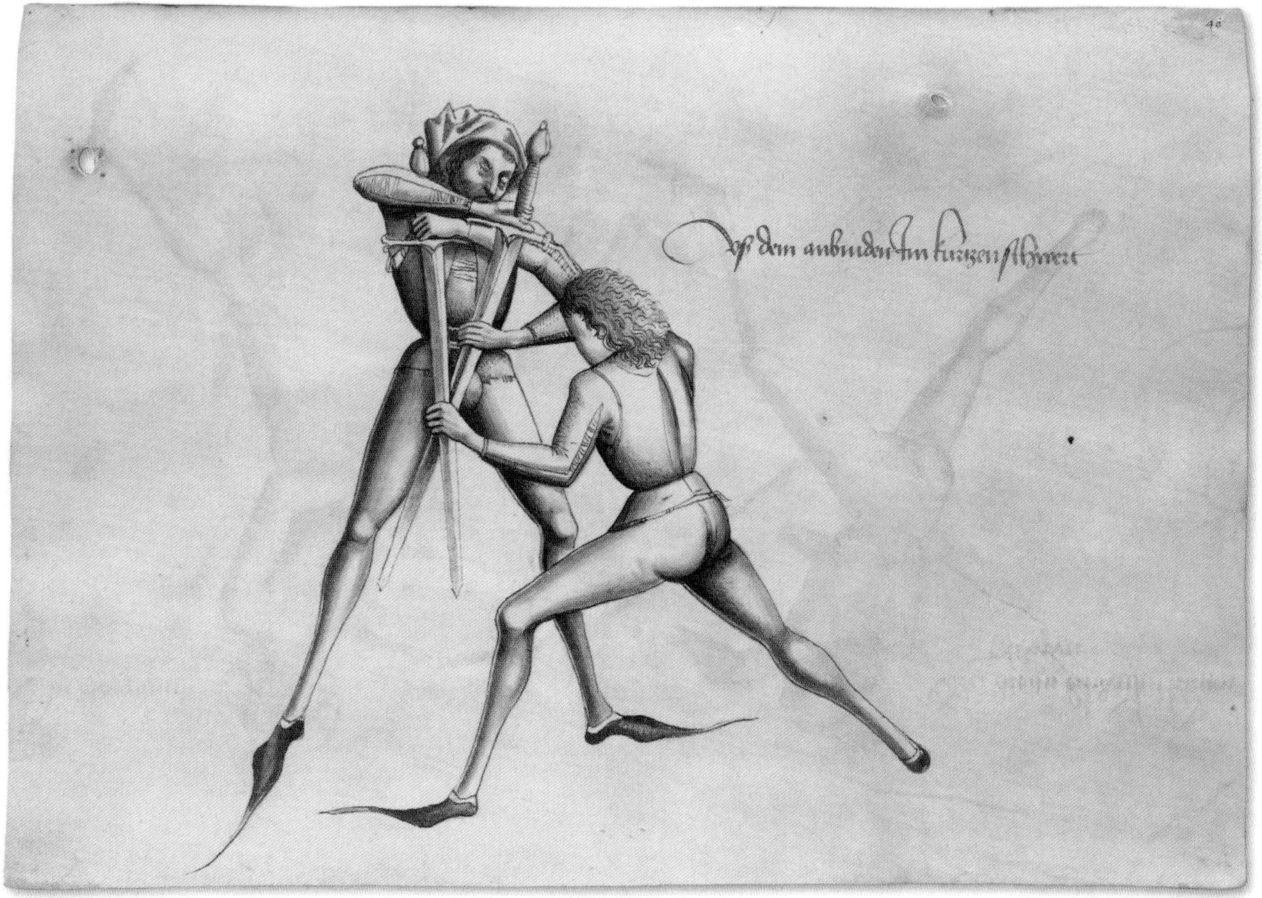

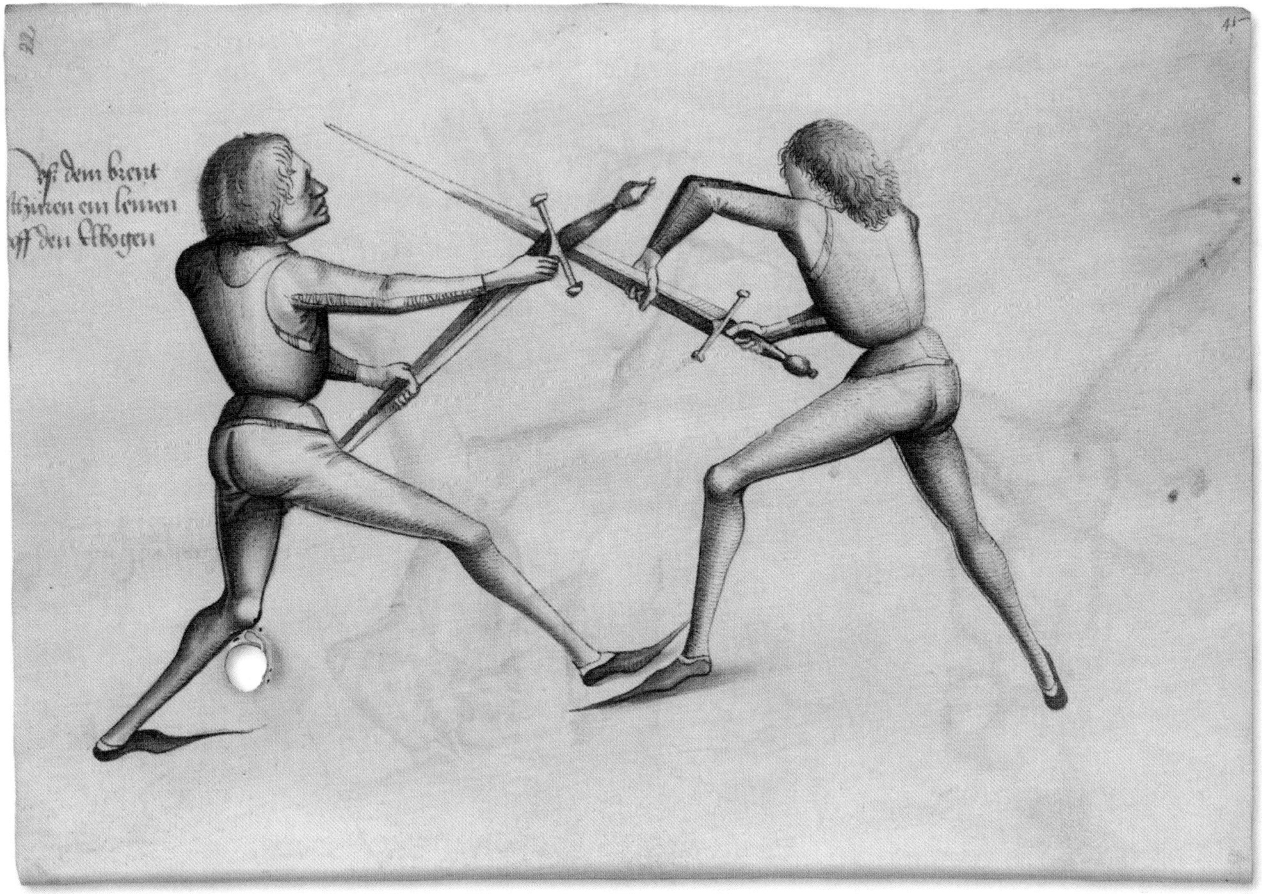

From the *Brentschüren*: paralysing the elbow.

A crisis stance against the thrust from above and below.

The work from the pressing.

Clinging to the gauntlet.

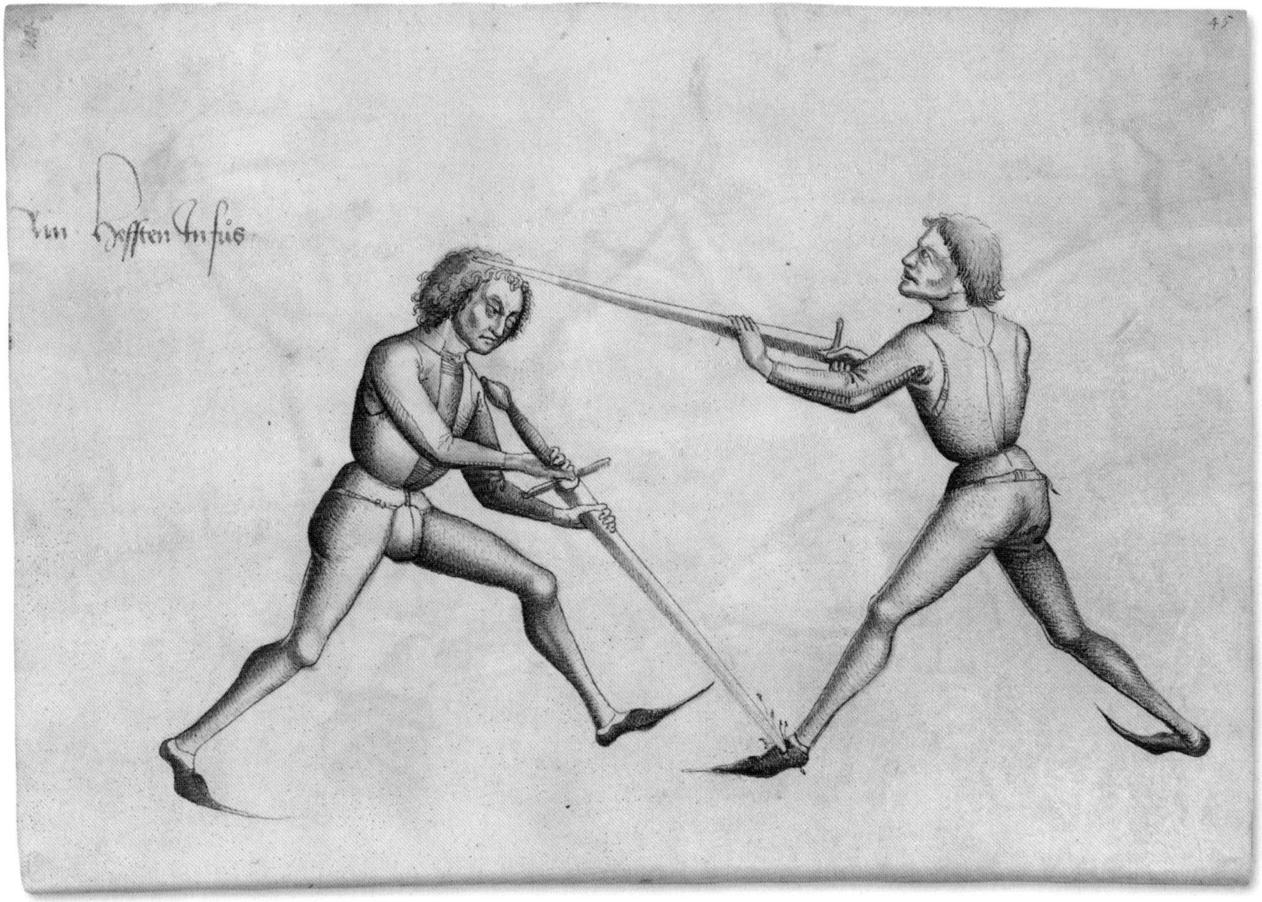

Clinging to the foot.

From the binding: wrenching the hollow of the knee.

Feinting a *Mord-schlag* and striking into the leg.

From the binding: wrenching at the neck.

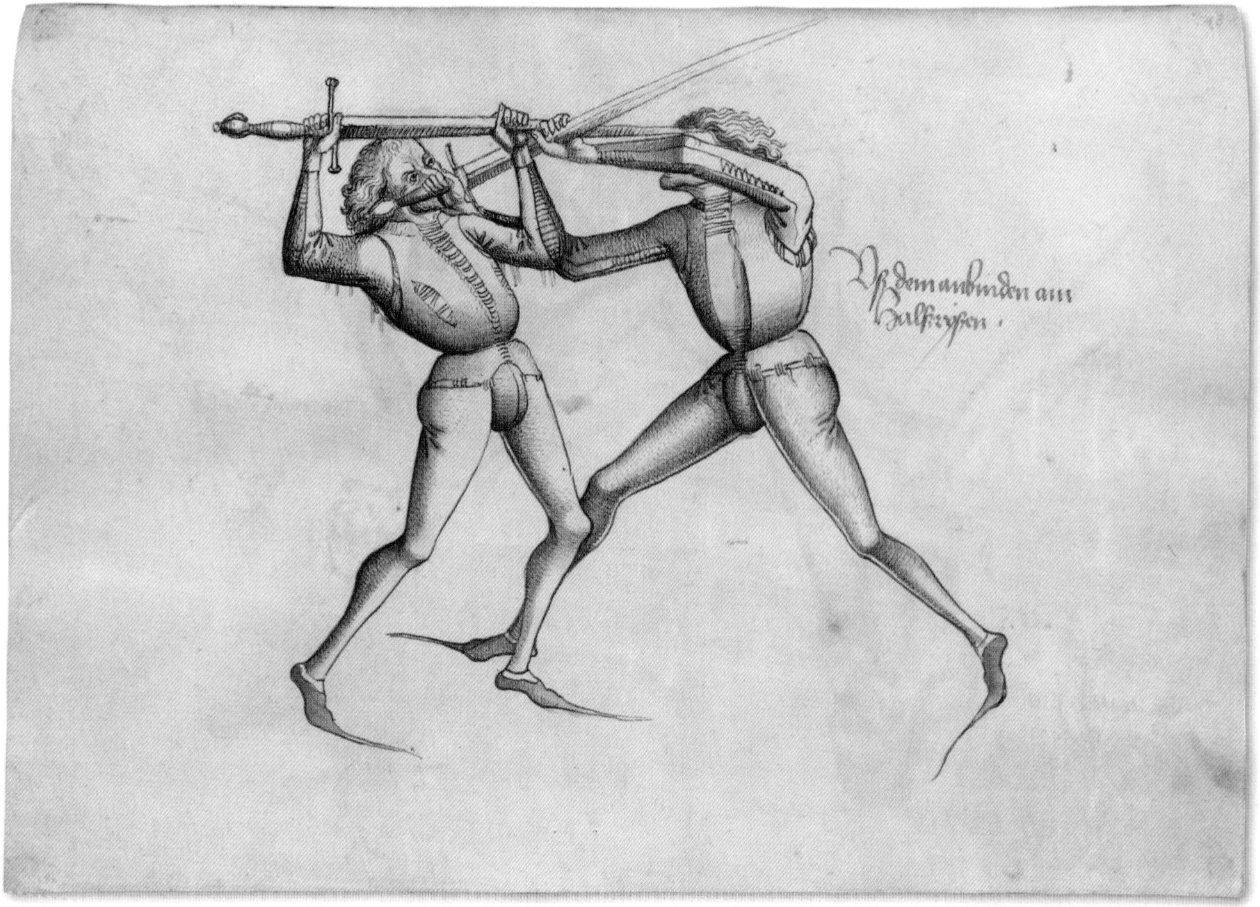

Here, they have seized the swords.

Taking the sword.

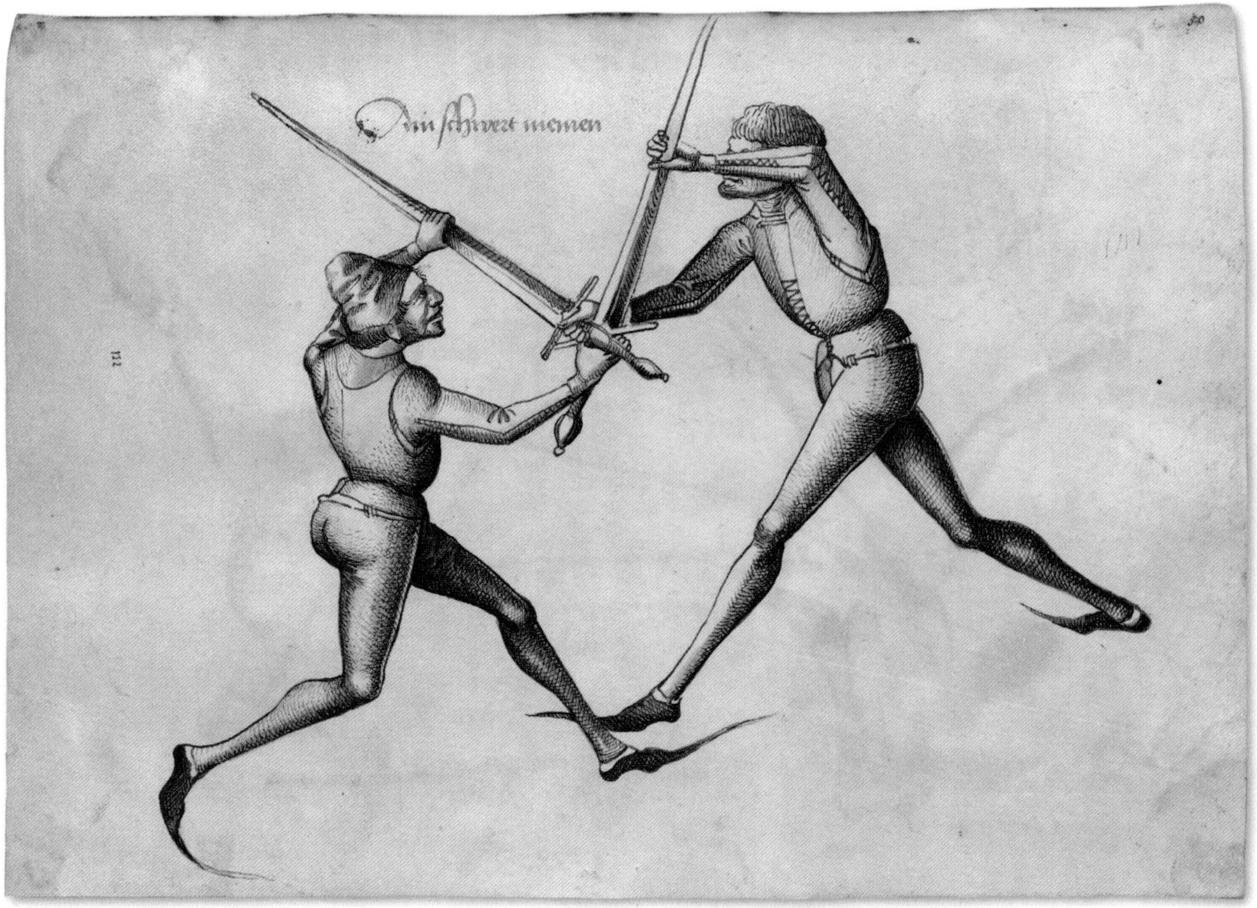

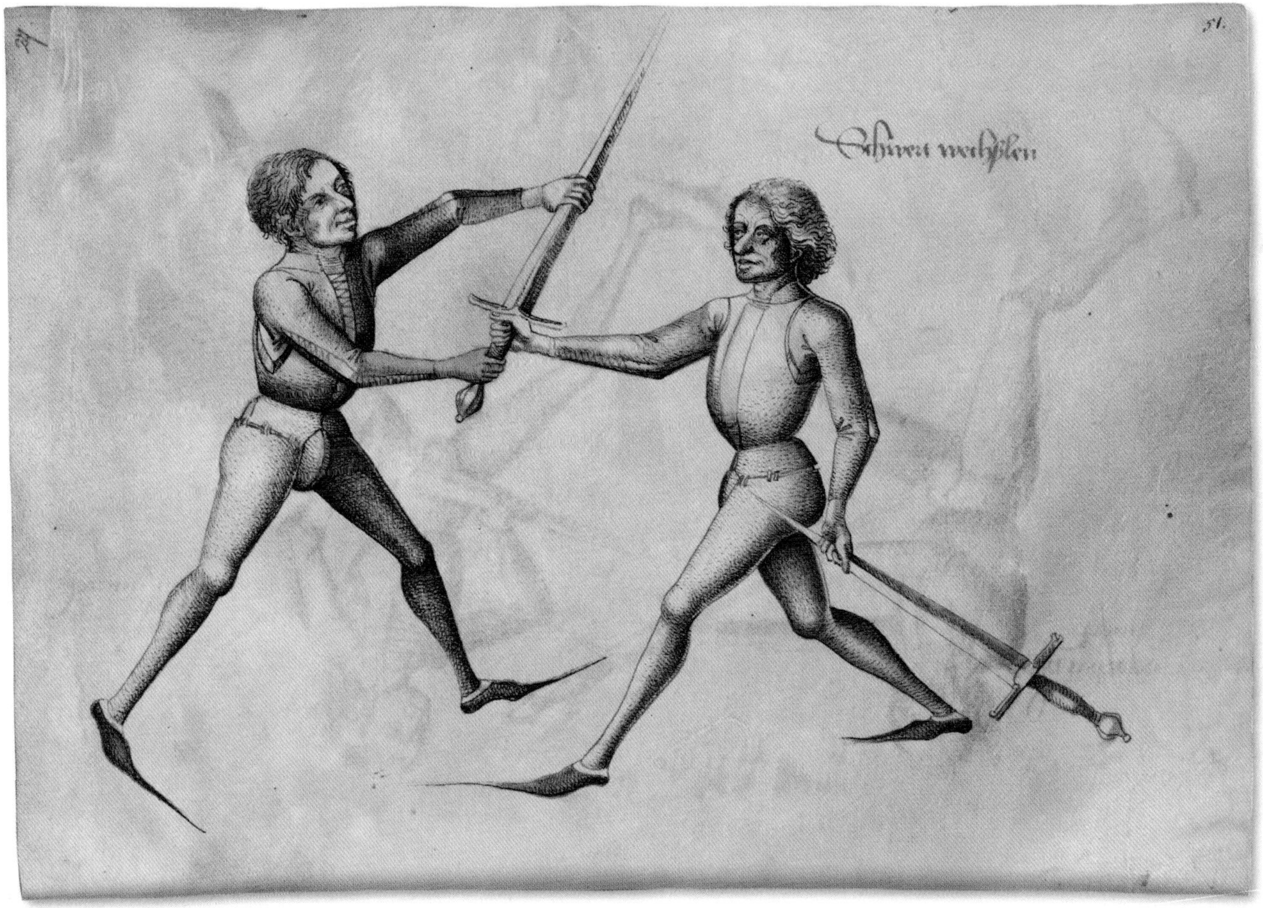

Changing the sword.

Another attempt.

—

This one strives to strike and wrench at the opponent.

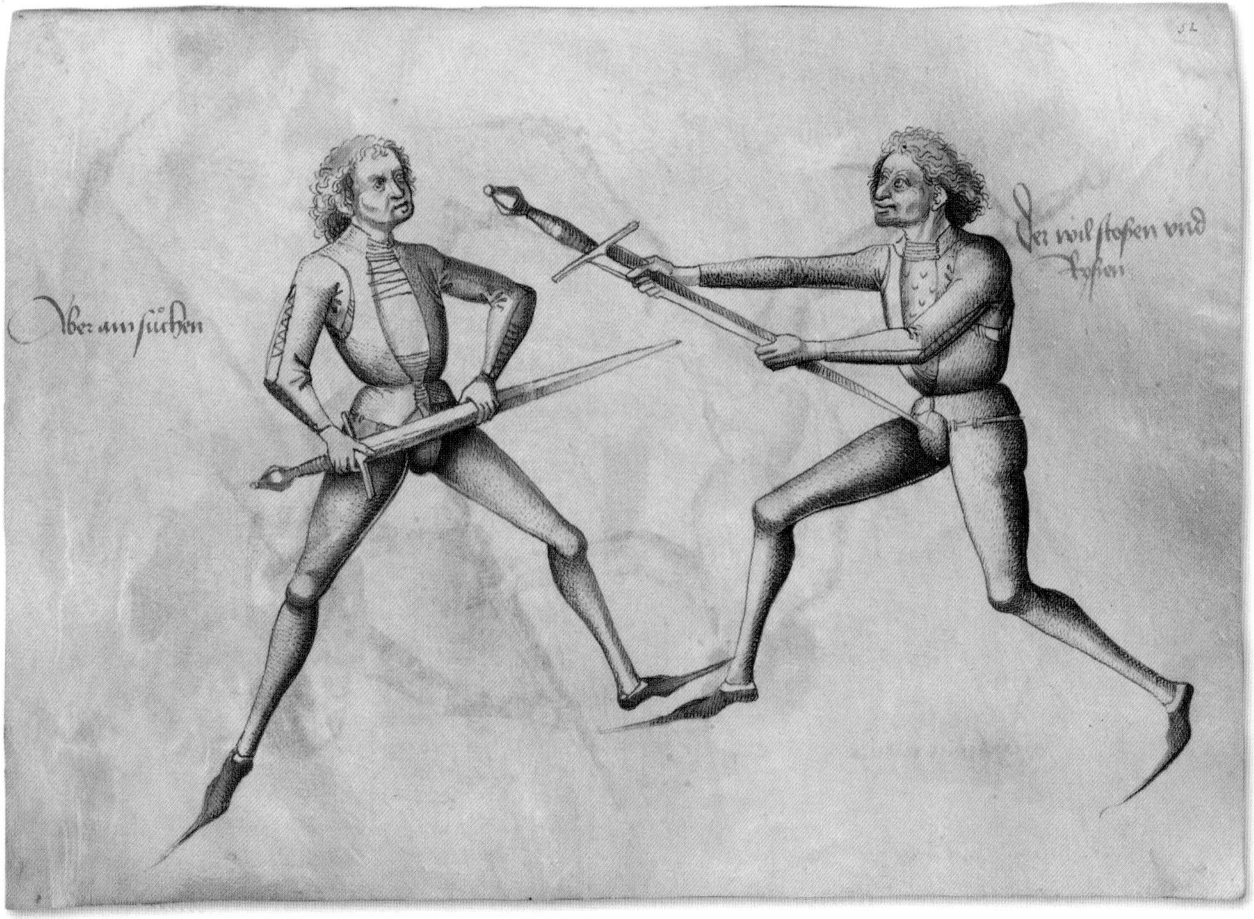

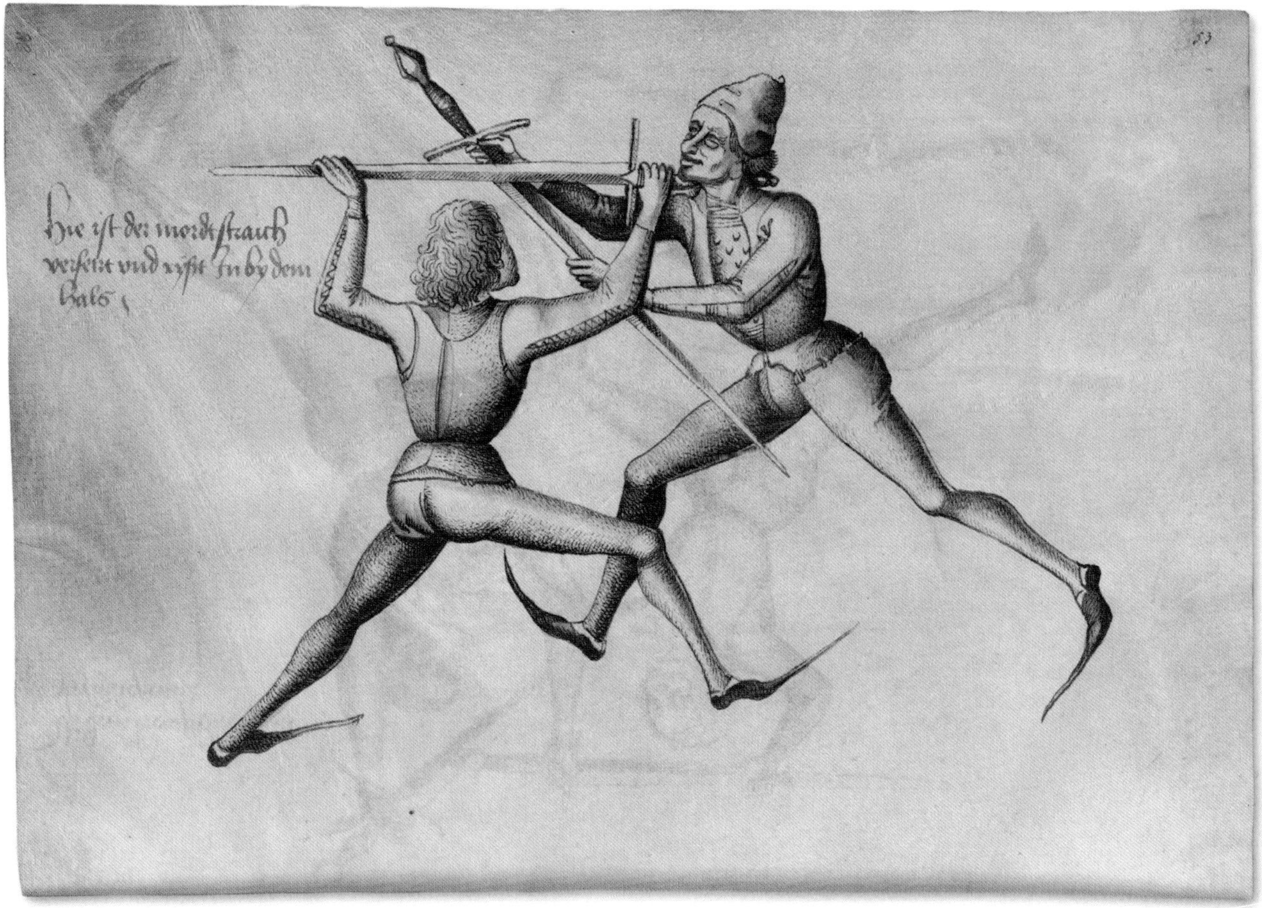

Here, the *Mordschlag* is displaced, and this one wrenches the other's neck.

From the *Brent-schüren* of the swords: enclosing.

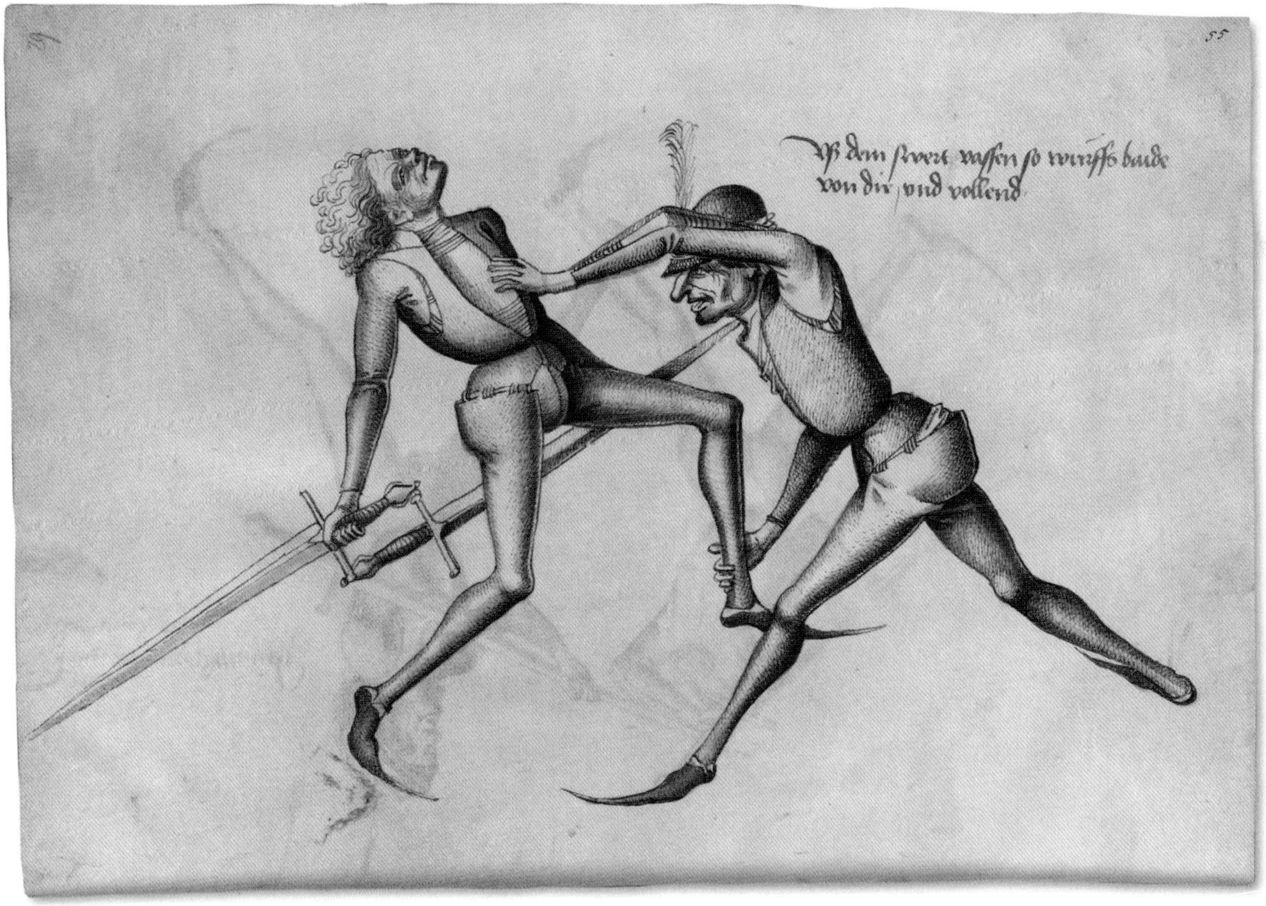

From grasping the sword: throw both away from yourself and bring it to an end.

From the *Mord-schlag*: turning around.

The break against turning around.

From the *Mordschlag*: running away from him, striking, hitting again and wrenching at him.

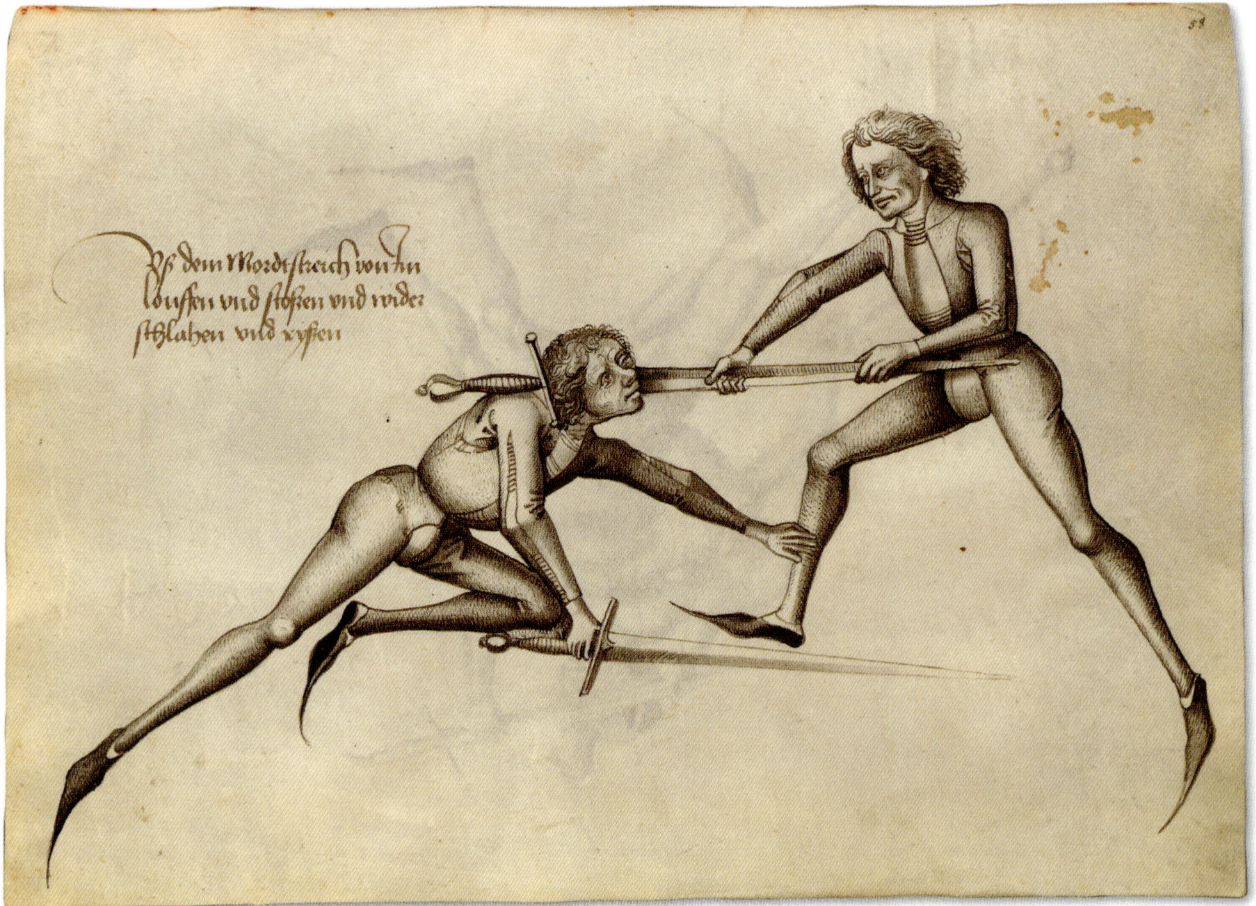

From grasping the sword: slipping through.

From the *Brent-schüren*: enclosing and throwing.

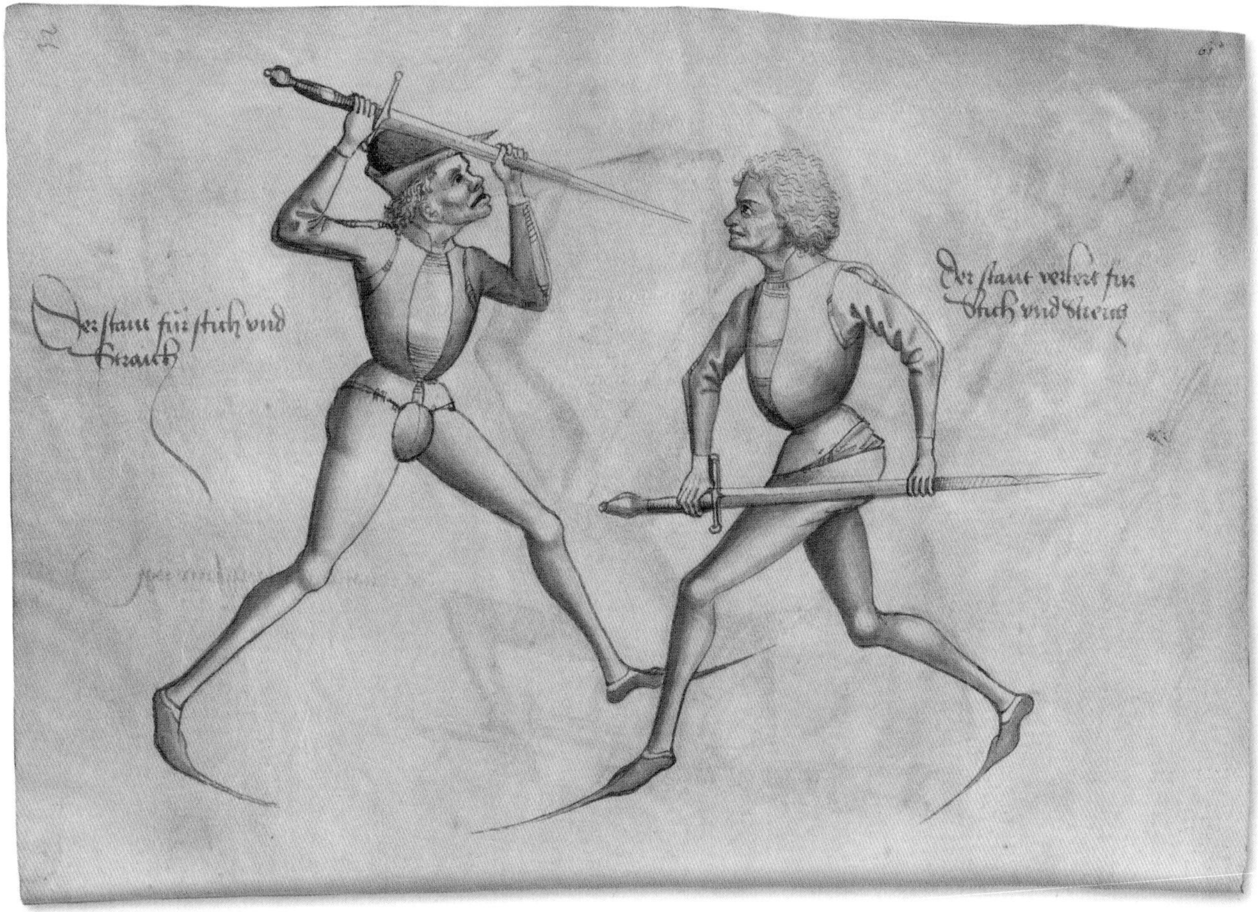

This one stands by to thrust and strike.

This one stands reversed in order to thrust and strike.

Another taking of the sword.

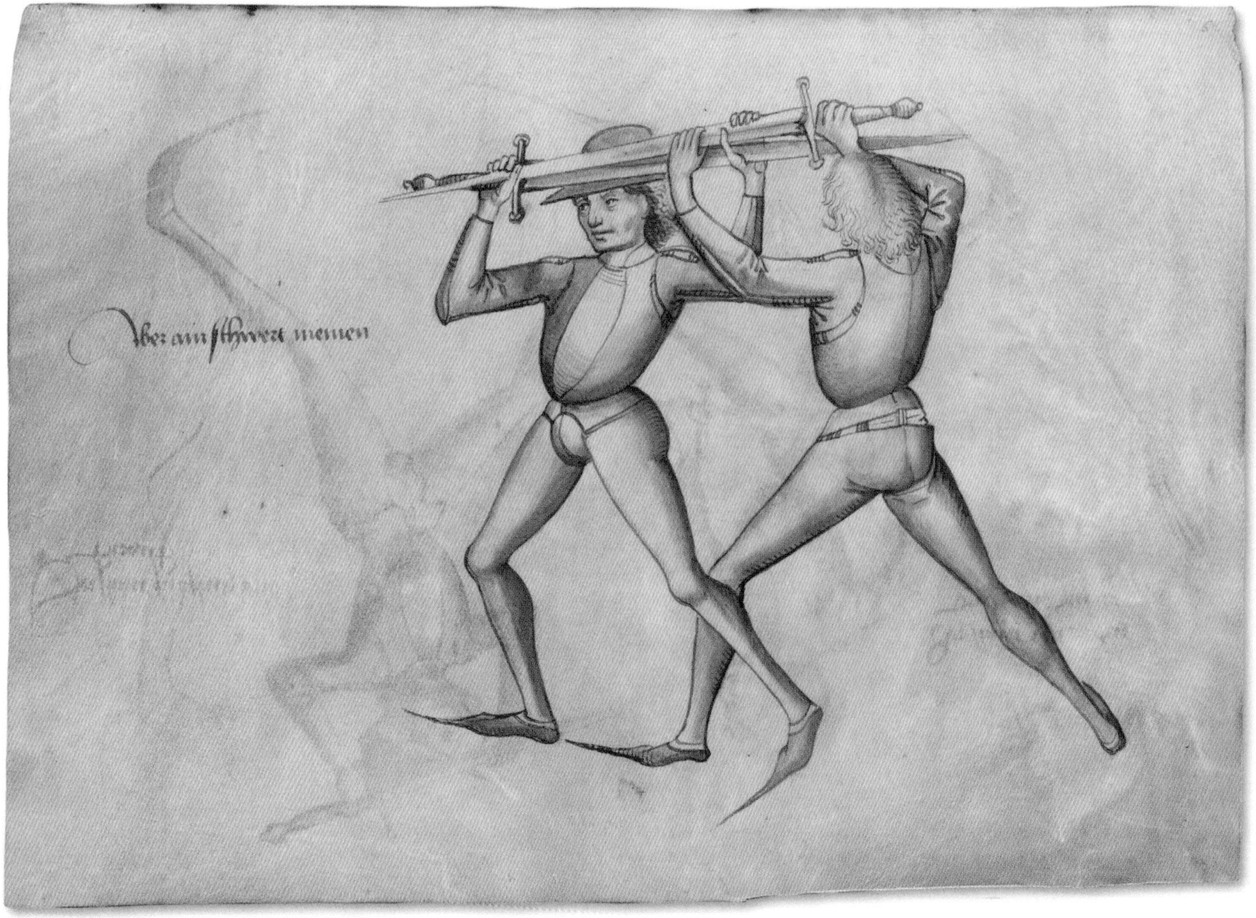

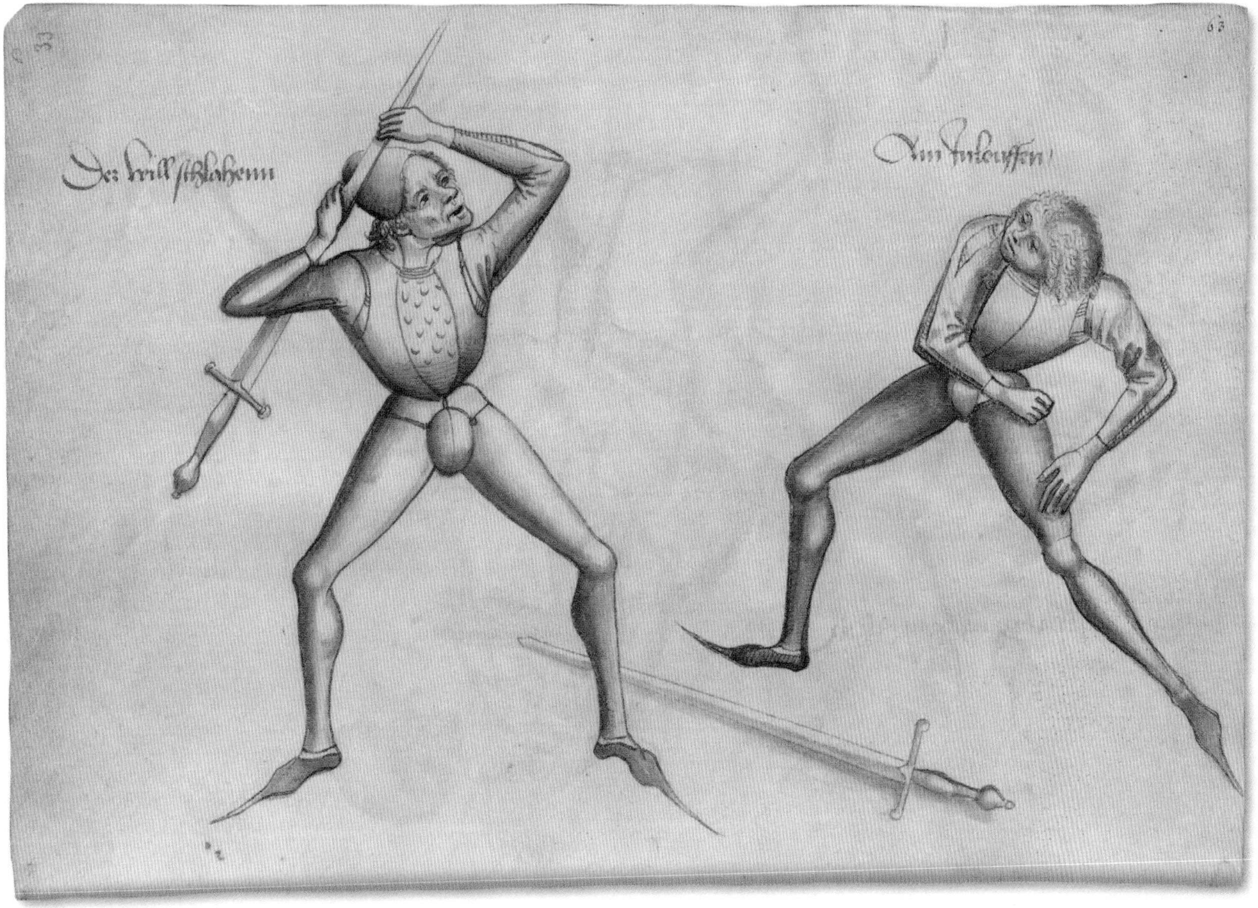

This one strives to strike.

Running in.

From running in: the throw.

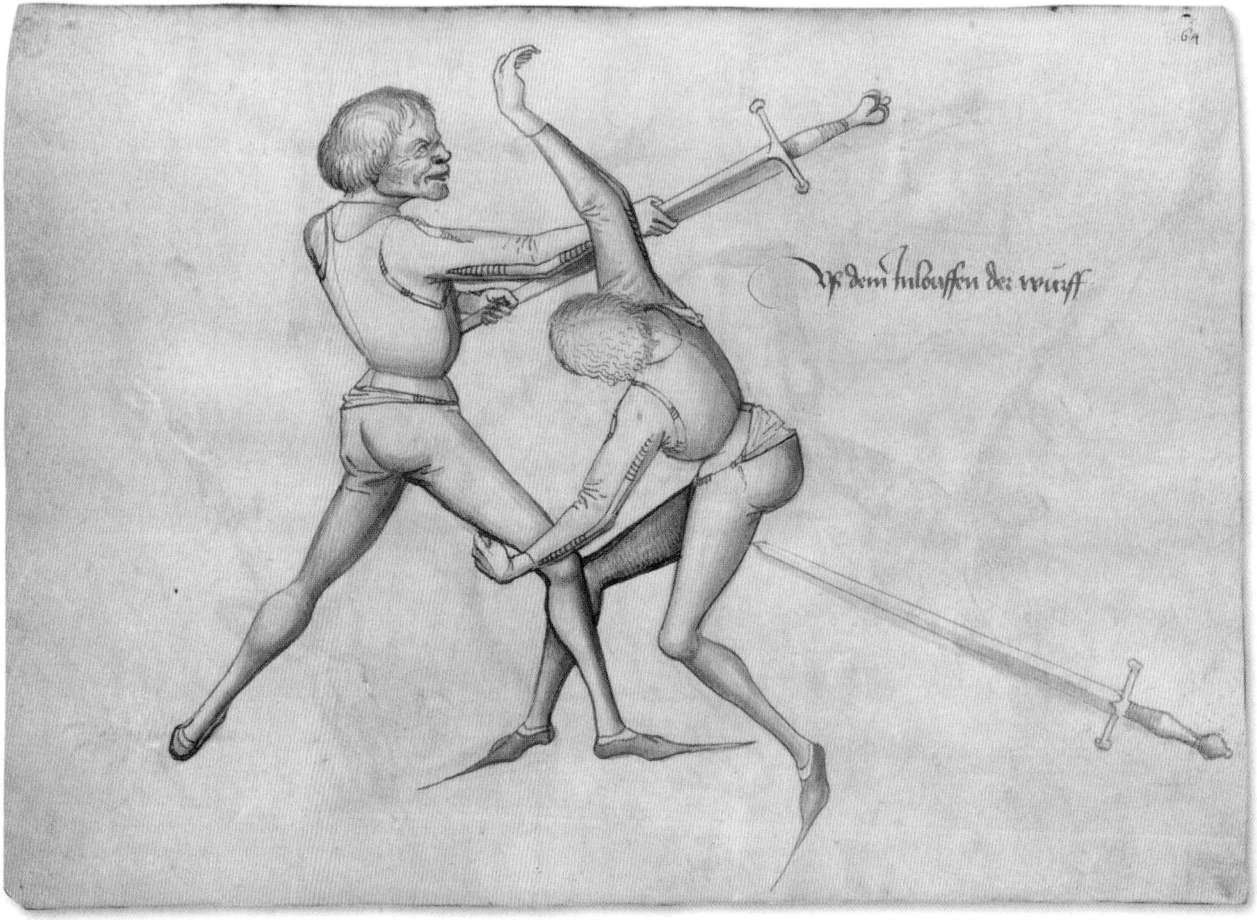

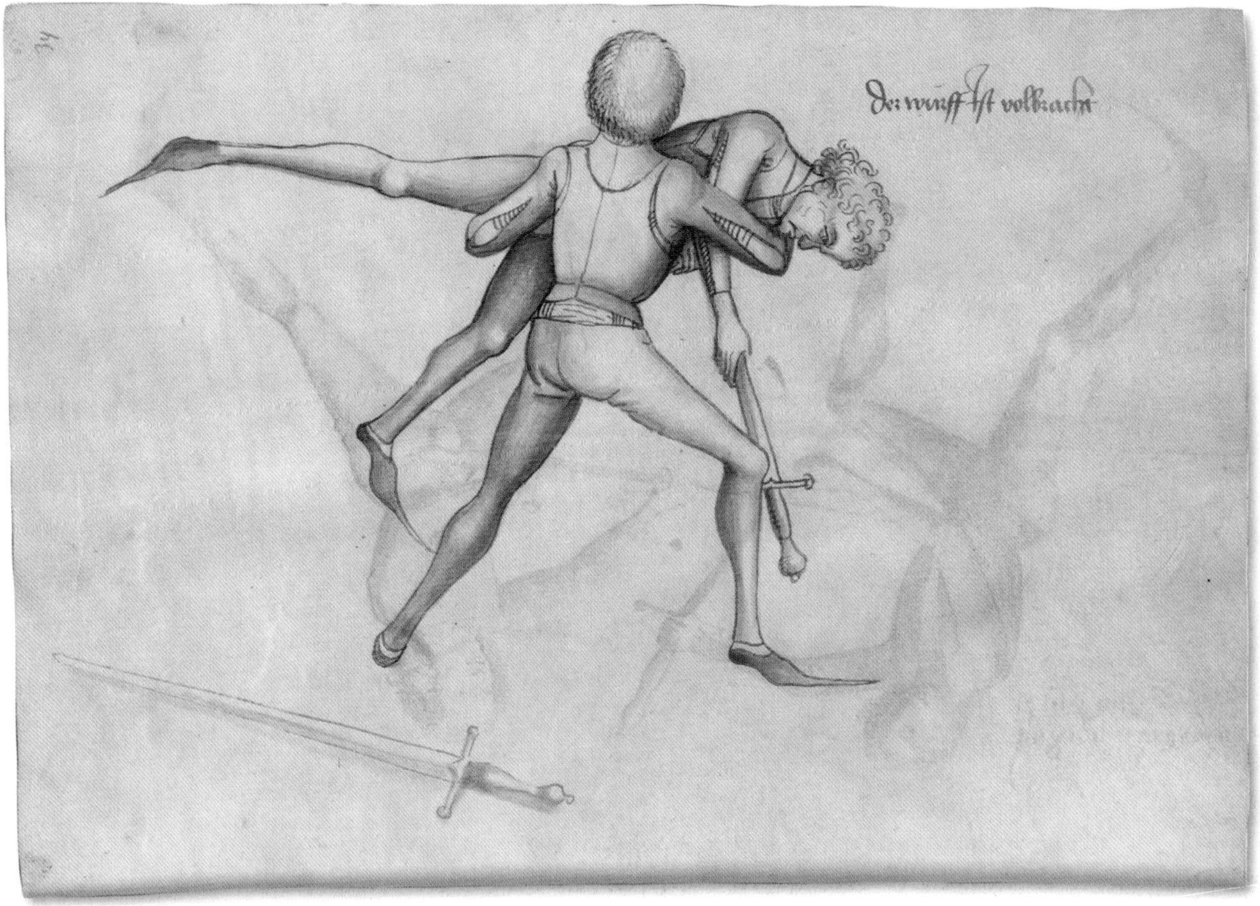

The throw is accomplished.

Against the thrust and wrenching the arm.

Hitting the face while stepping forward.

Here, he goes into the lists.

That one carries in his things.

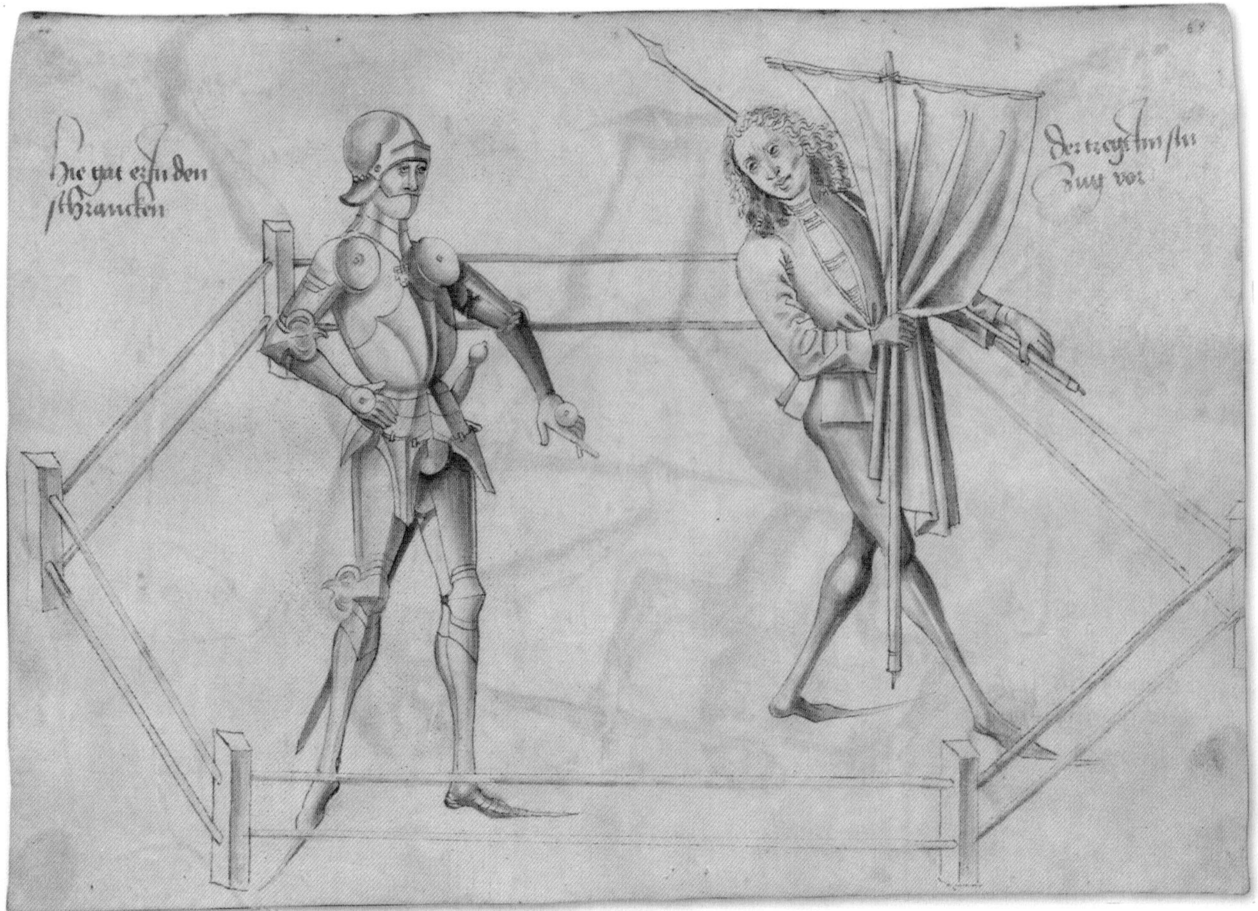

Here, both are in the lists and await the beginning. Each one has his coffin behind and his *Grieswart* in front of him.

Here, they have begun. This one stands displaced to defend against a throw.

That one strives to throw the spear.

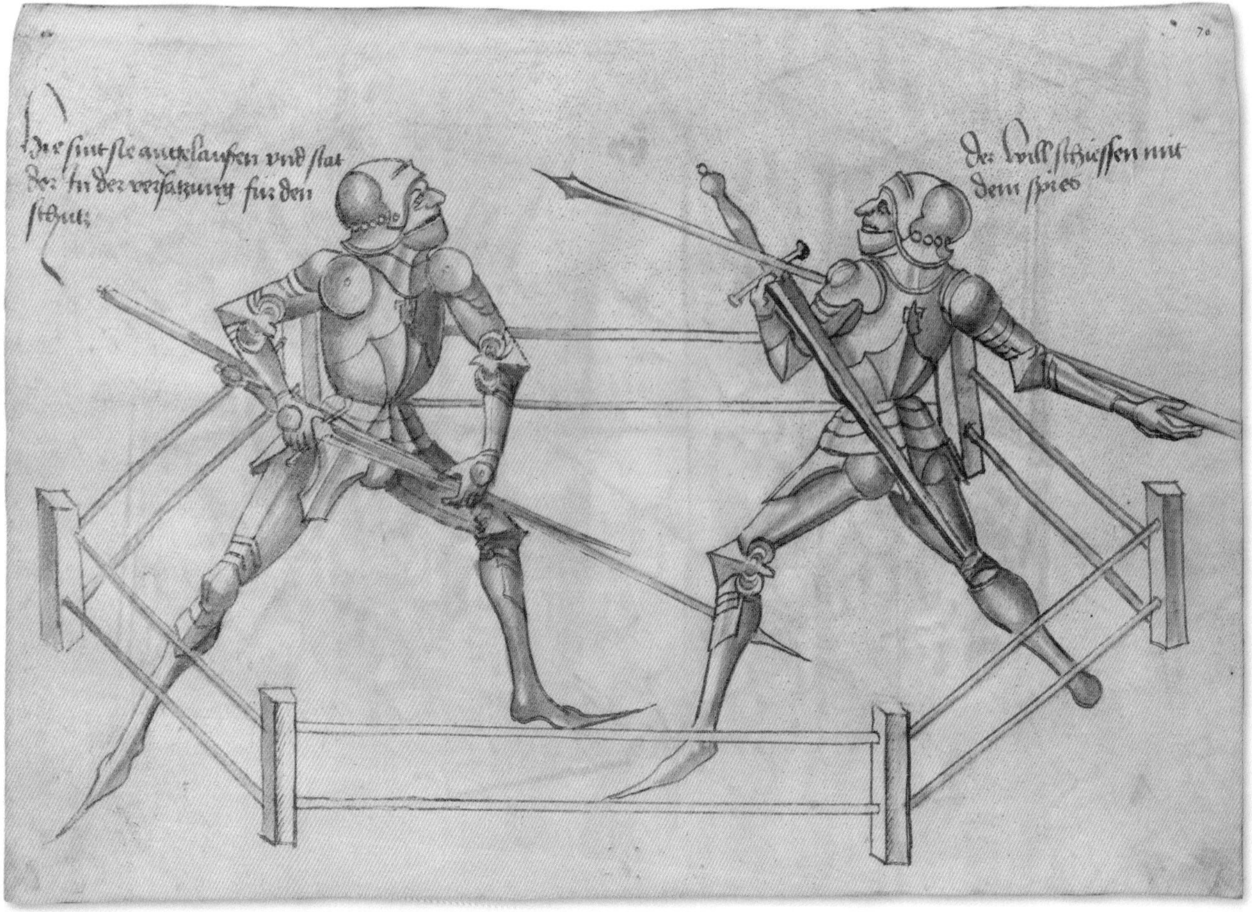

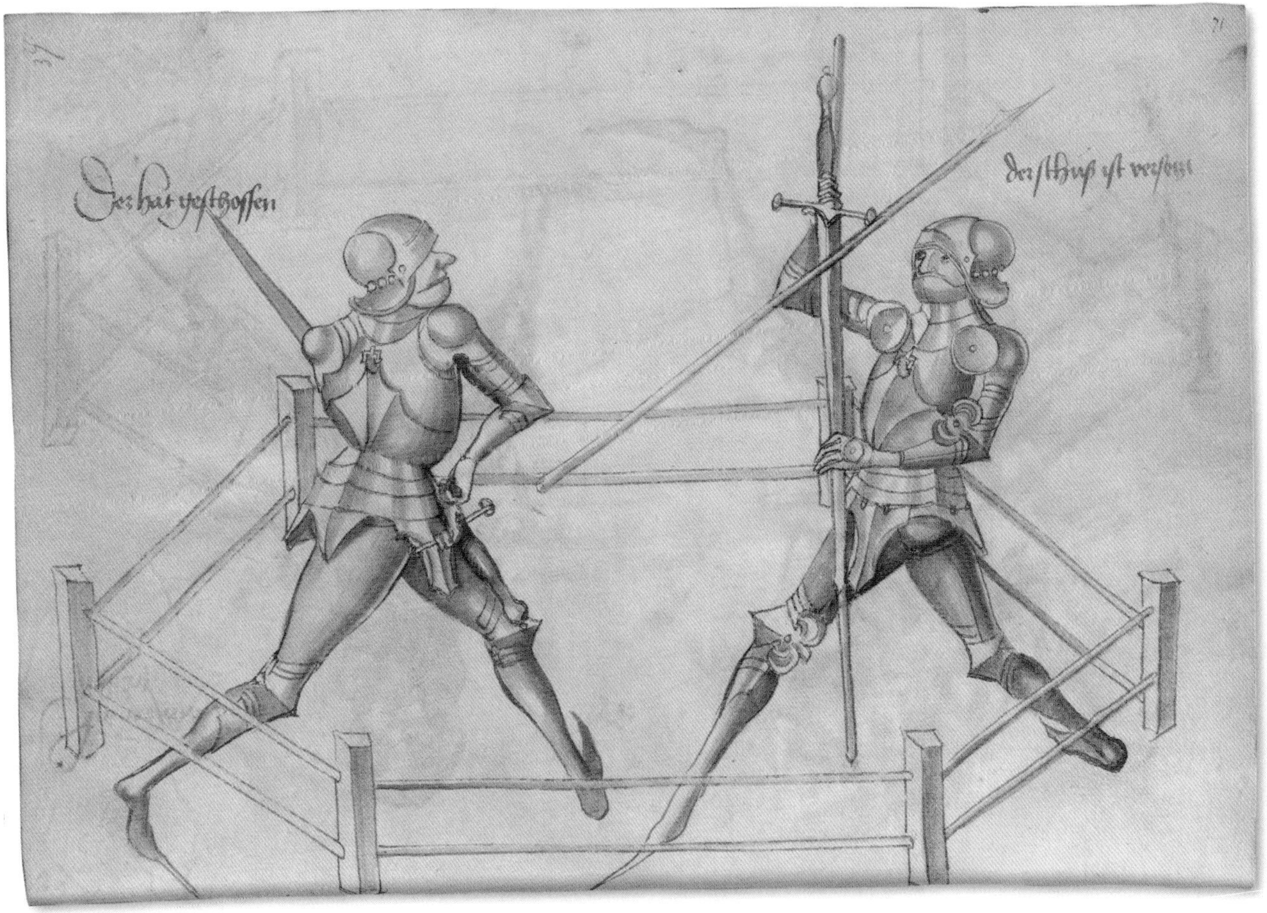

This one has thrown.

The throw has been displaced.

The displacement against the strike.

Running forward with the *Mordschlag*.

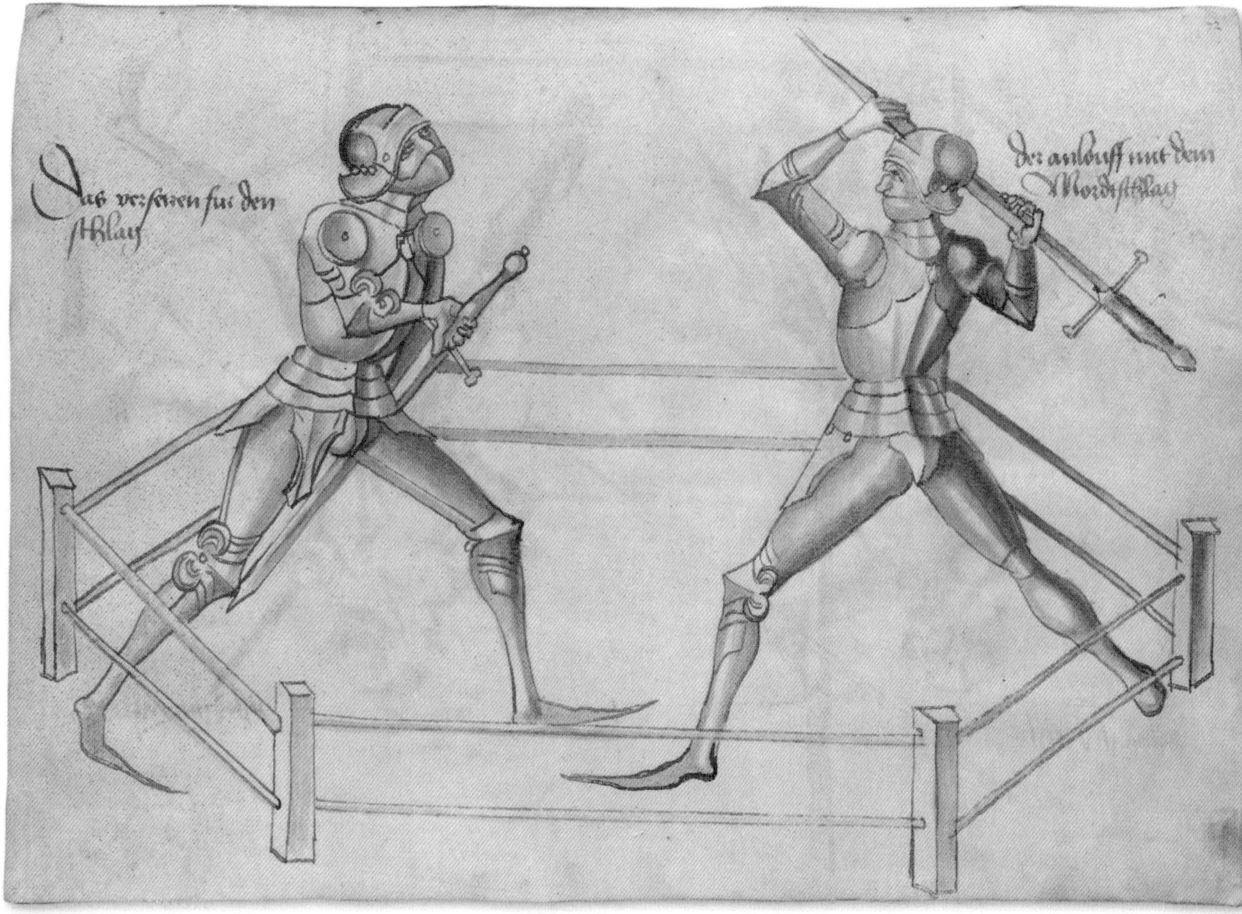

Pushing away from the displacement.

This one has accomplished the strike.

Stepping forward and throwing over the leg.

Break against a free strike: displace armed (with armoured hand), grasp over the opponent and fall onto the third foot.

Piece and break.

The thrust of doubt.

The correct posture when beset.

This one strives to work.

That one has the advantage and is unarmed.

Here, they wrestle for the sword.

Here, one has struck; the other one displaces him and makes an end.

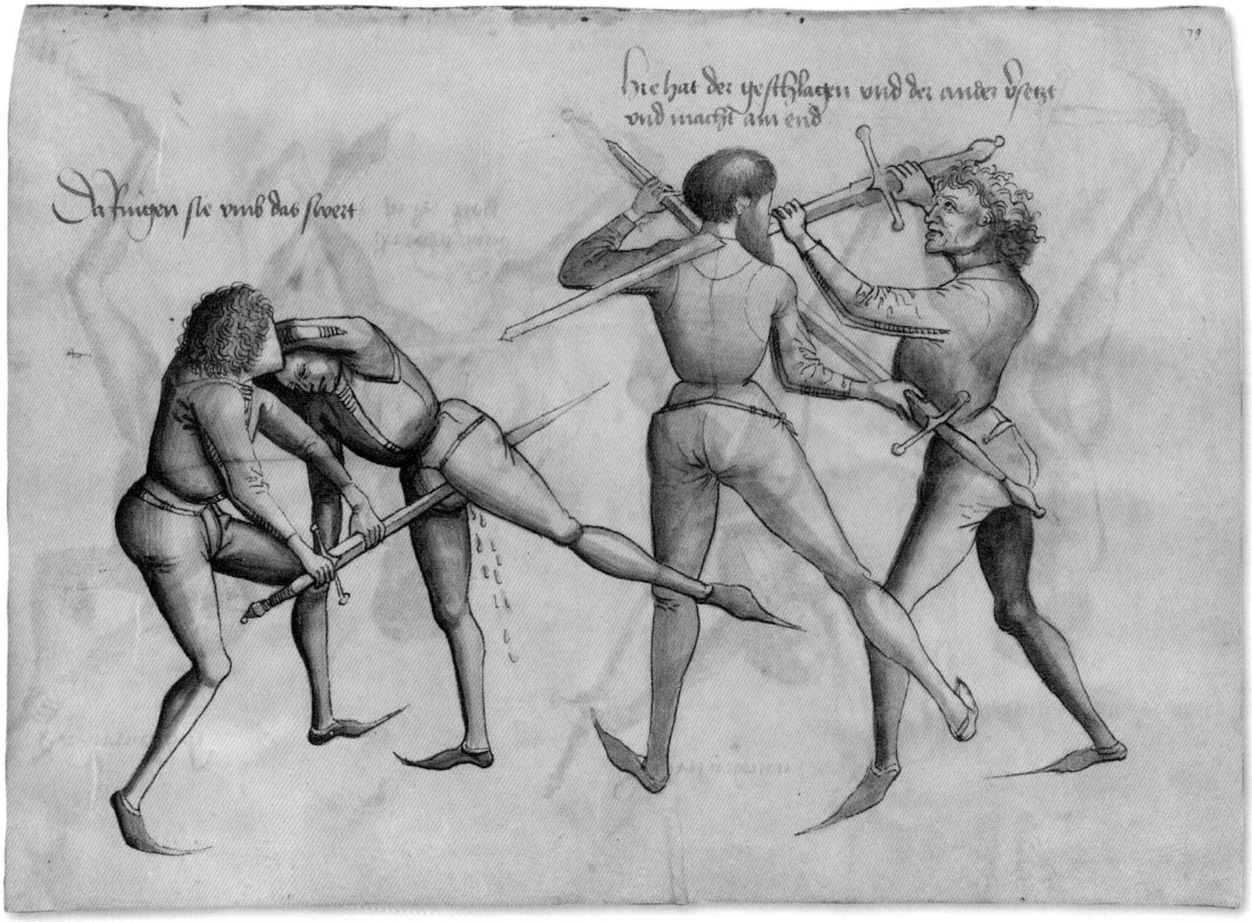

The first binding with the axe.

From the binding,
he has made
his opponent
defenceless.

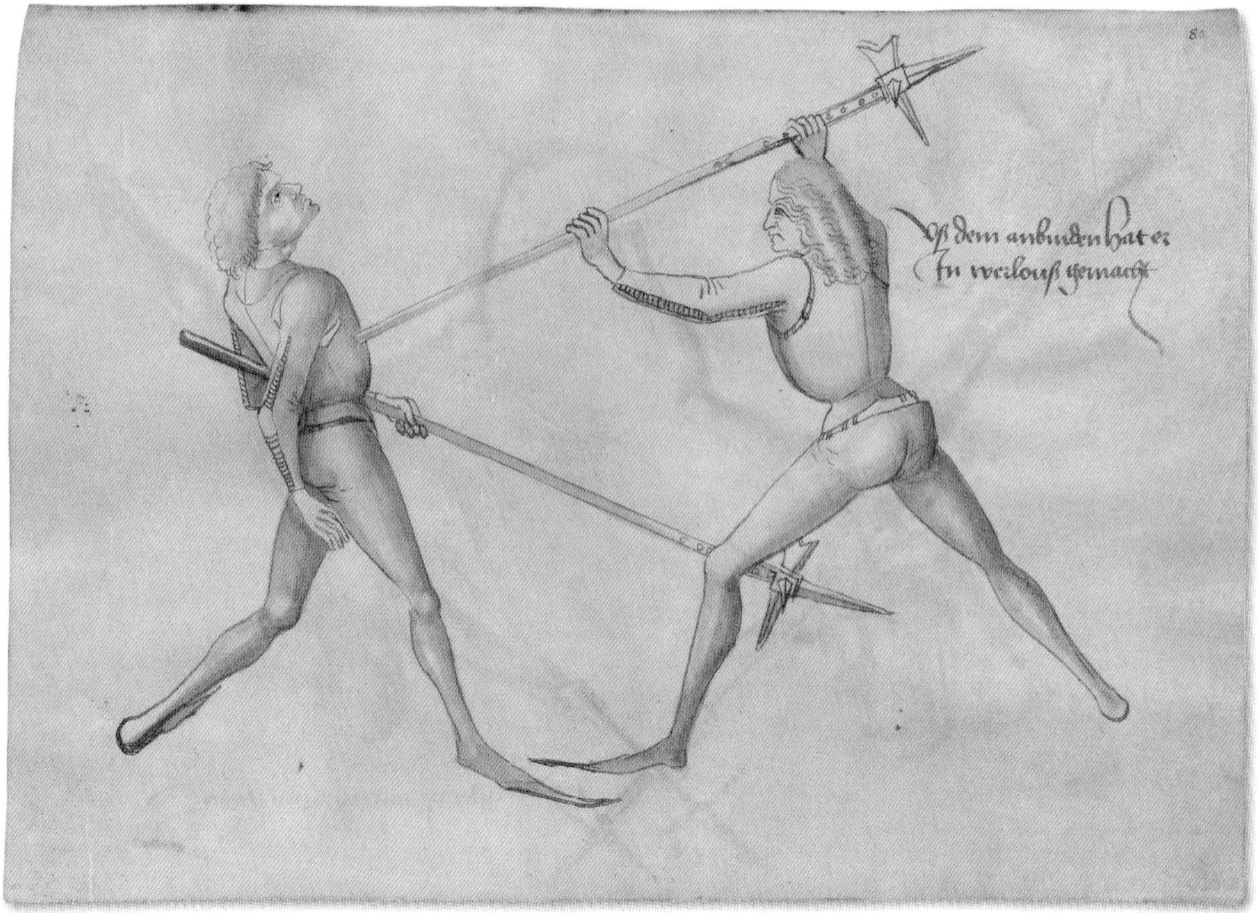

This one strives to strike freely.

That one strives to bind behind, make him defenceless and push away his strike.

Here, he has deflected the strike, and the piece is accomplished.

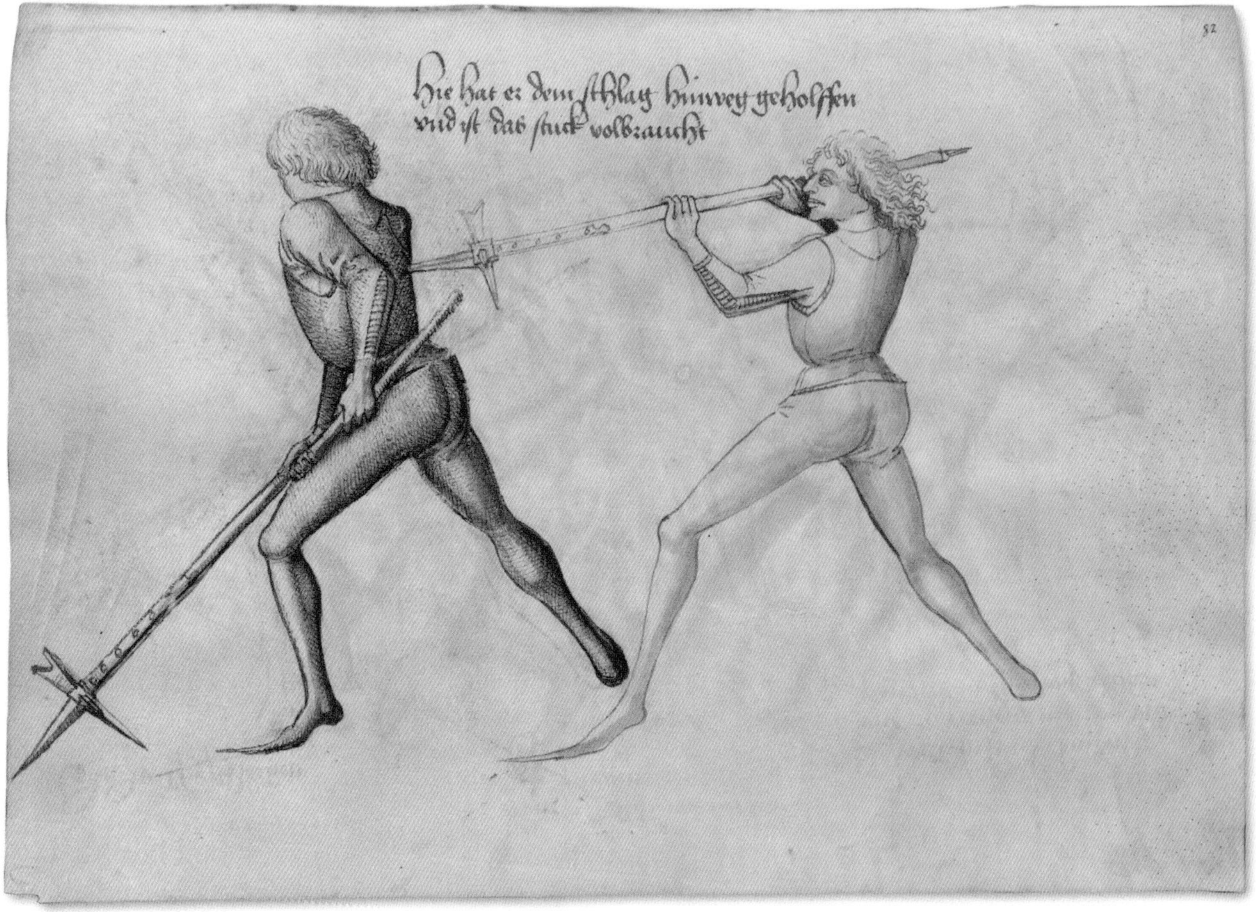

Displacing the feinted strike and wrenching after this.

From the binding as soon as both have struck *vom Tag*: one grasps the other one around the neck and wrenches at him.

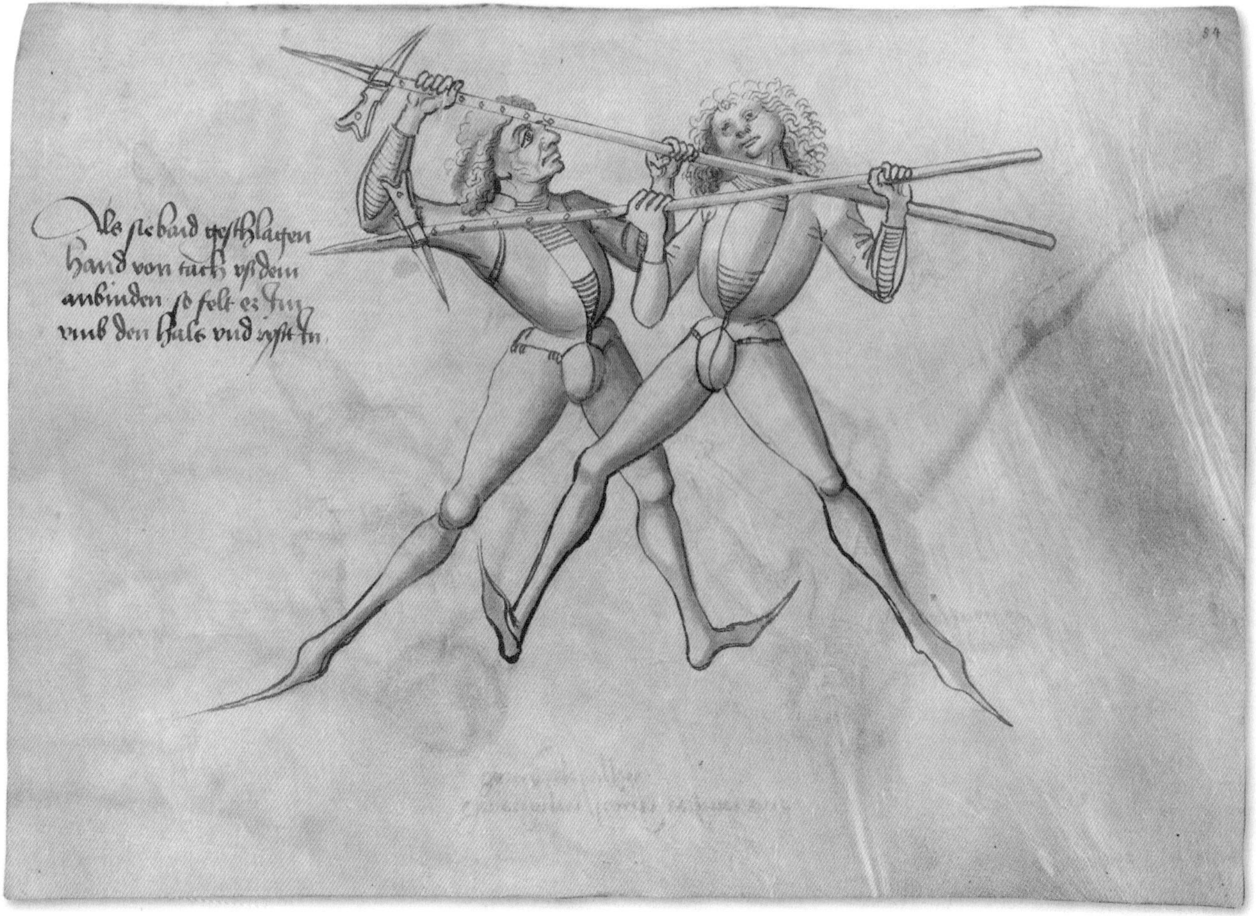

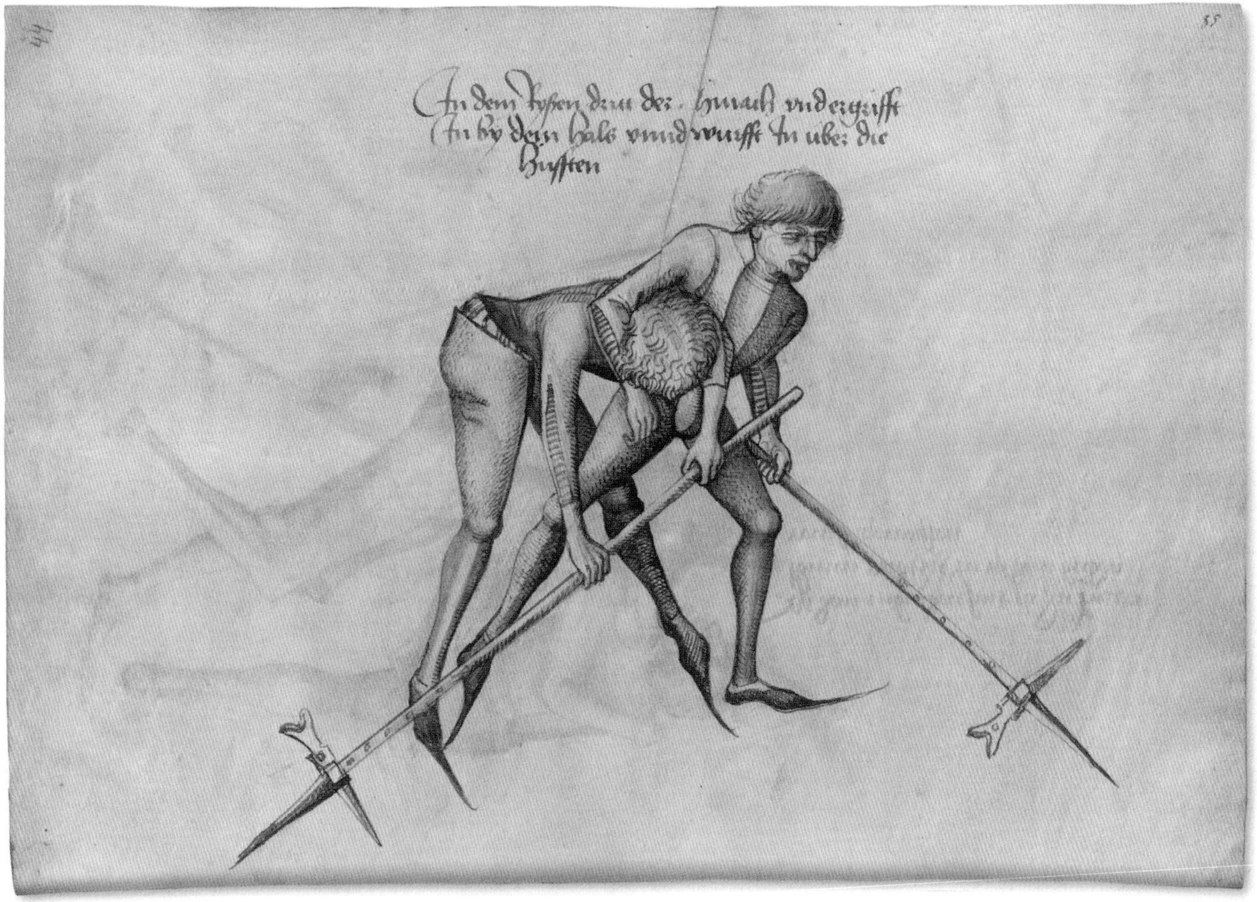

While wrenching at the opponent, one steps further, seizes the other one's neck and throws him over the hip.

From the binding: as soon as they have come to wrestling, one has thrown the other with the *Bubenwurf*.

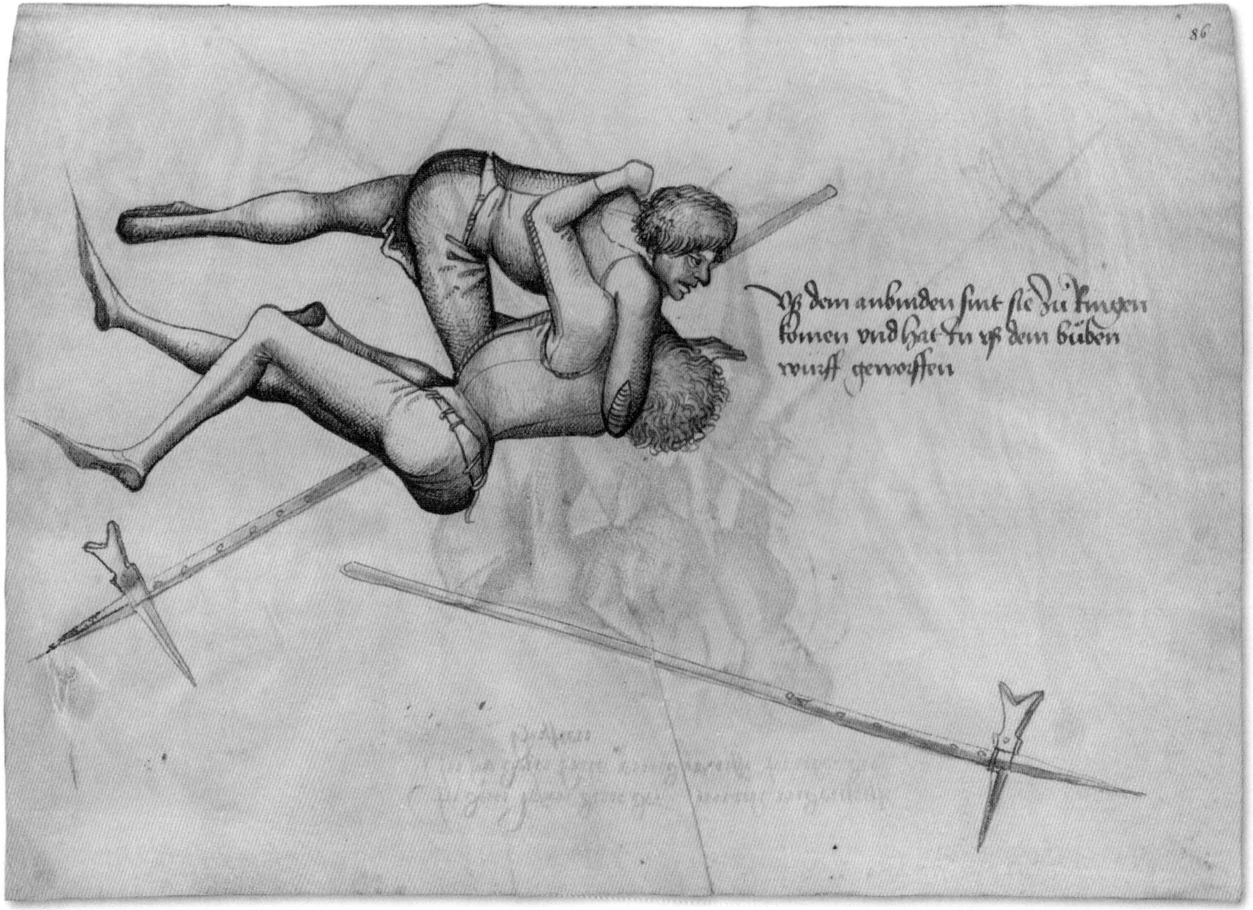

From the free strike and the bind: one has grasped around the other and strives to throw him.

This one has struck.

Here, one displaces with the shaft and the other strives to work.

Here, he has stepped forward and the piece is done.

One has struck the other in the feet.

One has displaced the strike and strives to continue working.

From the displacement: one strikes at the other one's neck and strives to wrench at him.

Here, one has displaced the thrust, strikes the other around the neck and strives to wrench.

He has thrust.

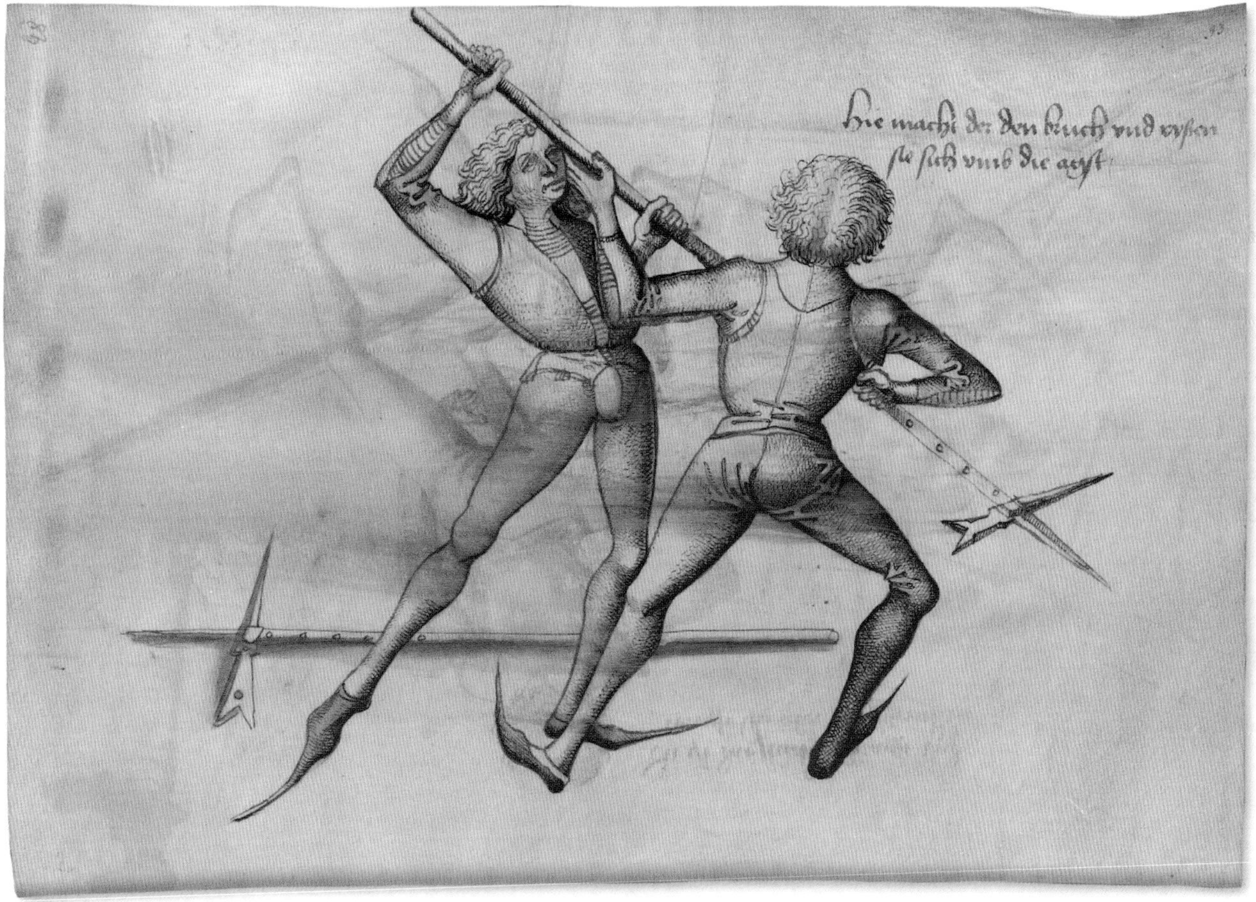

Here, one performs the break, and they quarrel for the axe.

Here, the piece is finished, and one throws the other over his leg.

From the binding: one falls behind the other one with his axe behind his elbow and shoves him away.

Here, he has let himself turn around completely and makes the break against the previous piece.

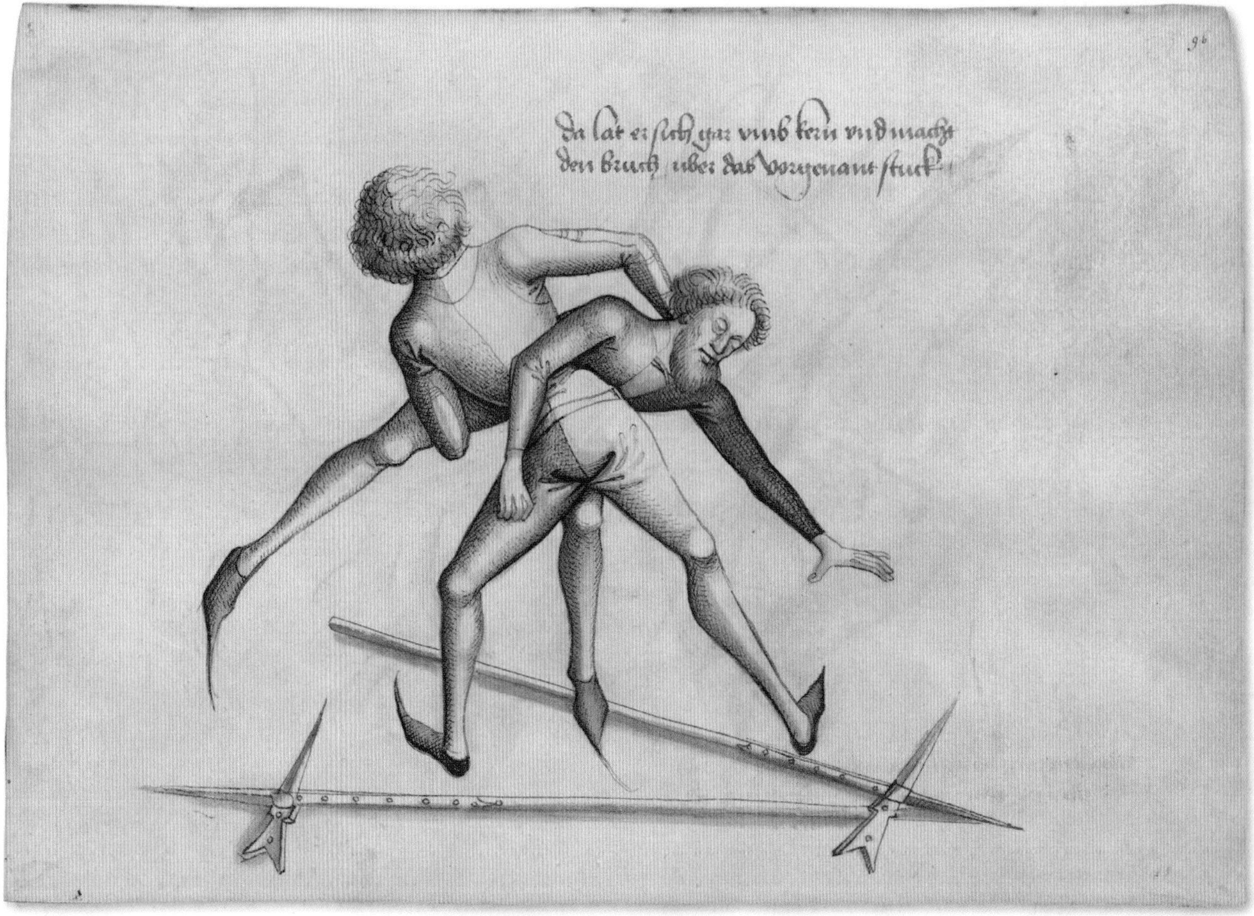

Another beginning.

Here, one strives to oppress the other with thrusts and strikes.

Here, one has displaced the thrust and strives to wrench at his opponent.

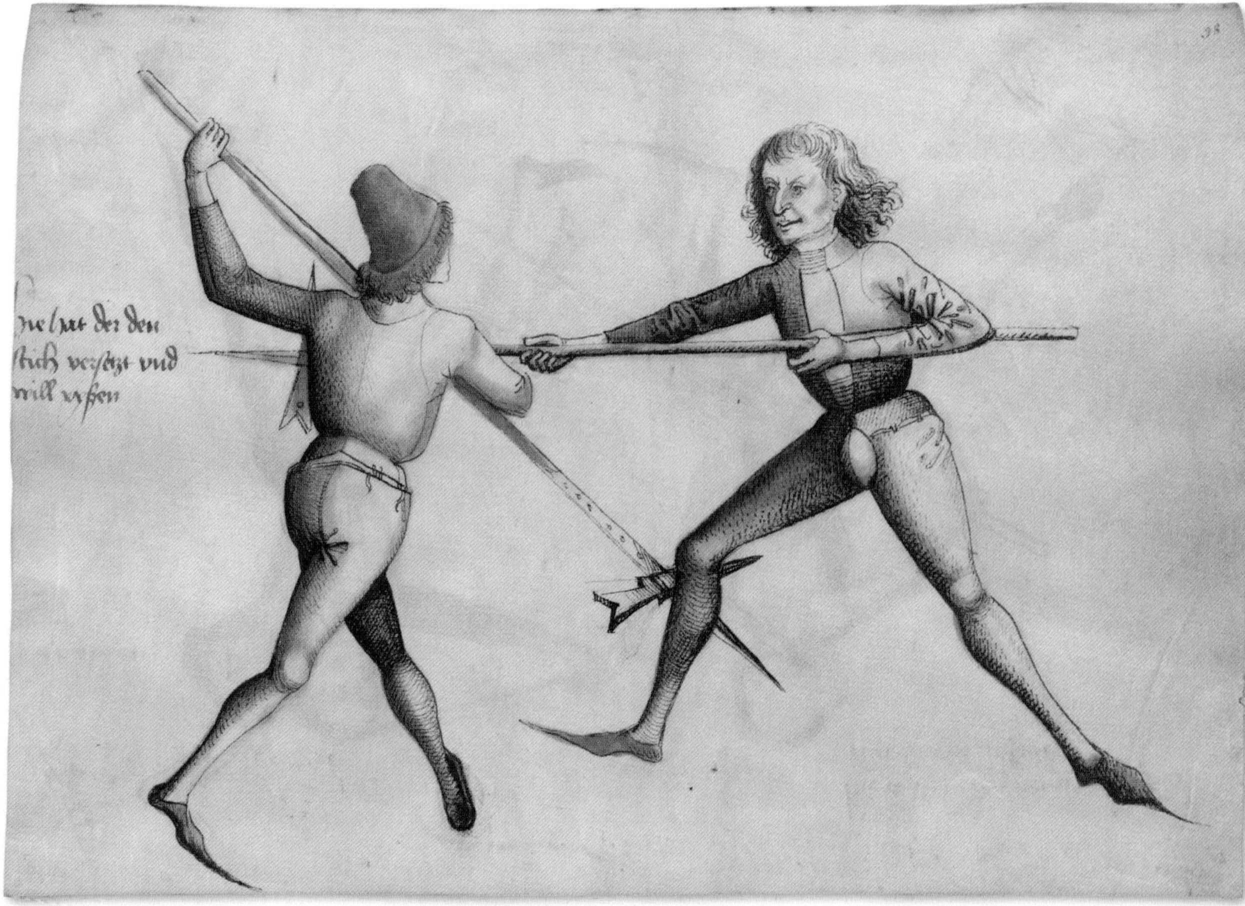

A binding from behind.

This one awaits the strike *vom Tag*.

That one strives to strike forcefully.

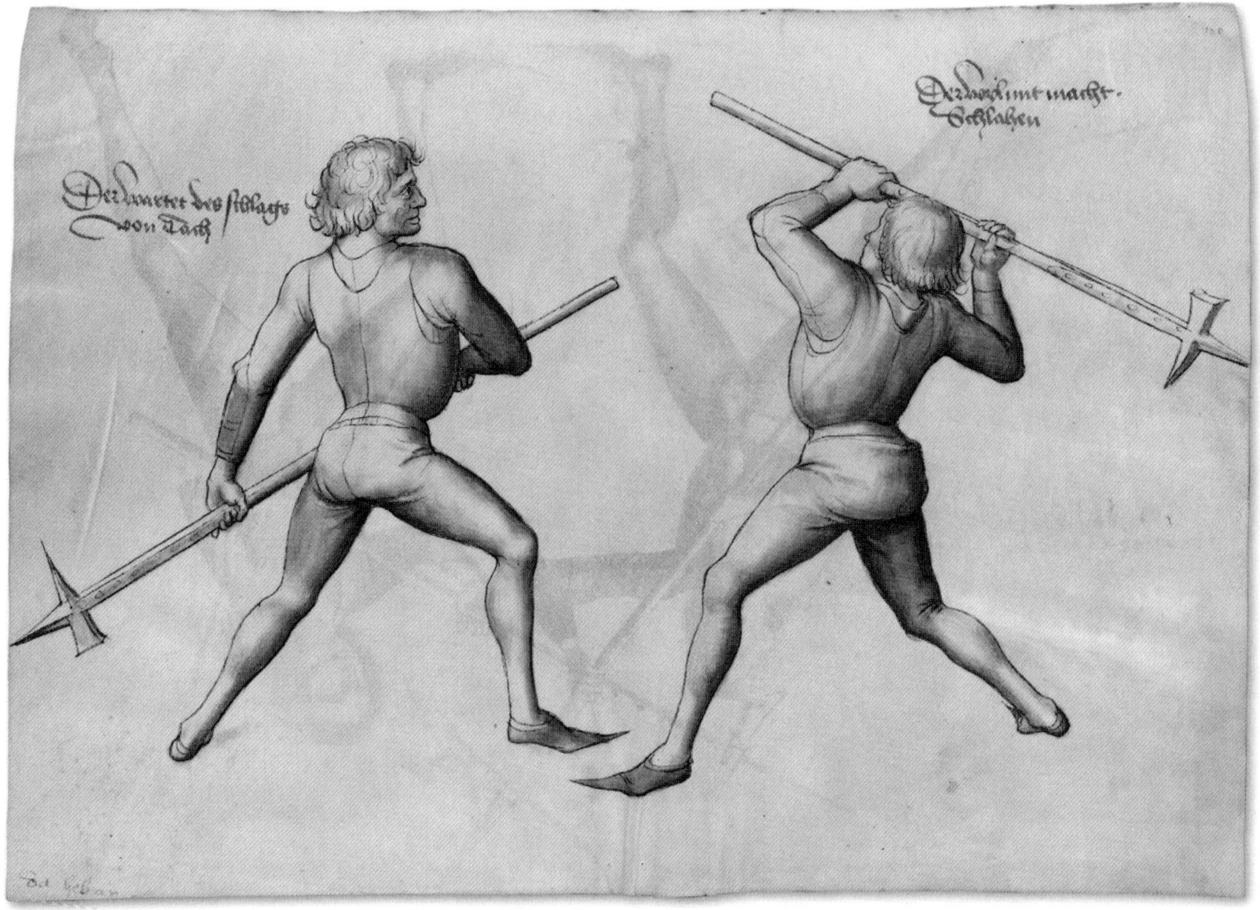

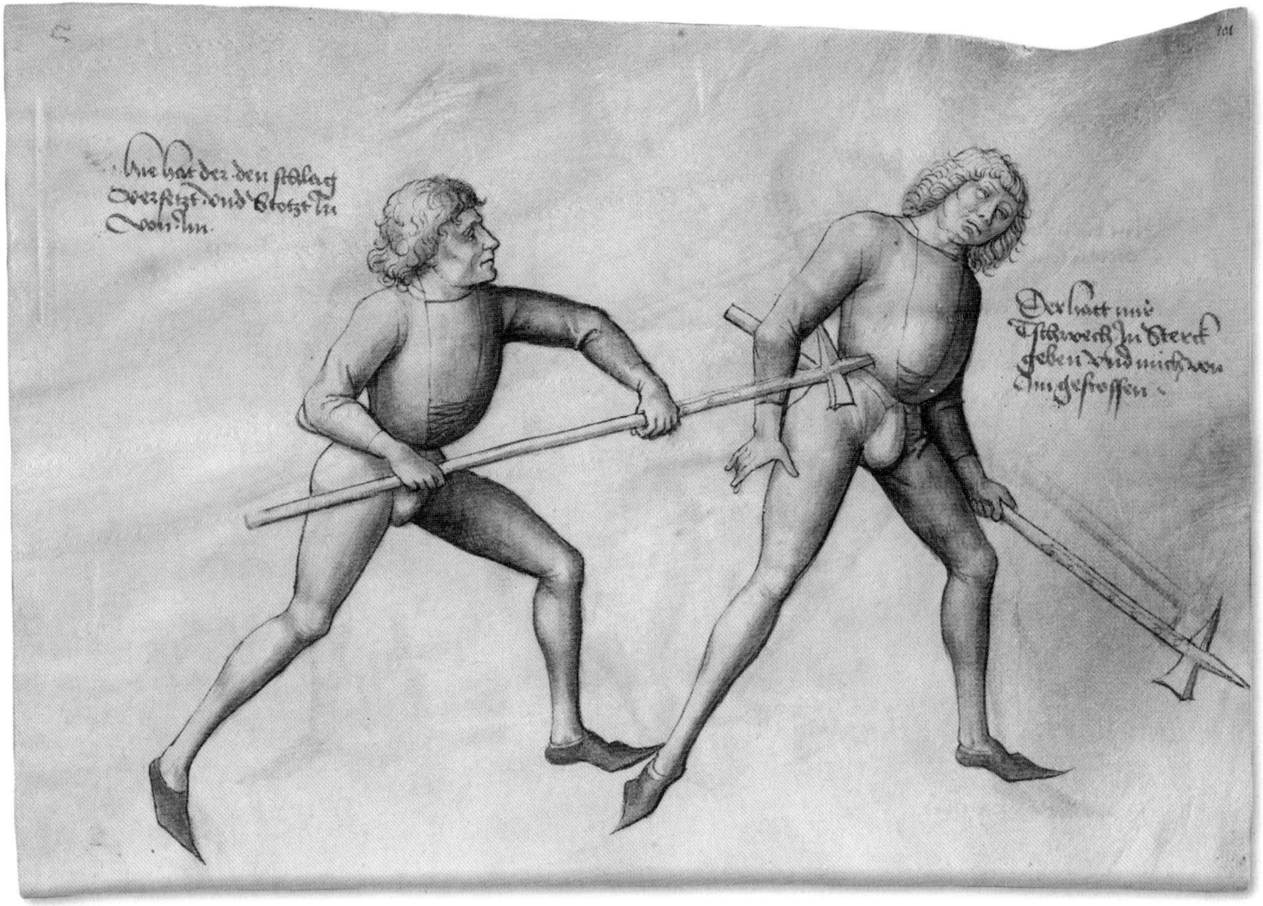

Here, one has displaced the strike and pushes the other away from him.

He has given me weakness in strength and pushed me away from him.

Here, I have lost my advantage.

Once one has pushed the other, he strikes the axe at his neck and throws him on his back.

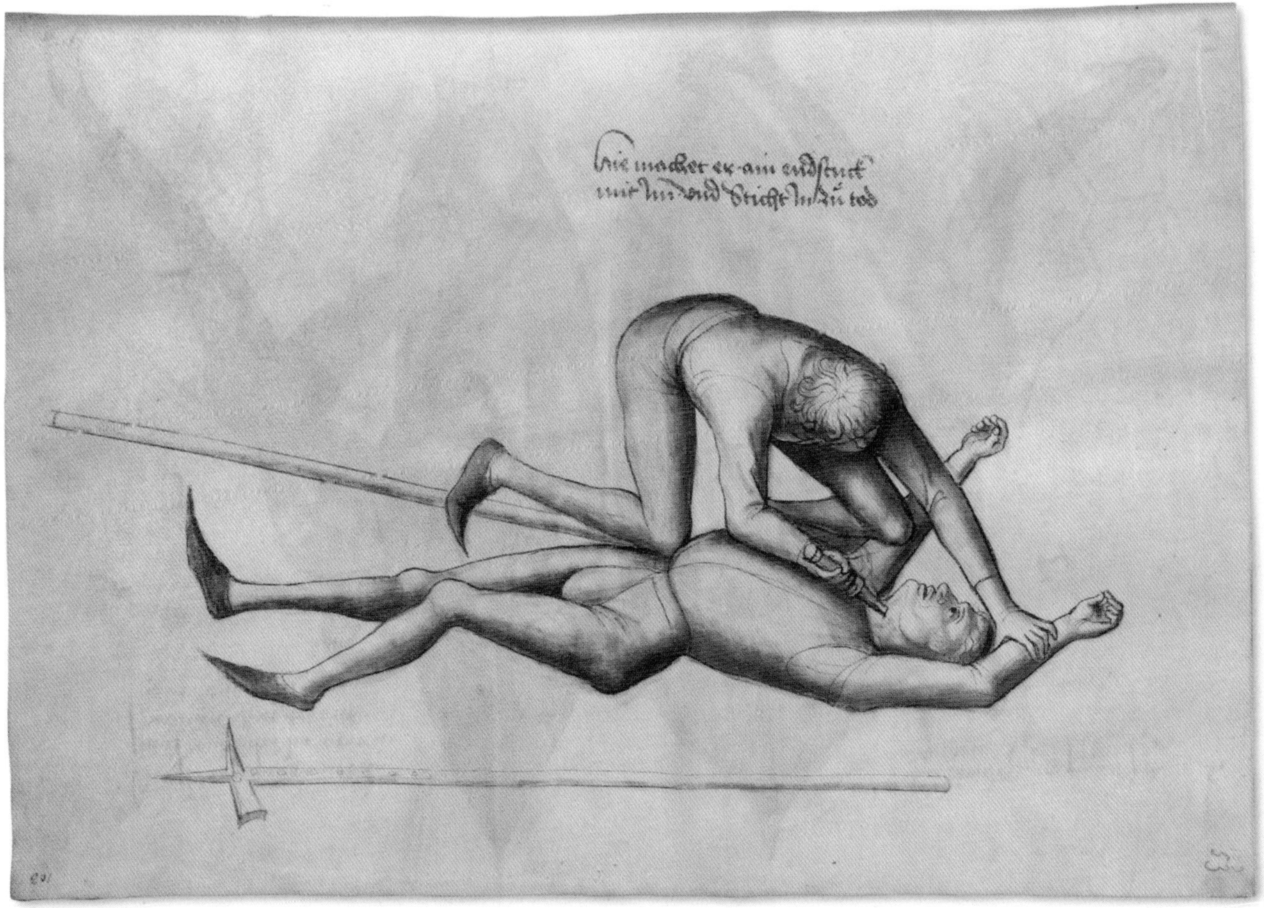

Here, he makes a final piece and kills him with a thrust.

This is the beginning of the shield. I stand advantageously. May God give us good fortune and bless us.

Here, I stand according to Franconian law.

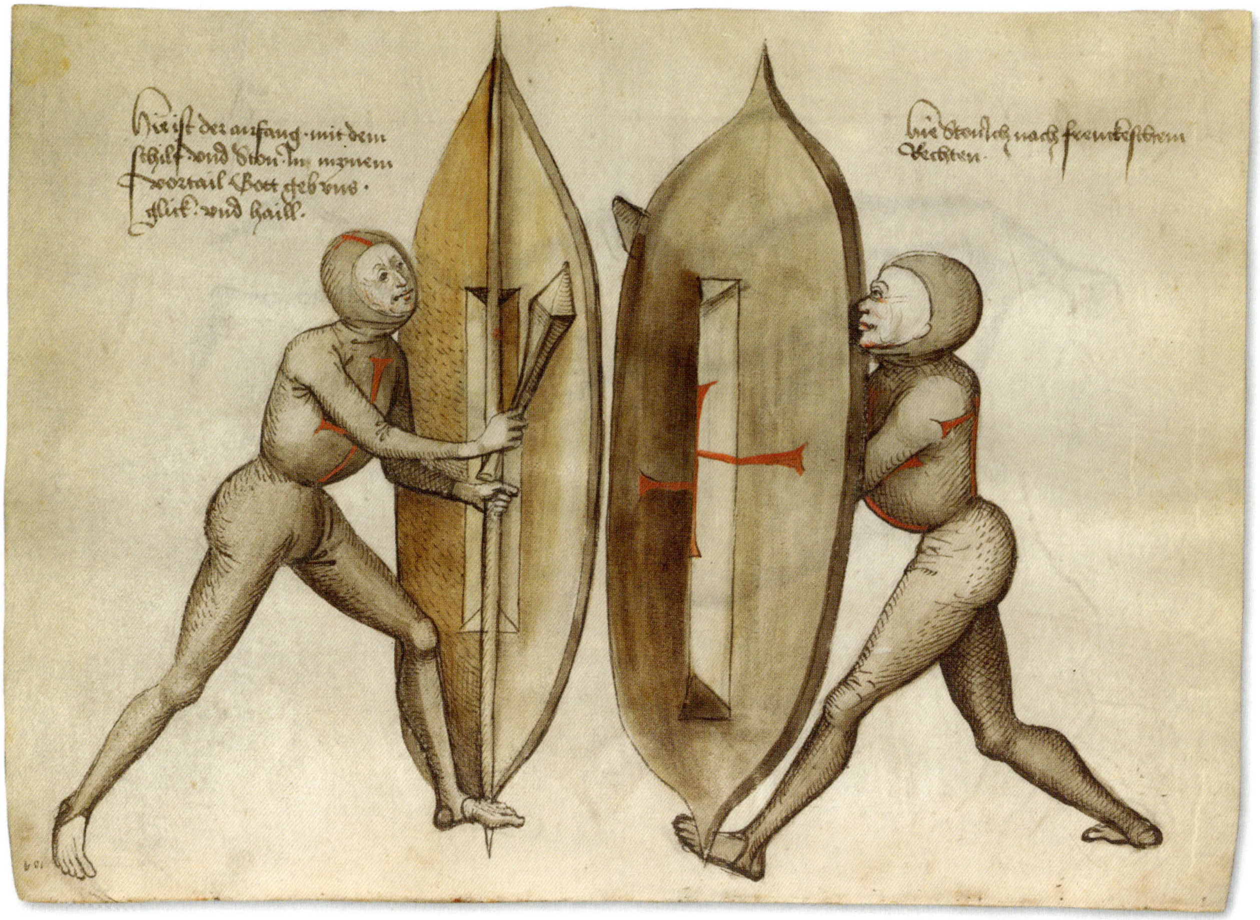

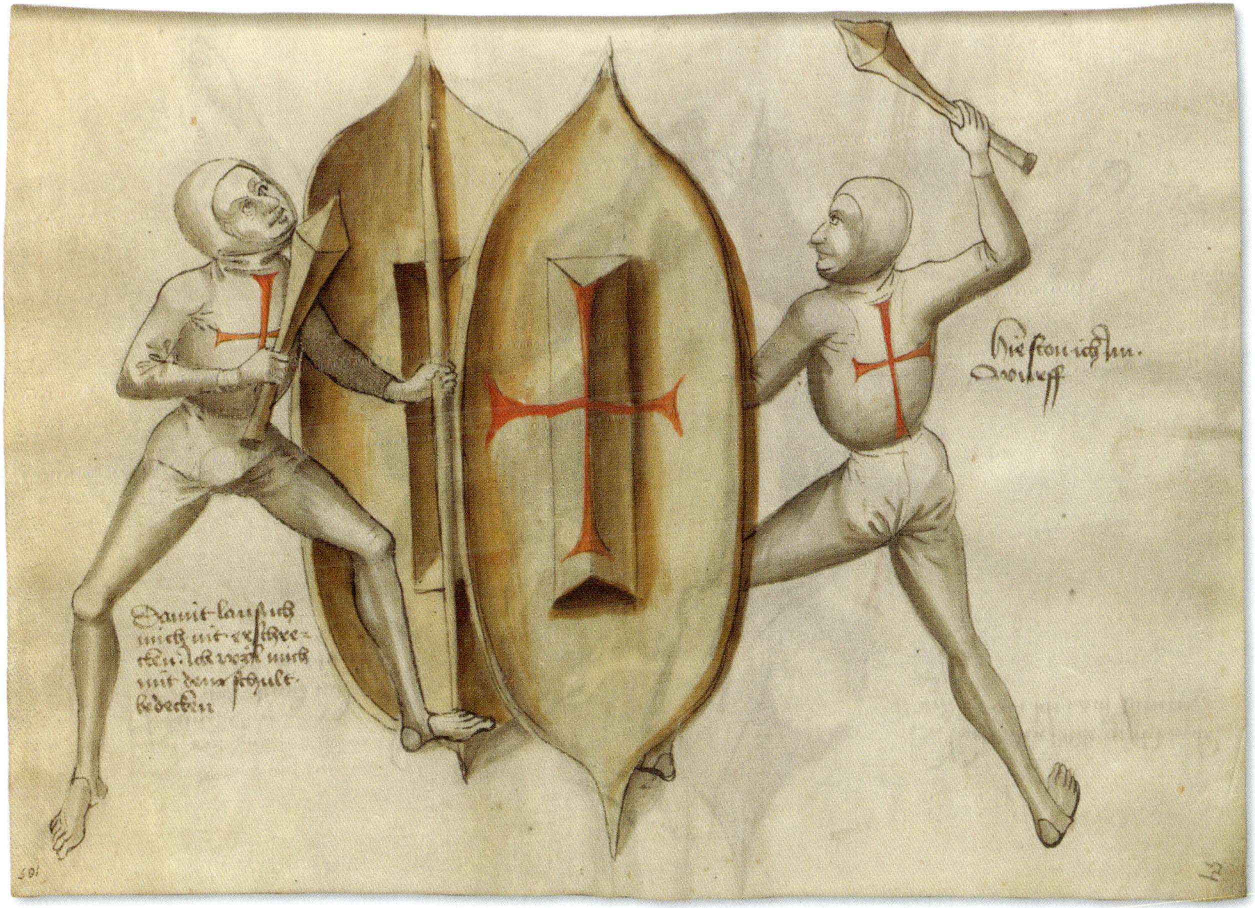

This does nor frighten me. I strive to cover myself with the shield.

~

Here, I stand, [ready] for a throw.

In this instant, you have found me uncovered.

Here, one has turned the shield and hits the other's head.

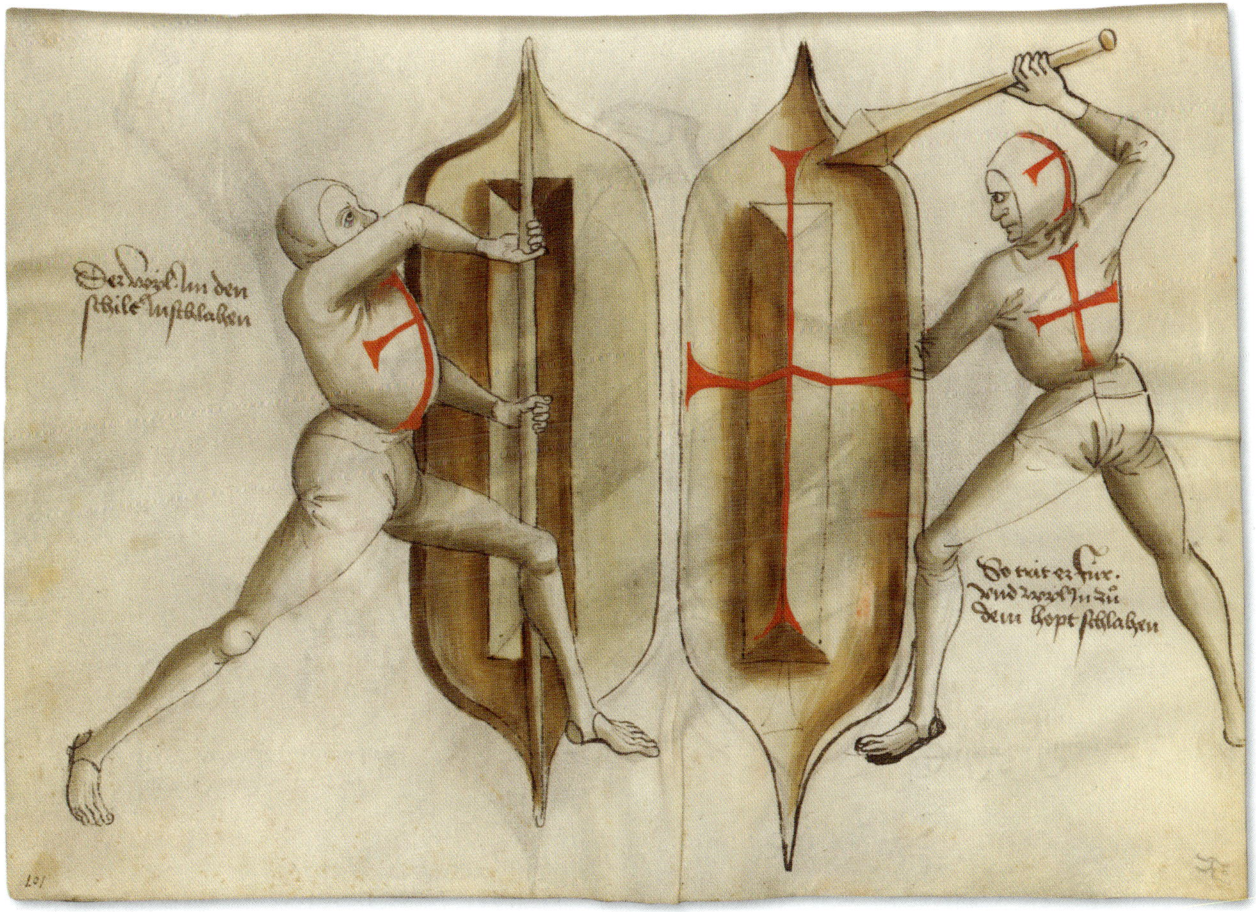

One strives to strike against the other's shield.

Thus, he advances, striving to strike at his opponent's head.

I have been found uncovered, and I fear I shall receive a thrust.

~

One has been bound behind the other's shield.

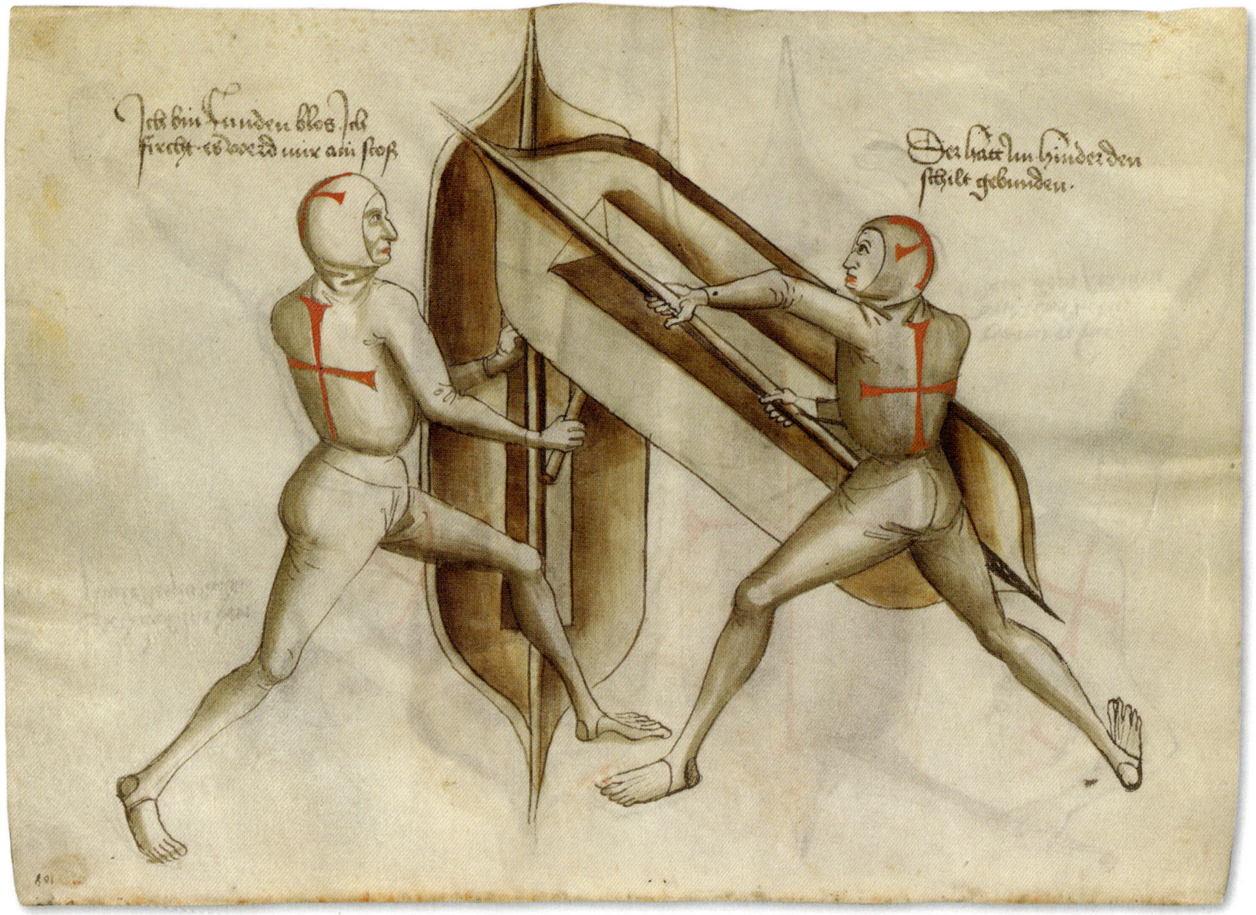

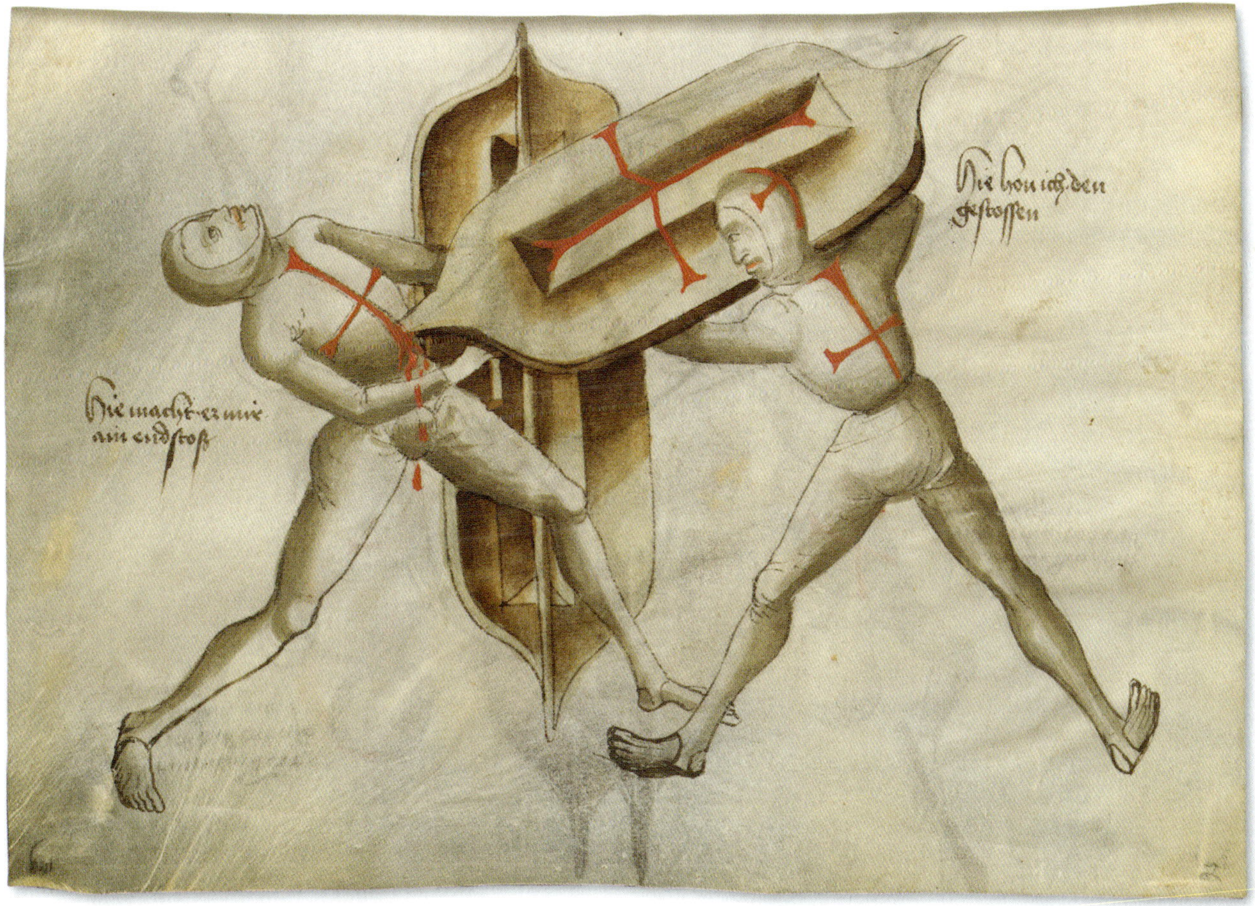

Here, he puts an end to me with a thrust.

~

Here, I have thrust into him.

With my step, I have rendered you harmless and strike you confidently.

~

He has kicked my shield to the side, therefore I will throw him.

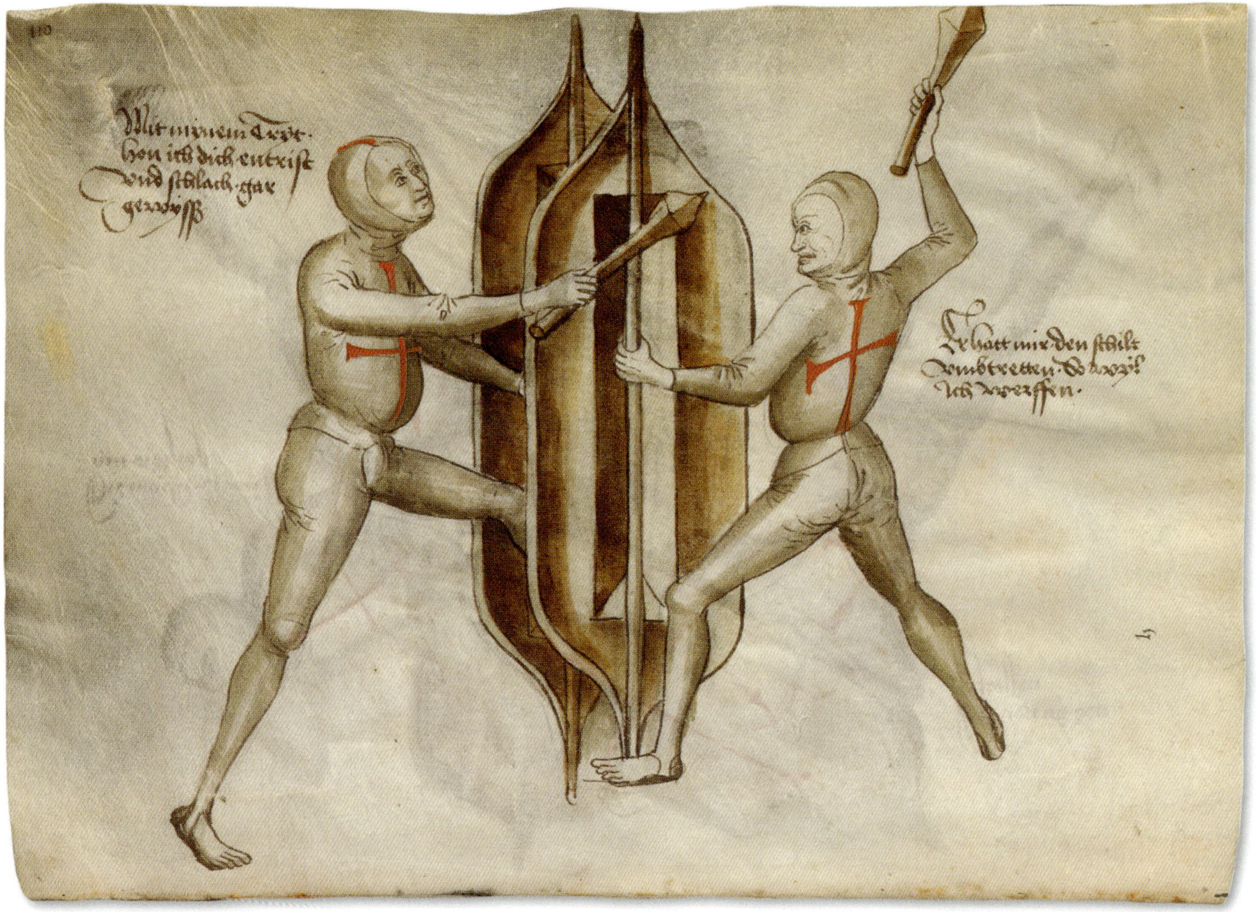

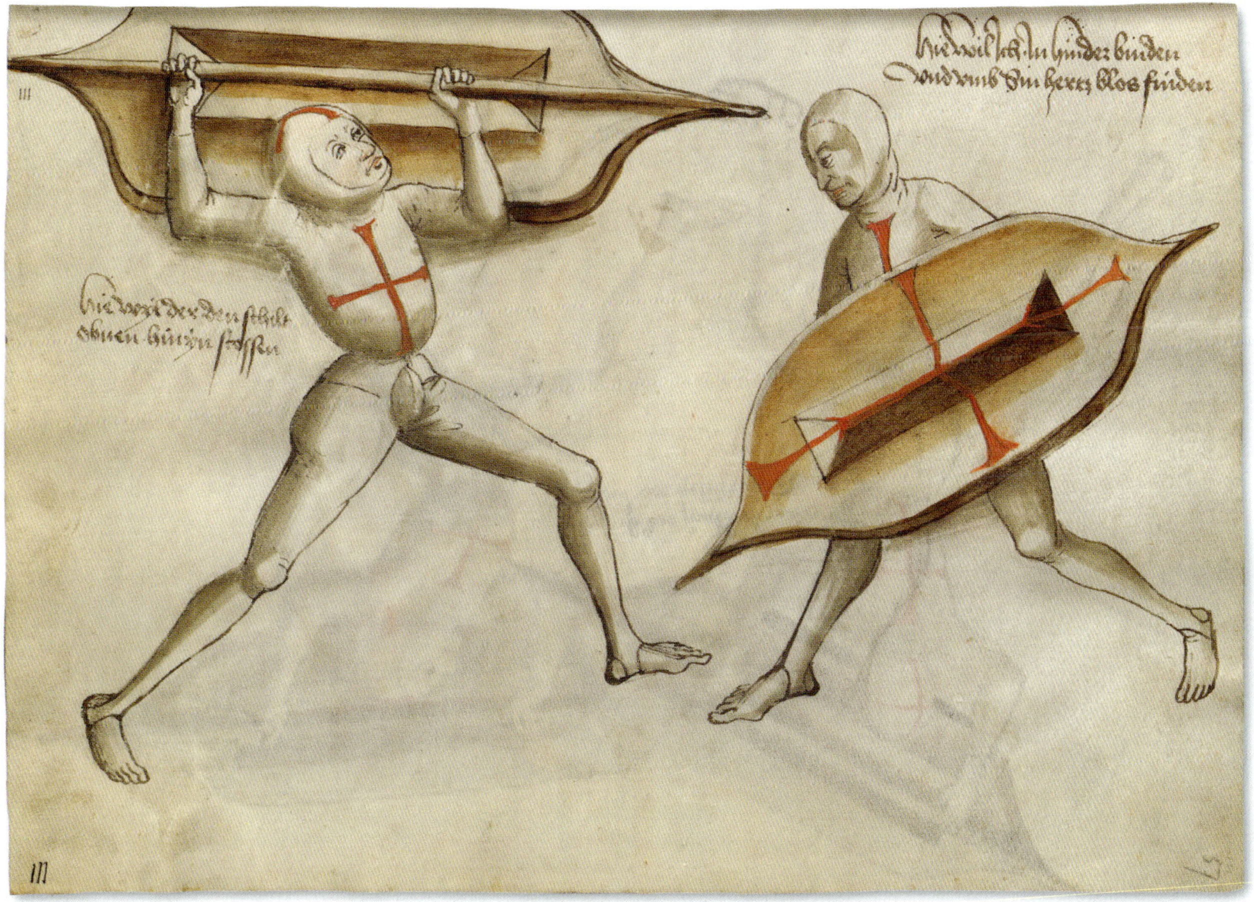

Here, he strives to thrust the shield above into the other.

Here, I strive to bind from behind in order to find his heart uncovered.

Here, the piece is finished, as described above.

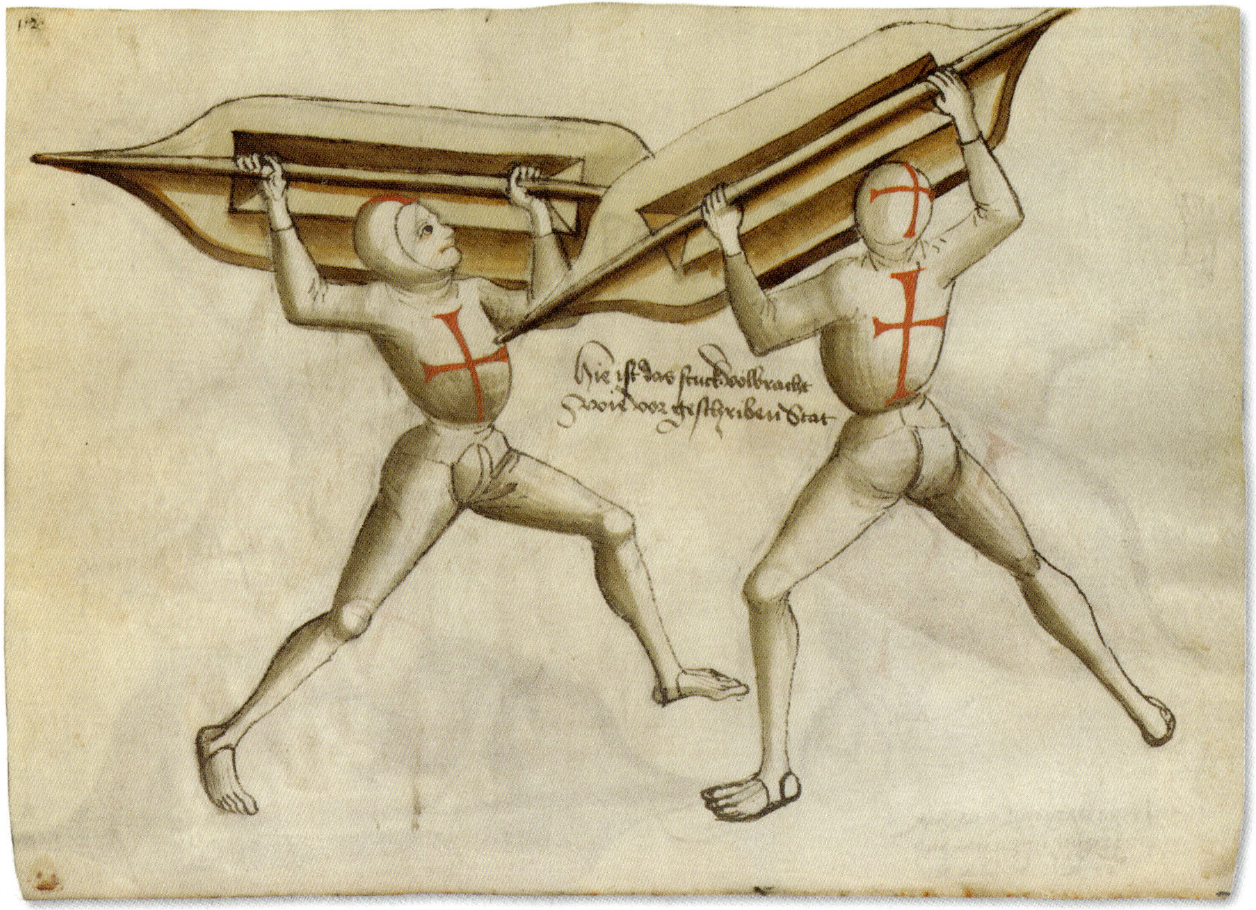

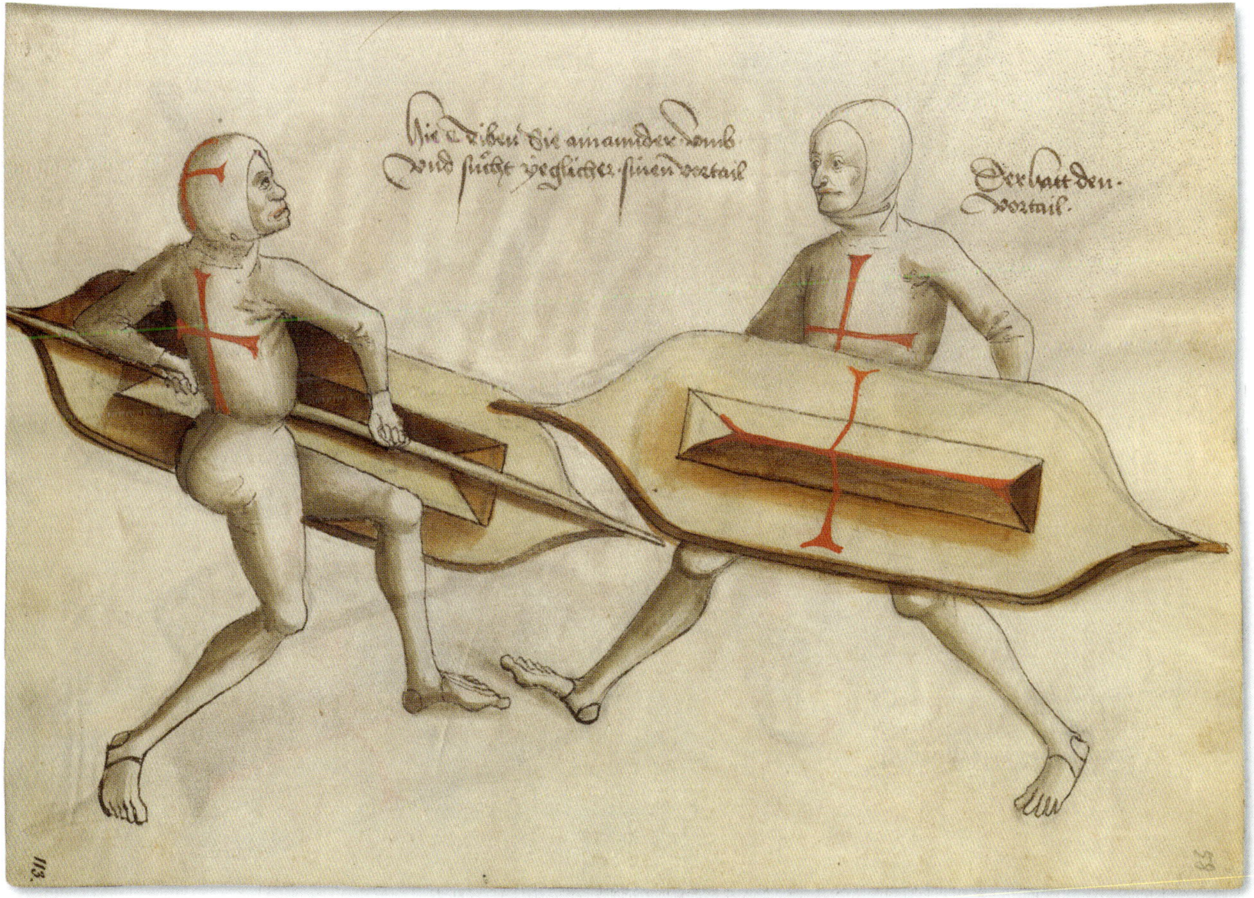

Here, they drive each other around and seek their advantage.

~

This one has the advantage.

Here, from the binding, one has struck the other's shield out of his hands. He intends to thrust where he may find him uncovered.

But the other grabs with the arm, throws away the shield and pushes him away.

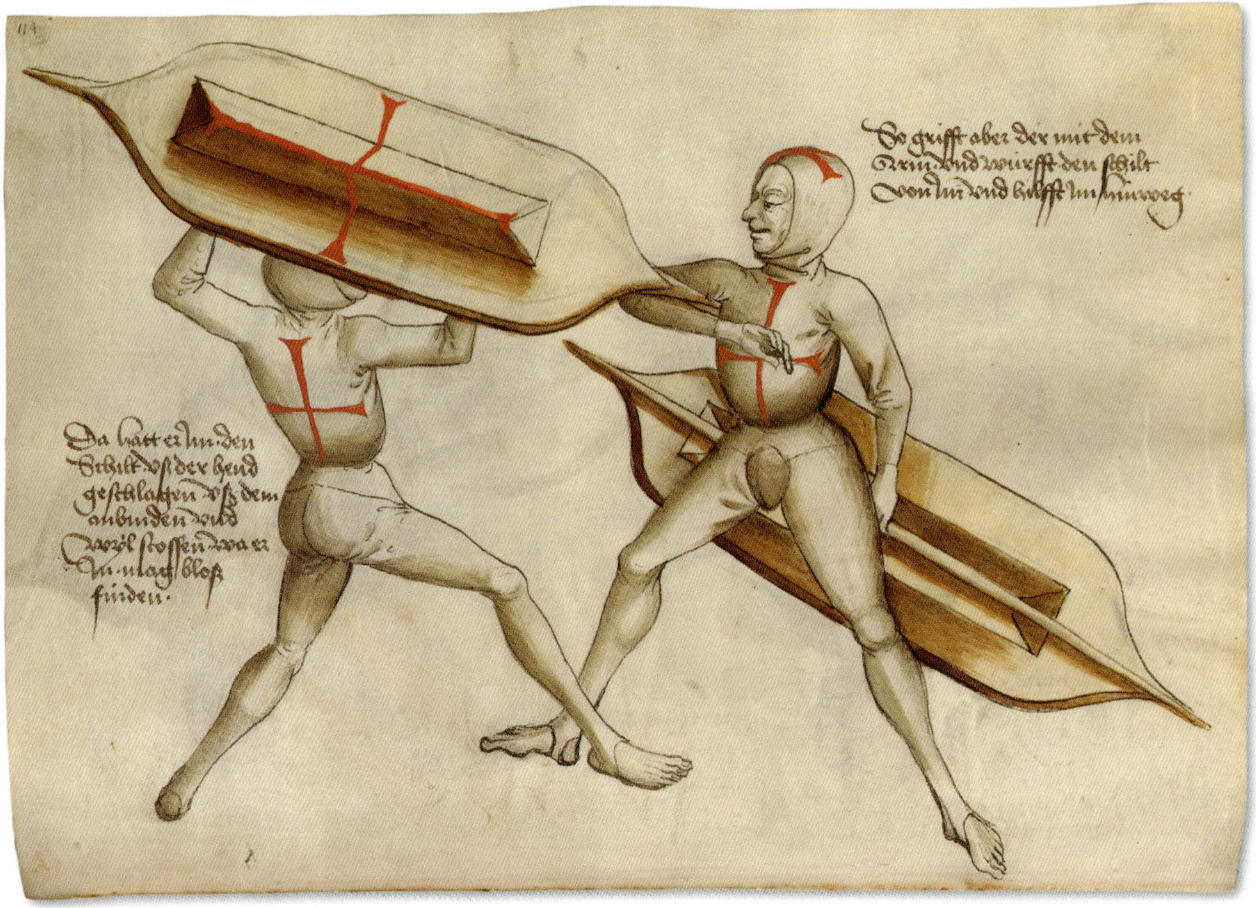

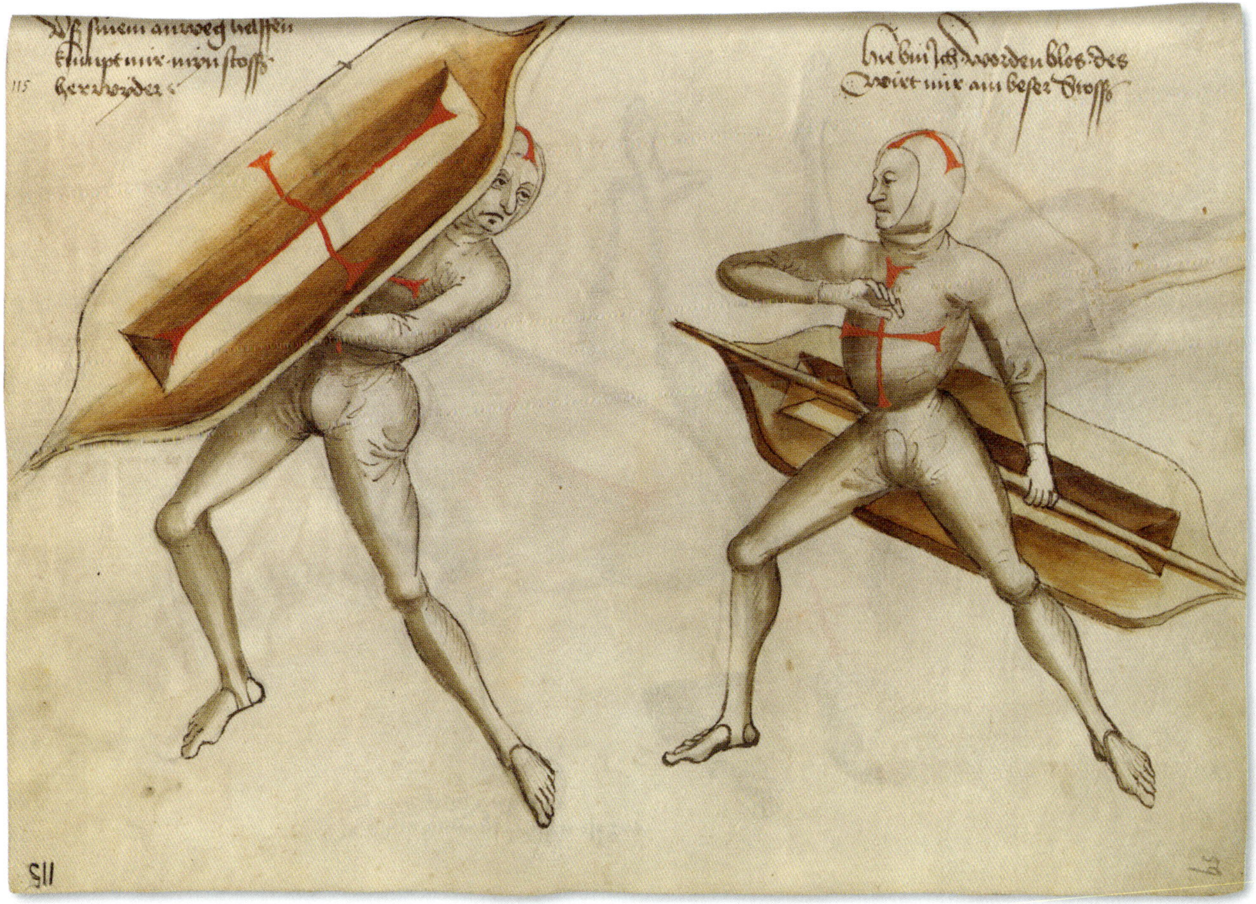

After he has pushed away, I follow with another blow.

Here, I became uncovered, thus I'm in for a bad blow.

From the pushing away, the piece has come to a complete end.

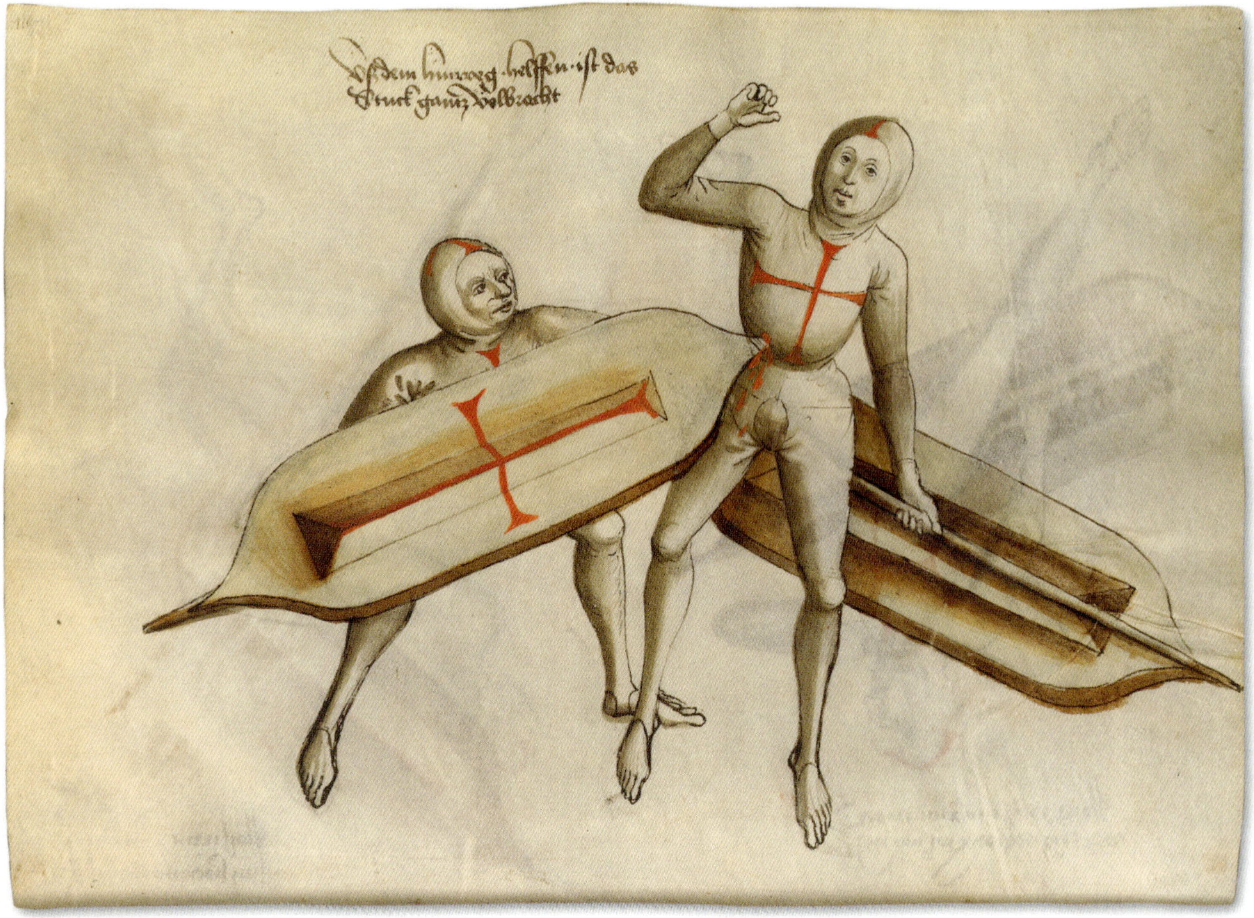

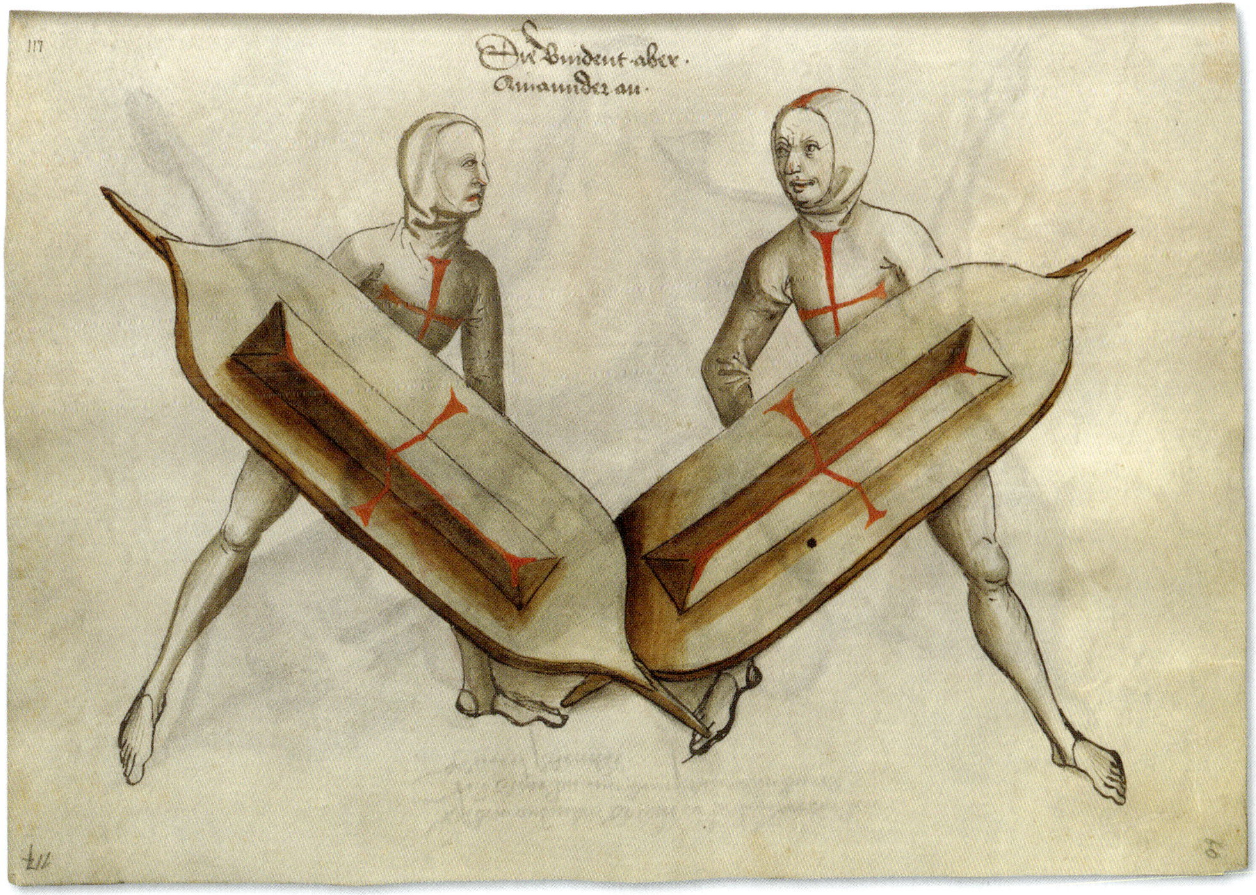

They are binding again.

From the binding: one has bound behind and thrusts down with the shield through the other's upper leg.

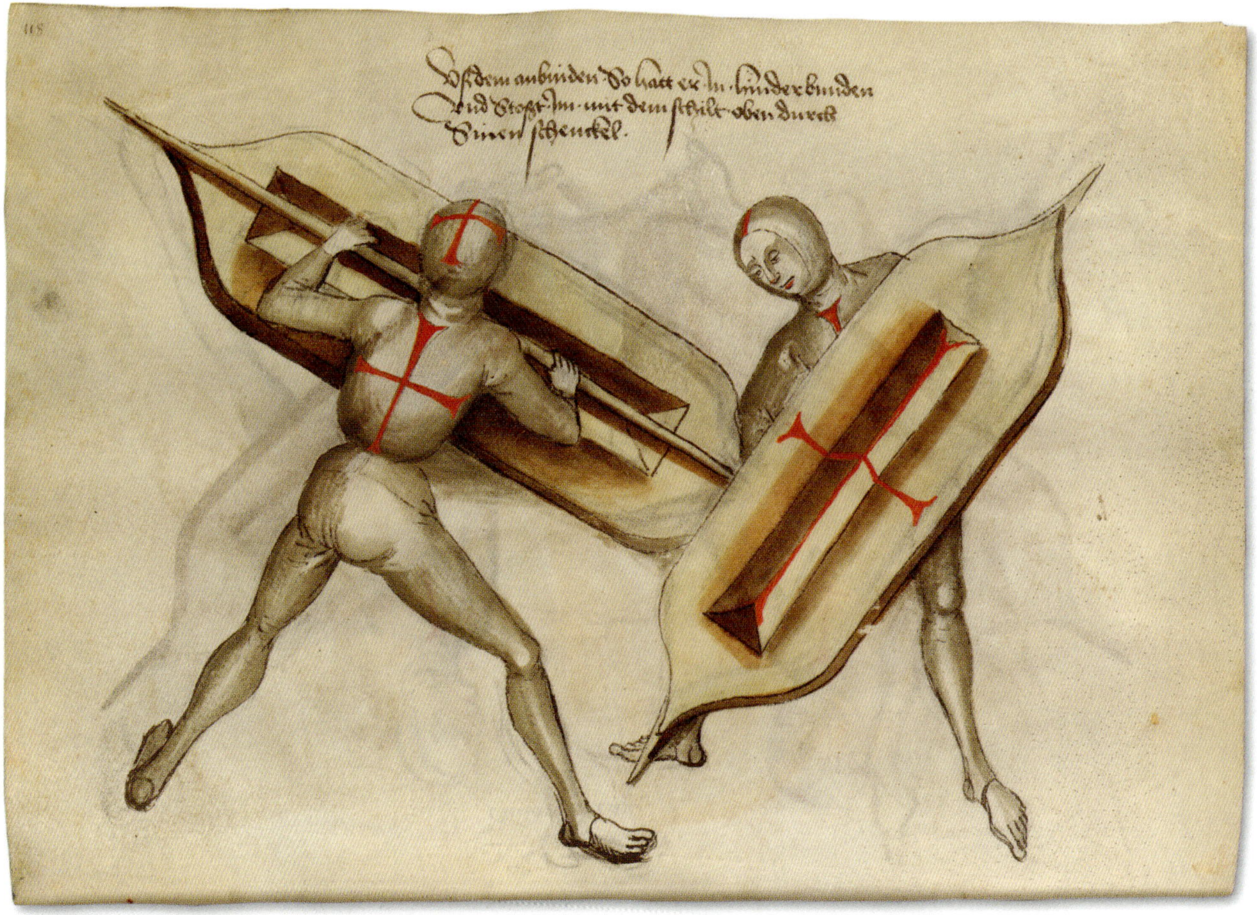

Here, he stands in guard and observes the opponent.

One strikes the hook behind the other one's shield and wrenches backwards.

This one wrenches backwards, as is written before.

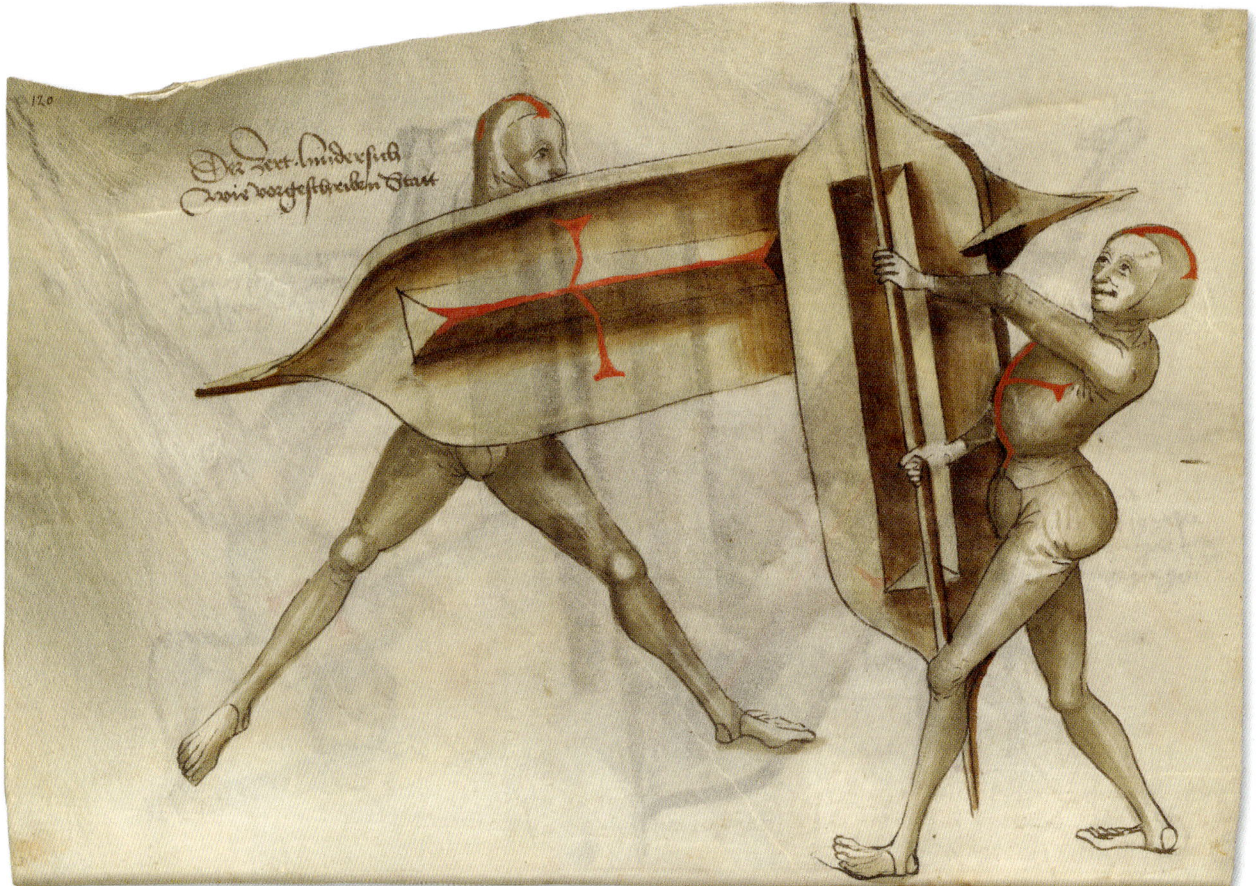

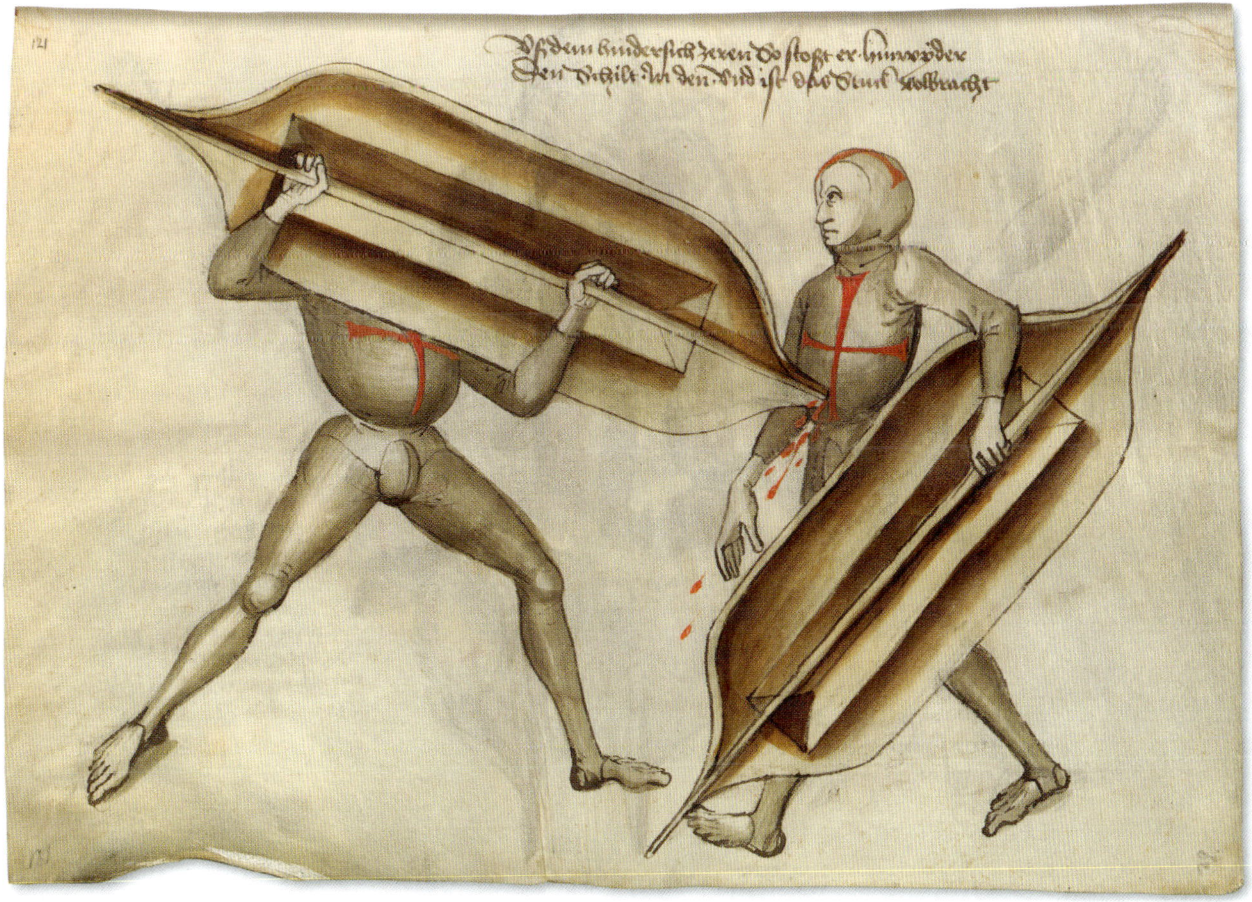

From wrenching backwards: one thrusts his shield again into the other one, and the piece is finished.

Here, one throws his arm upwards and grabs over the other's arm. He drops the shield and steps forward. He tosses his left arm around his neck and throws him over his hip.

Here, one turns the shield around and strikes crossed-armed towards the other's head.

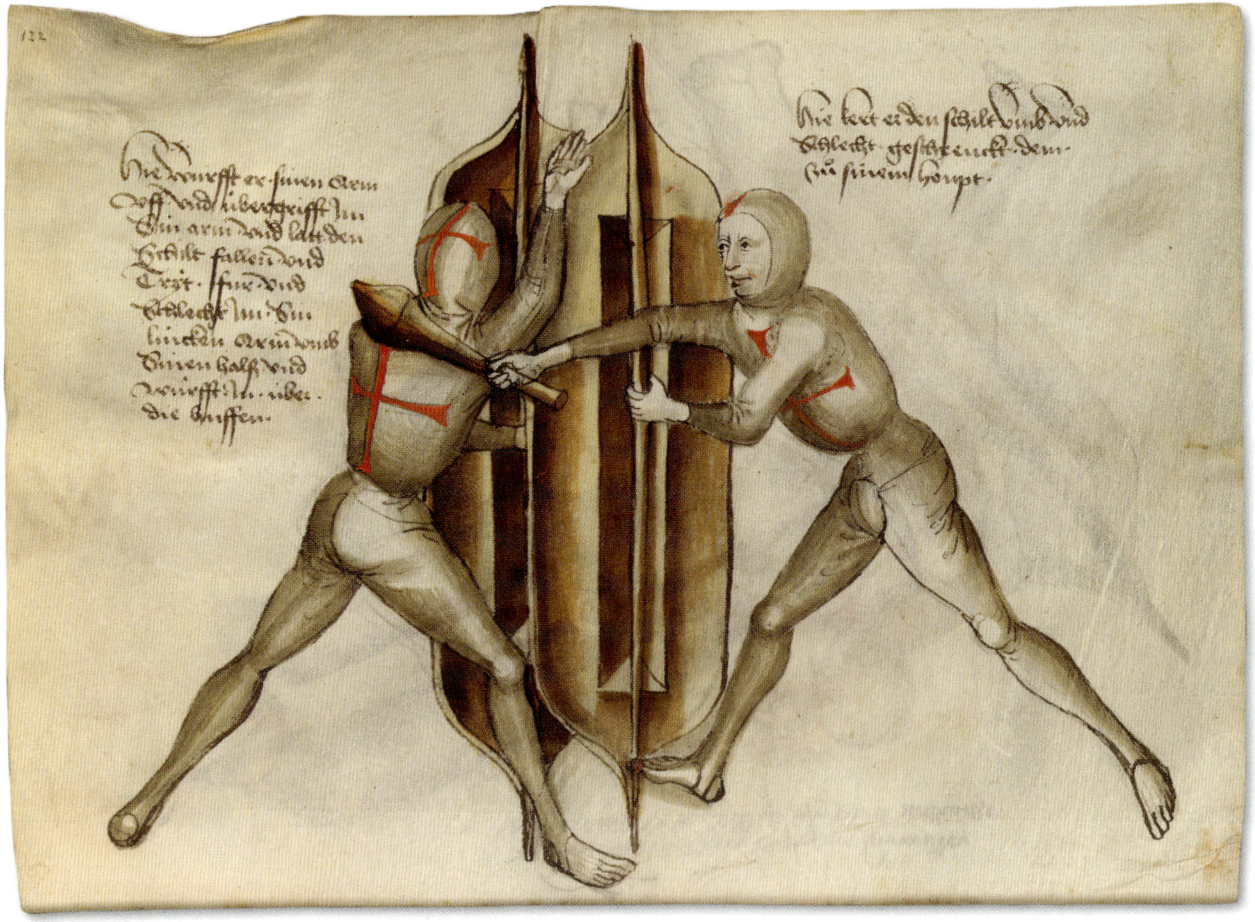

Here, the piece is brought to an end, as is written above.

This one is completely defenceless and is being thrown.

That one has struck his shield between opponent and shield. He releases the shield and intends to throw the other one.

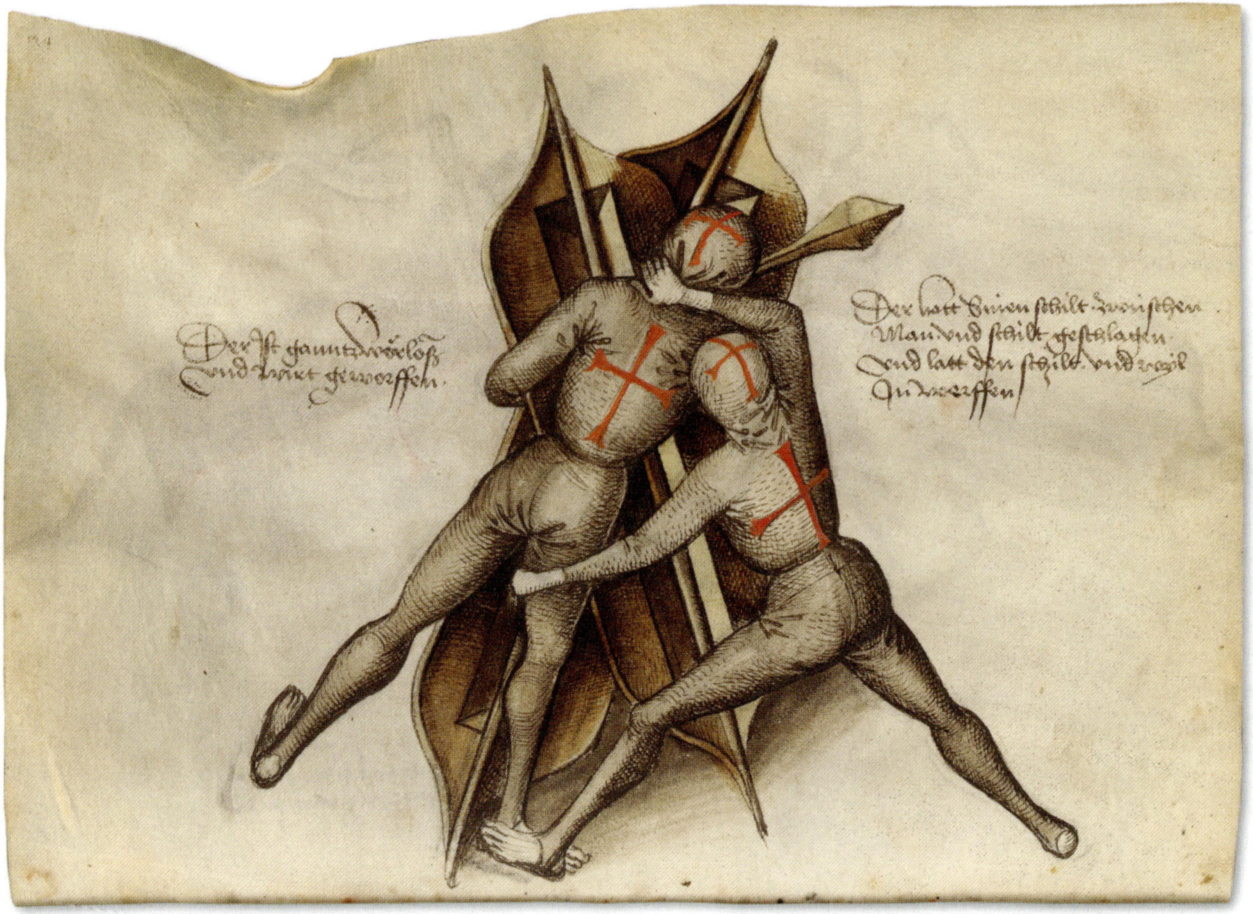

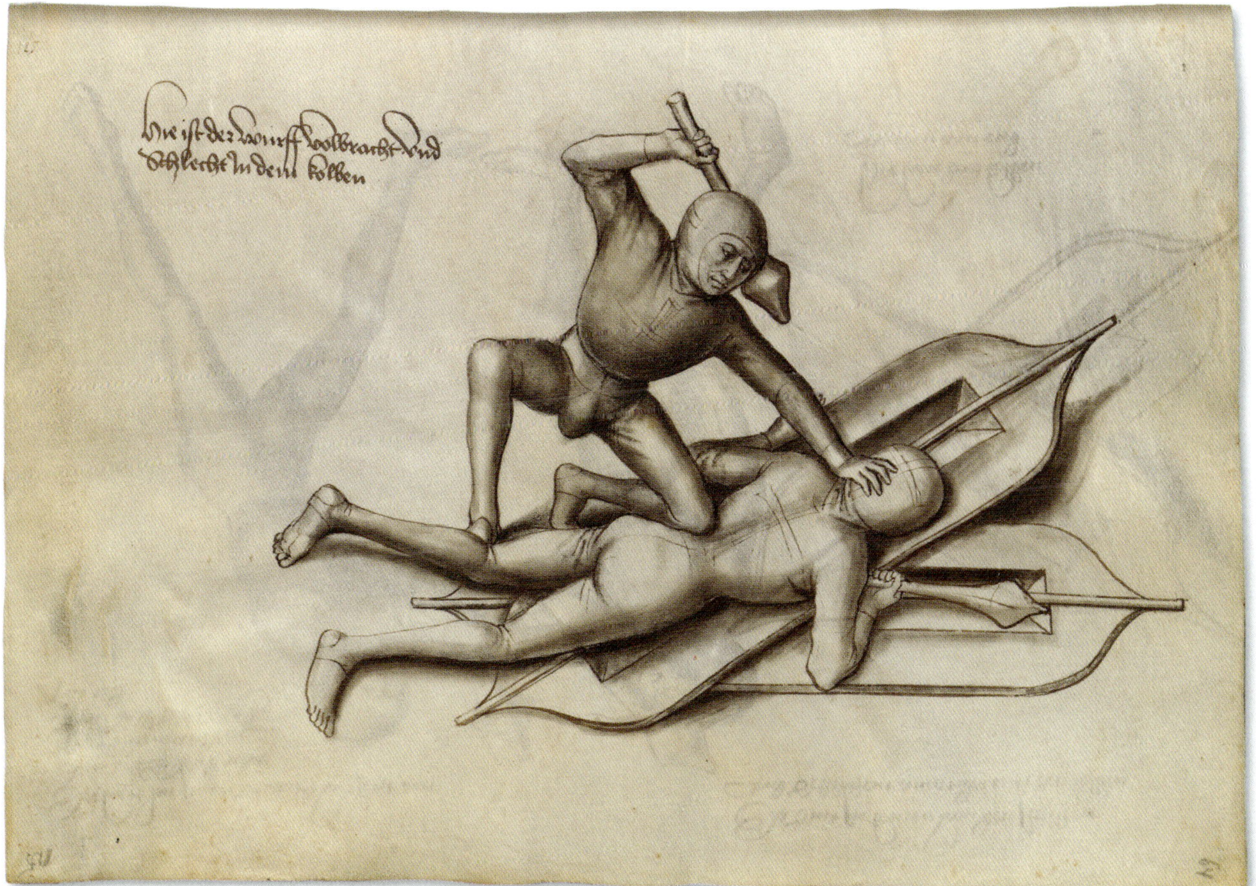

Here, the throw is completed and one strikes the other one with the club.

Here, they have abandoned their shields and strike each other with clubs.

~

One has displaced the strike with the club. He grabs the other's arm from above and beats him to death.

~

Here, the fencing with the clubs comes to an end.

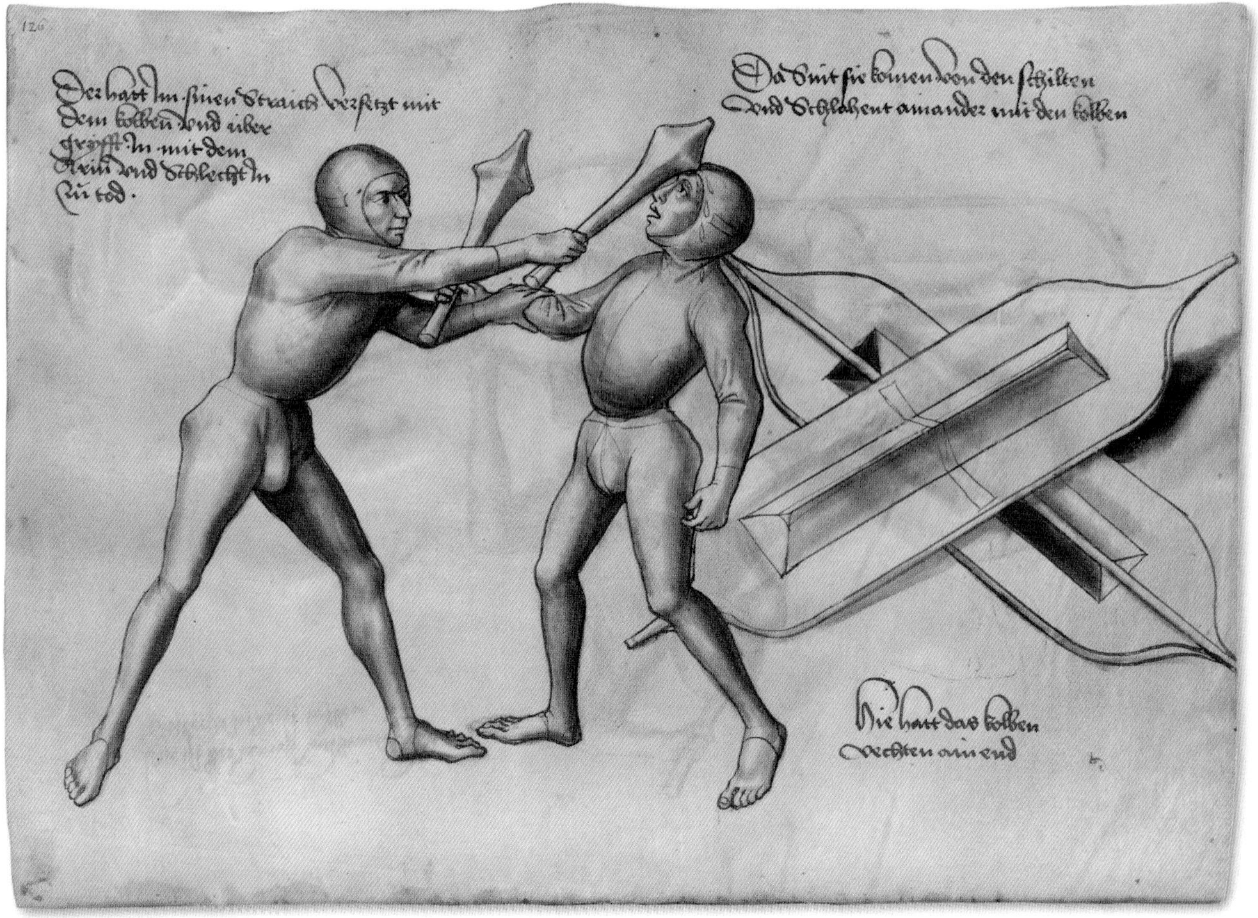

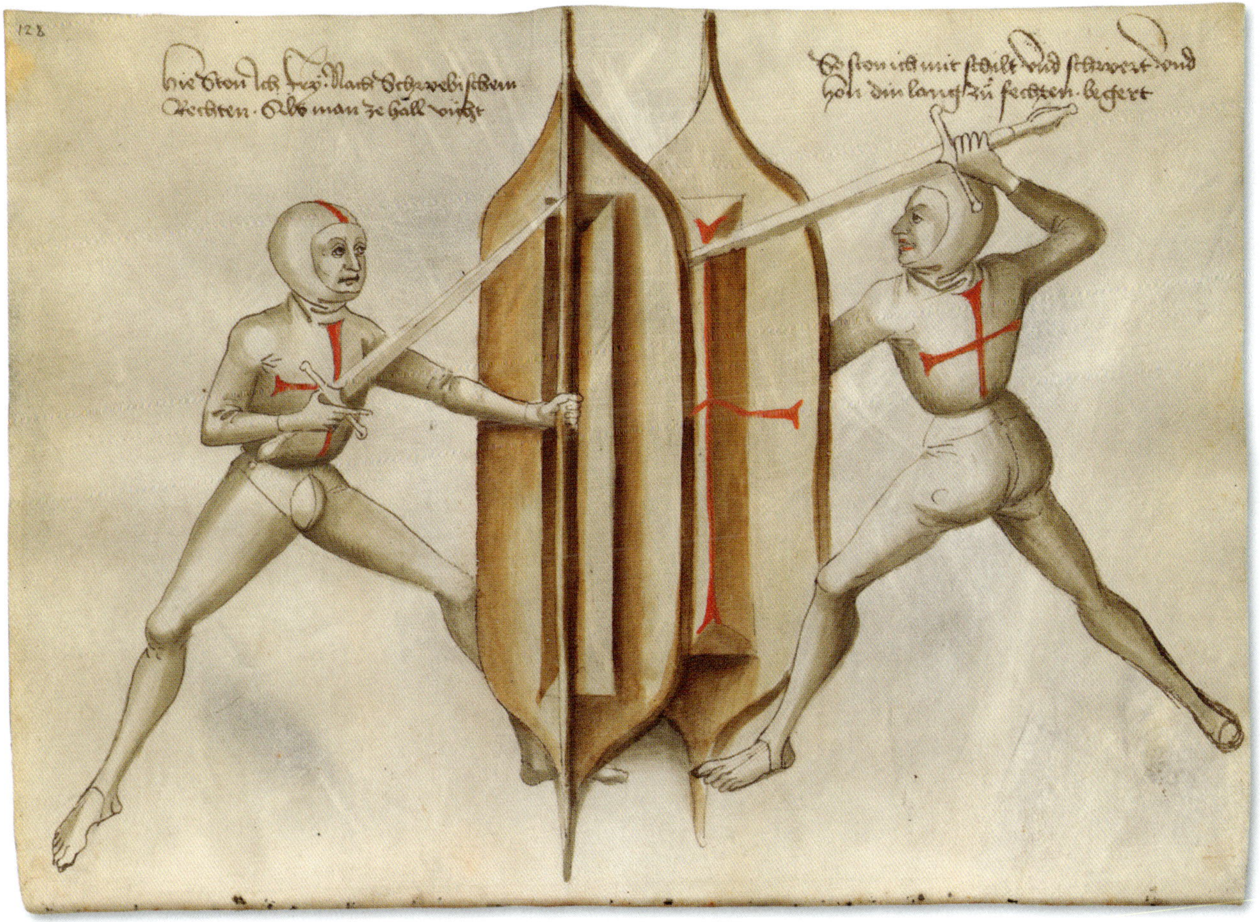

Here, I stand freely according to Swabian law, as one fights in the city of Hall.

So I stand with shield and sword. I have hoped for a long time to fence with you.

This one strives to turn the shield and strike.

So he brings his sword between the shields and tests how strongly the other stands.

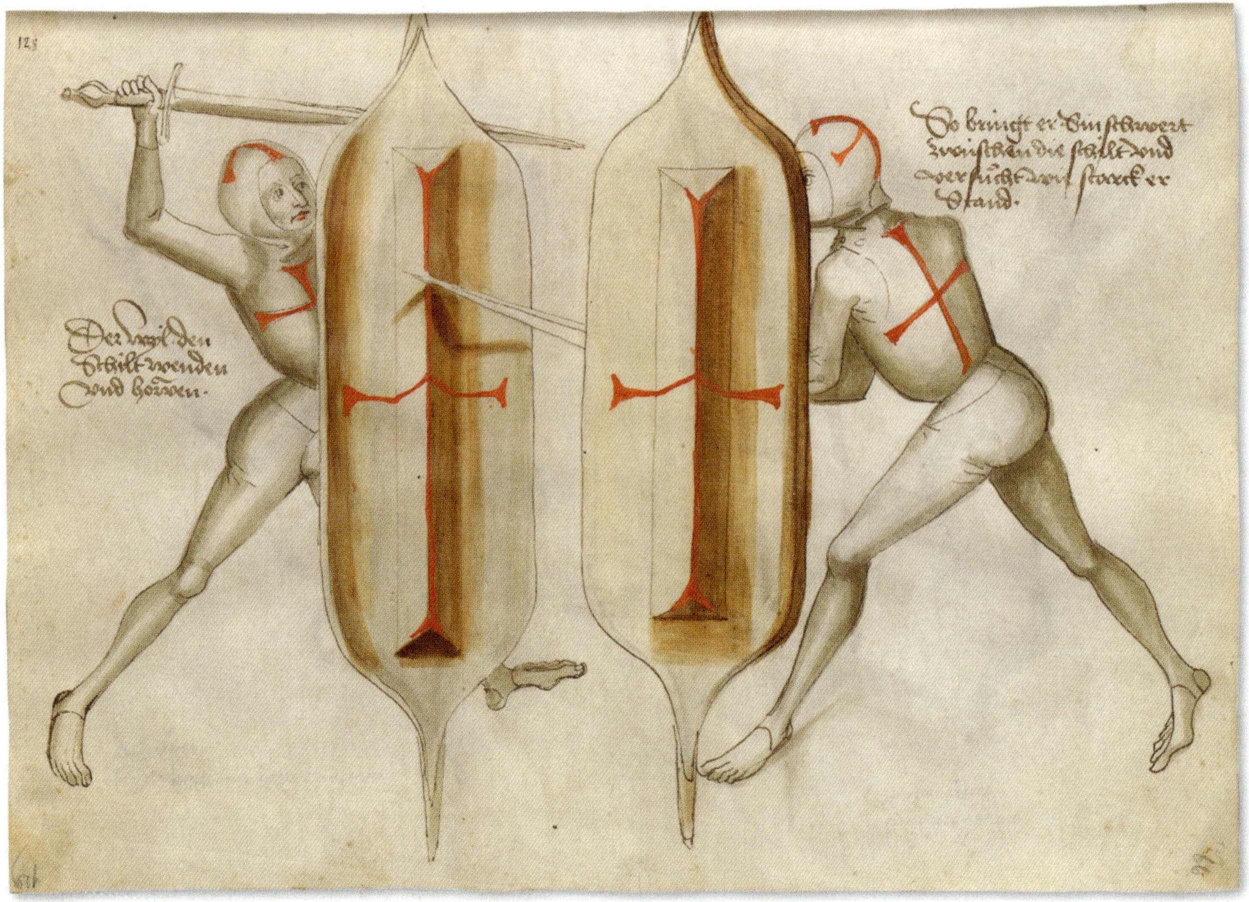

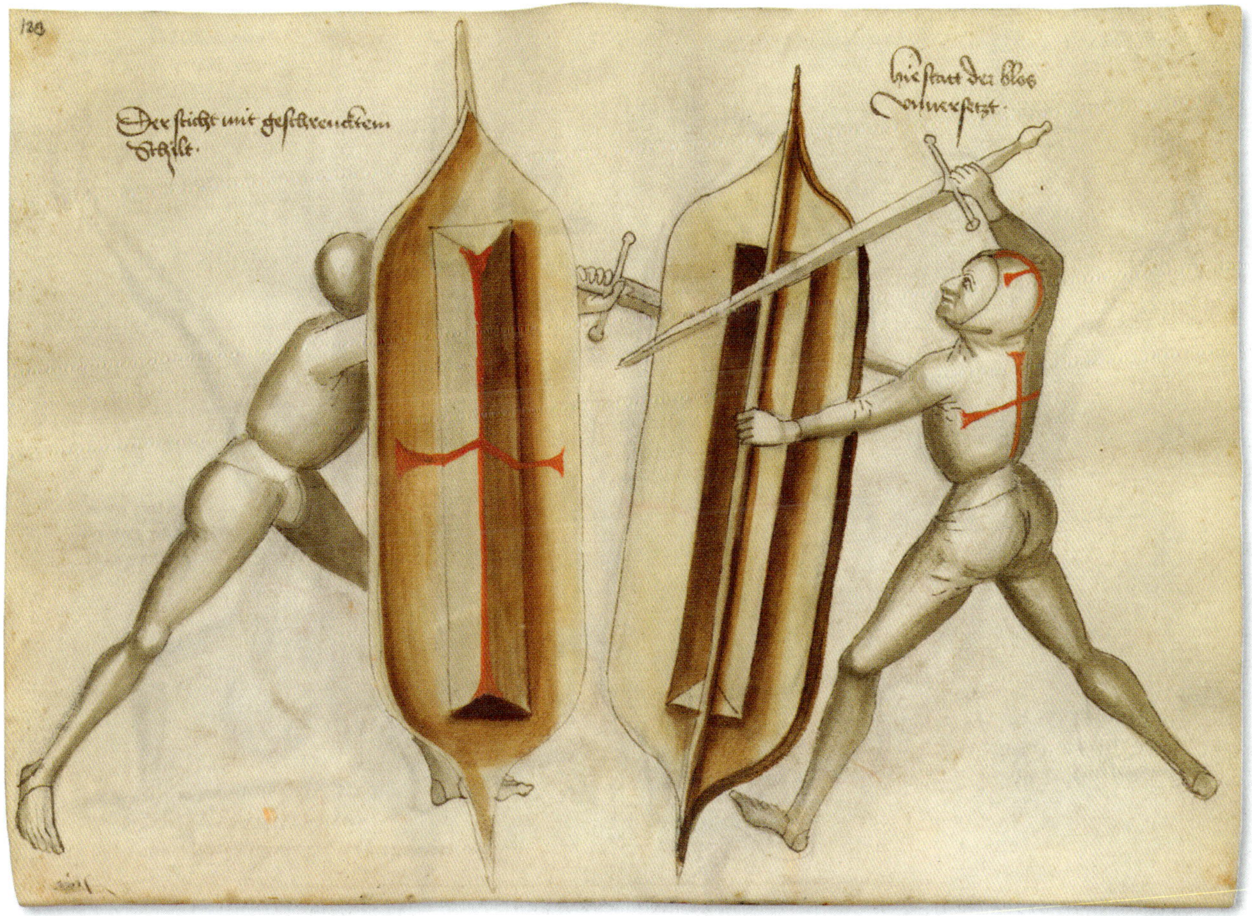

This one thrusts with clasped shield.

Here, he stands uncovered without displacement.

The two of us stand like the two before, but we have turned around so that each has assumed the other's position.

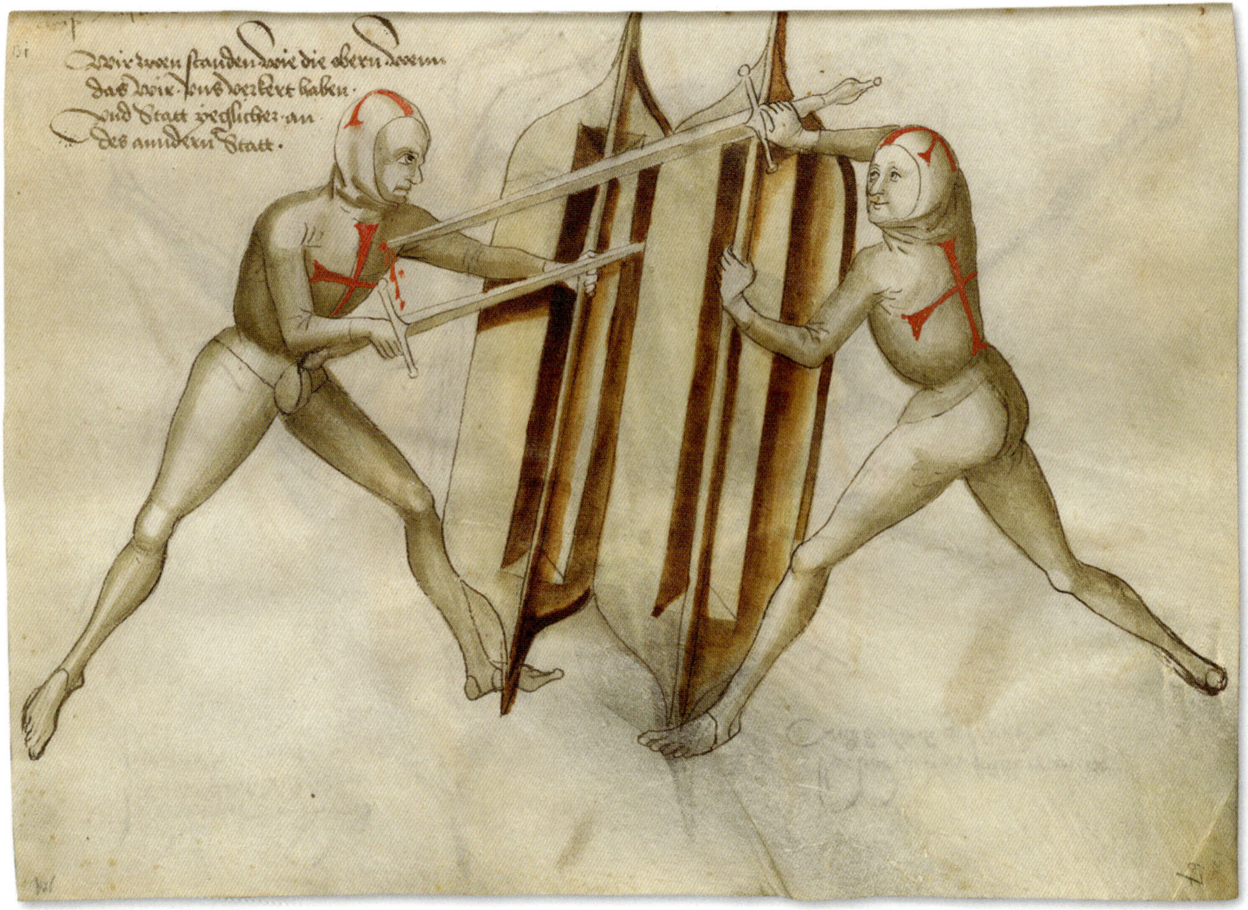

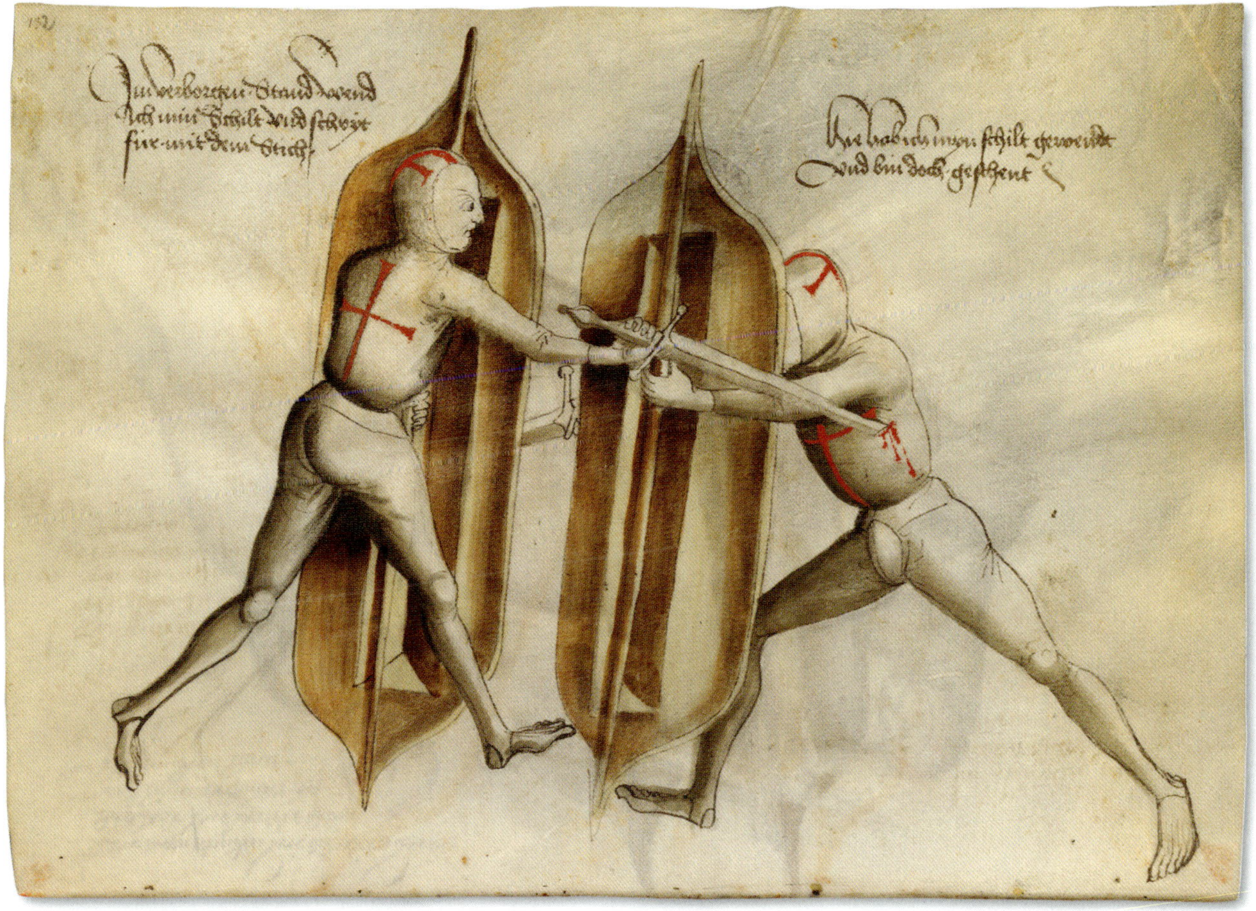

A secret stance: I turn my shield and step forwards with a thrust.

Here, I have turned my shield, and yet I have been injured.

So I have twitched my shield and stepped away from his strike. Thus I break his piece.

I intended to thrust crossways behind his shield.

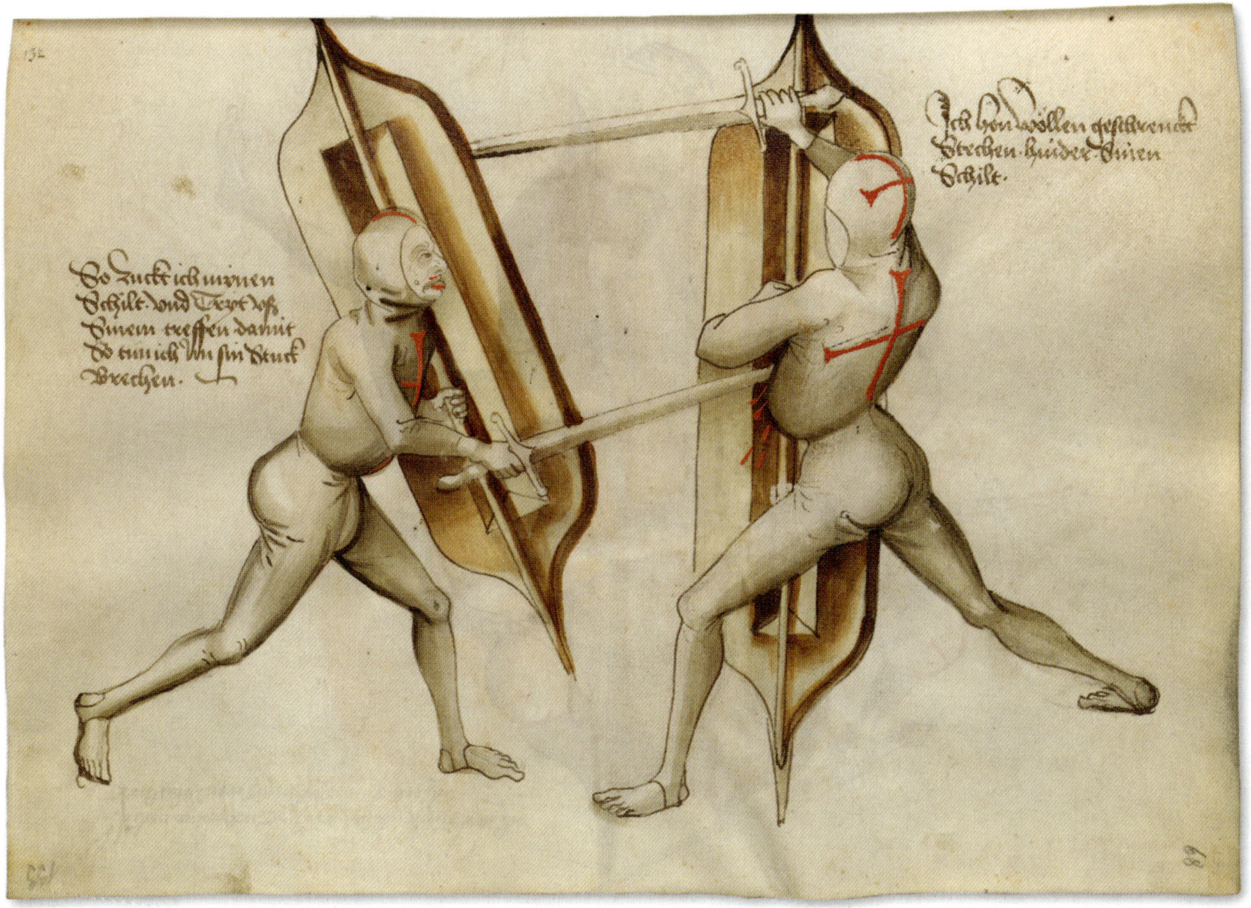

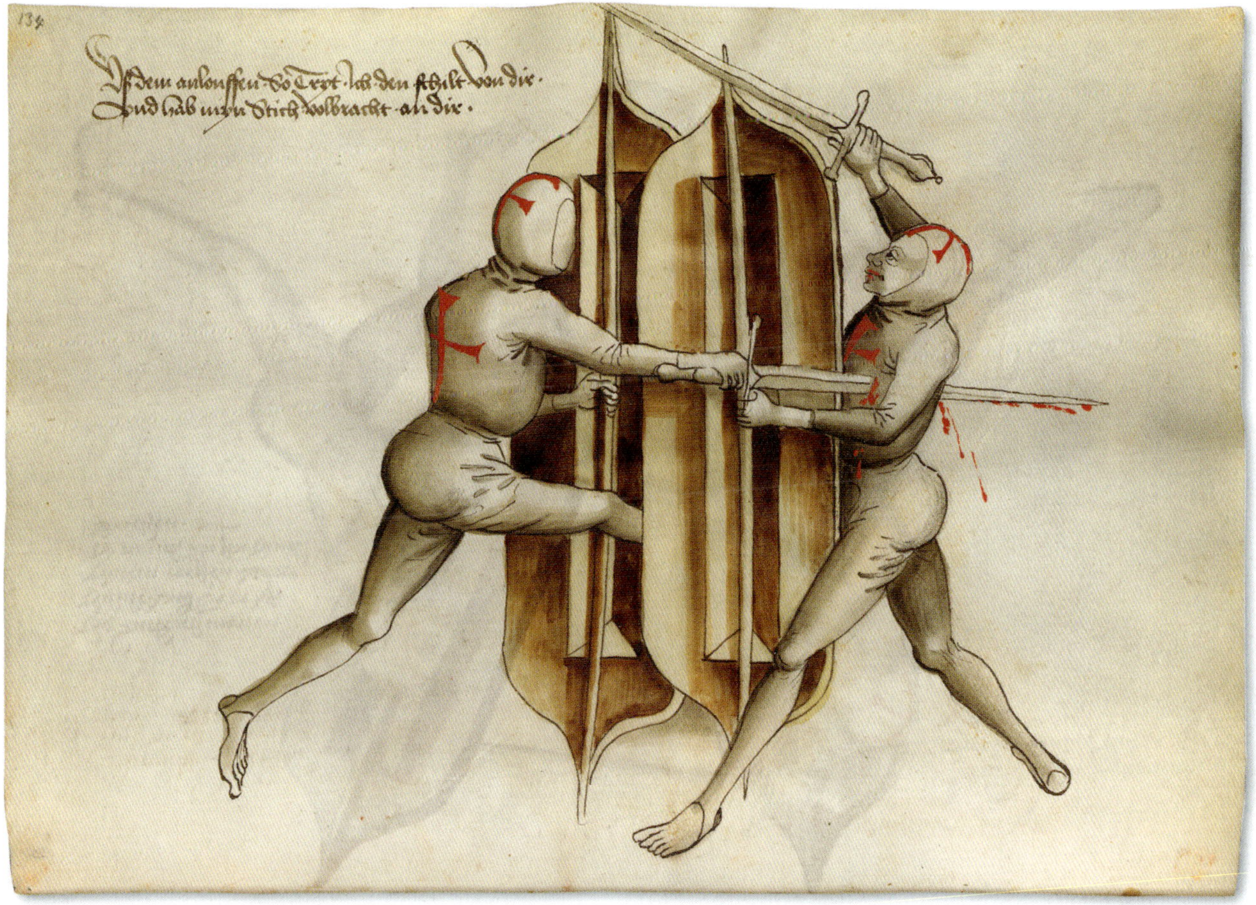

From the running in: So I kick the shield away from you, and my thrust to you is completed.

Here, I displace with strength in order to thwart your strikes and thrusts.

~

This one has turned his shield; note that.

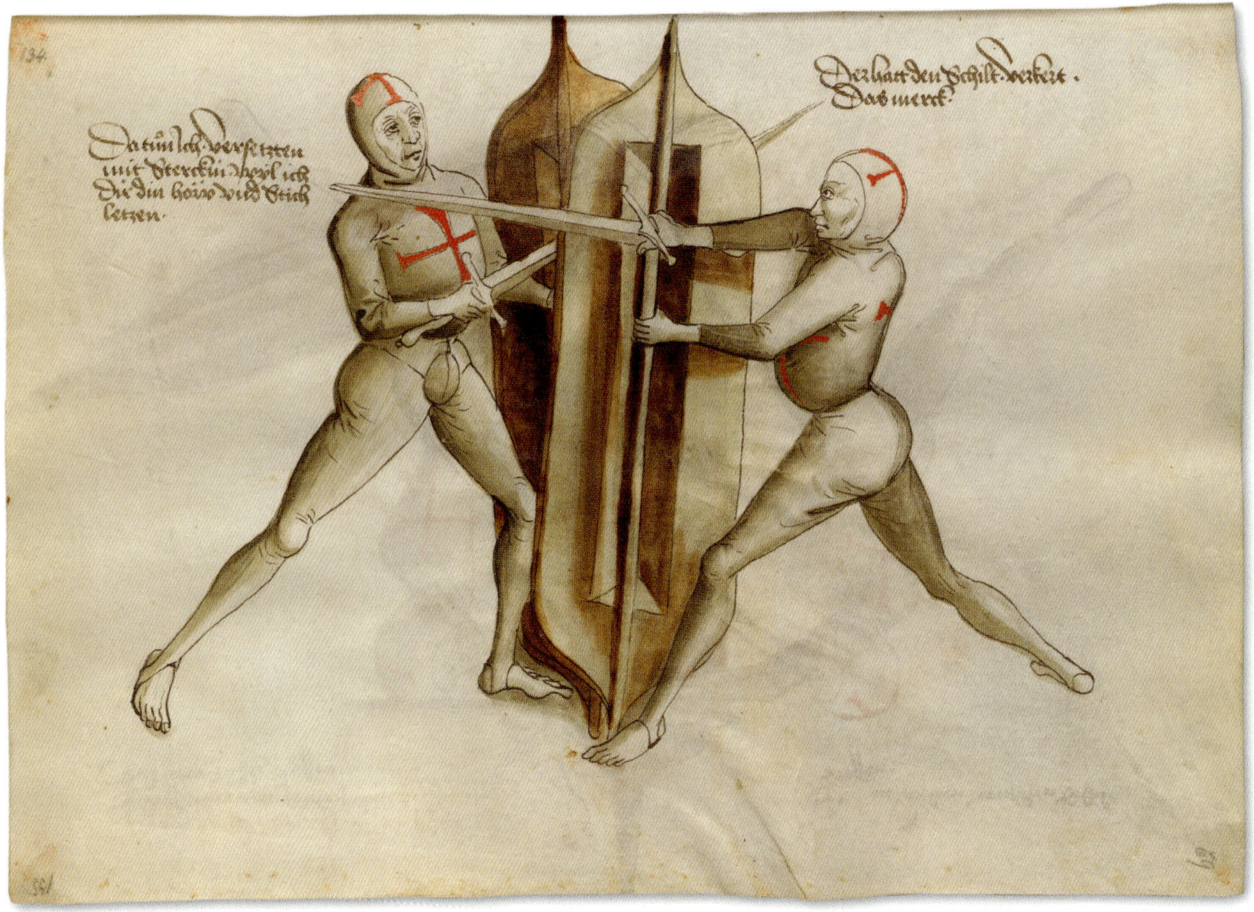

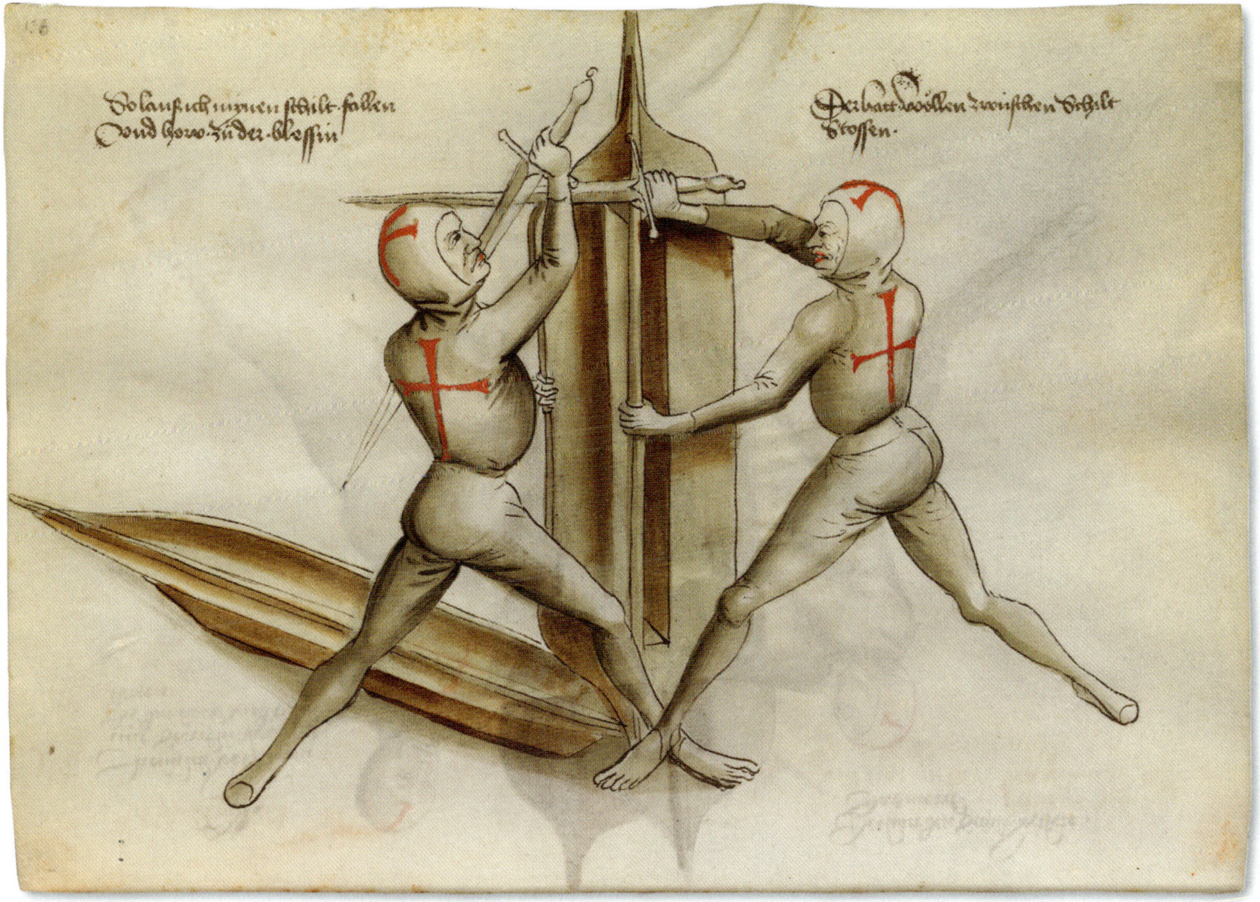

So I drop my shield and strike to the opening.

This one intended to strike between the shields.

This is another ending.

This one has also dropped his shield, grabs the other by the elbow and thrusts the sword through him.

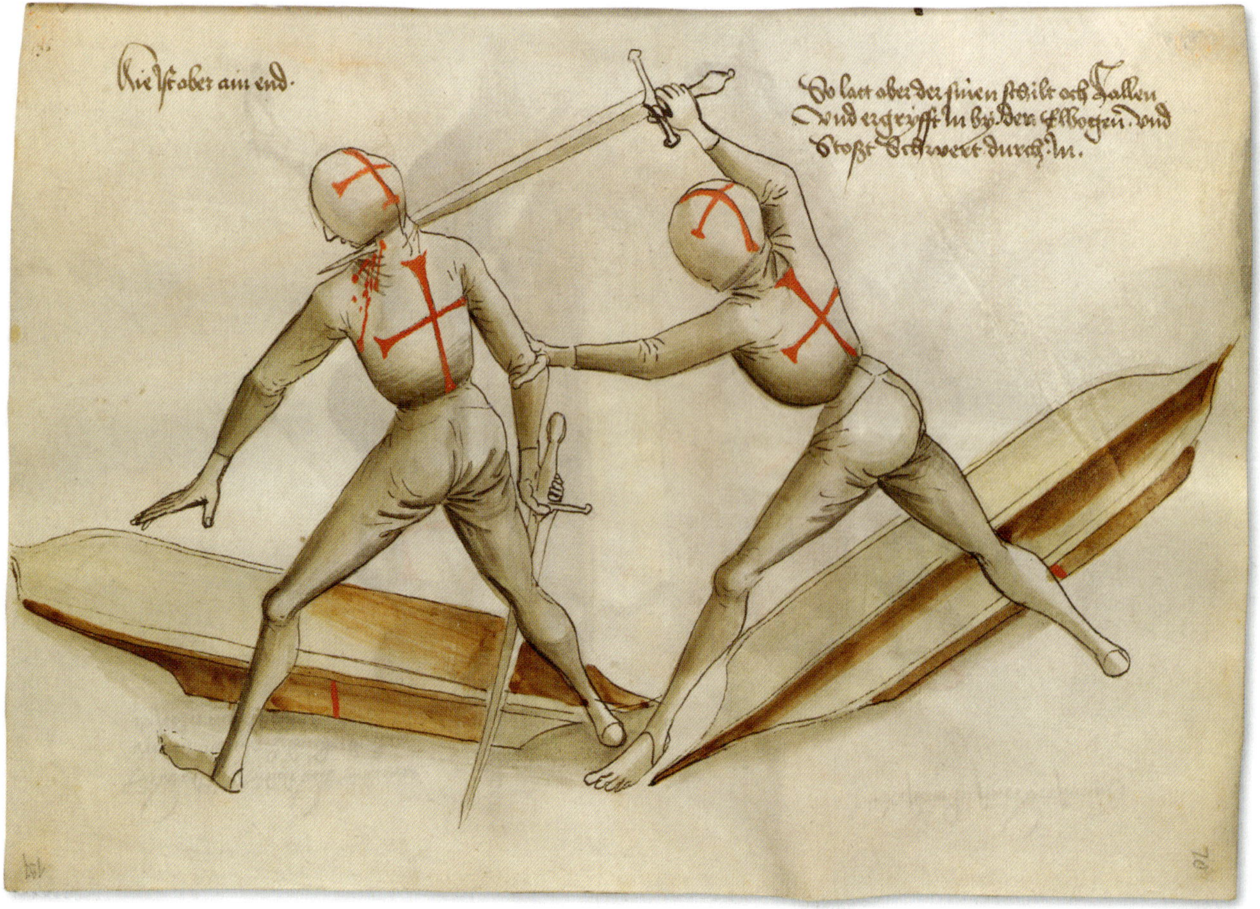

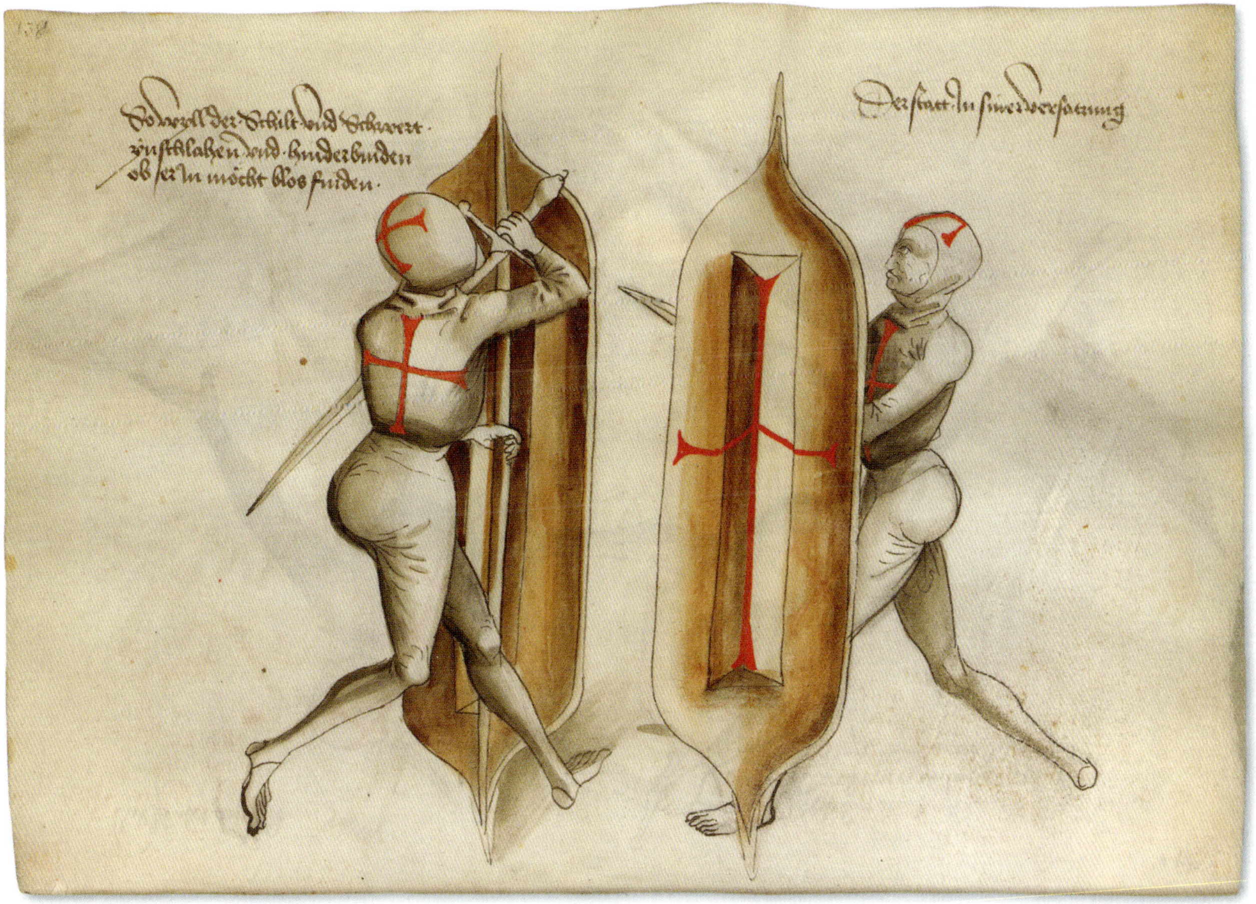

So this one strives to strike with shield and sword in order to bind behind in case he finds the other uncovered.

That one stands in a displacement.

He has twitched away from my hit. With this, he deceives me.

~

Here, I have twitched away from his hit so that he has missed. In this way, I have travelled after him with this piece.

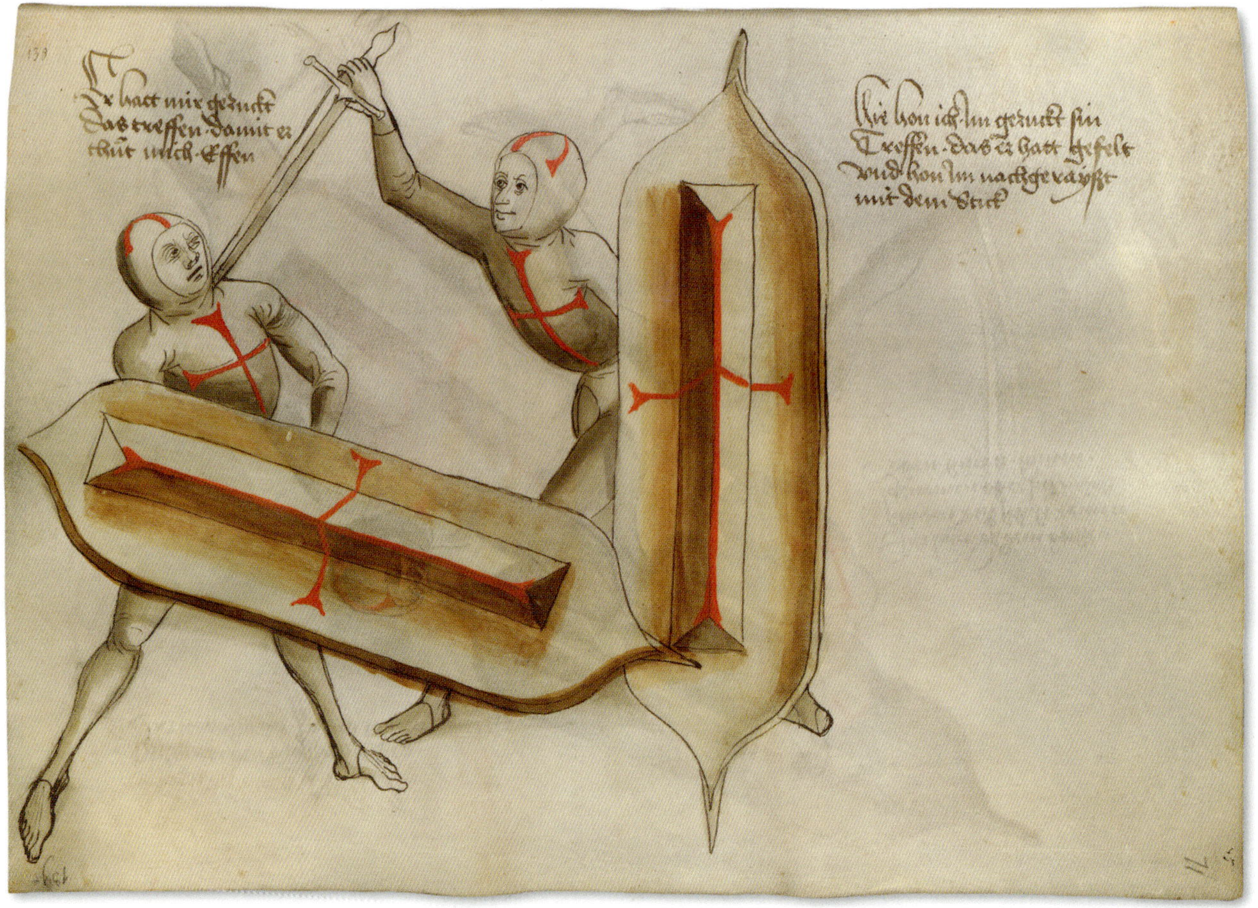

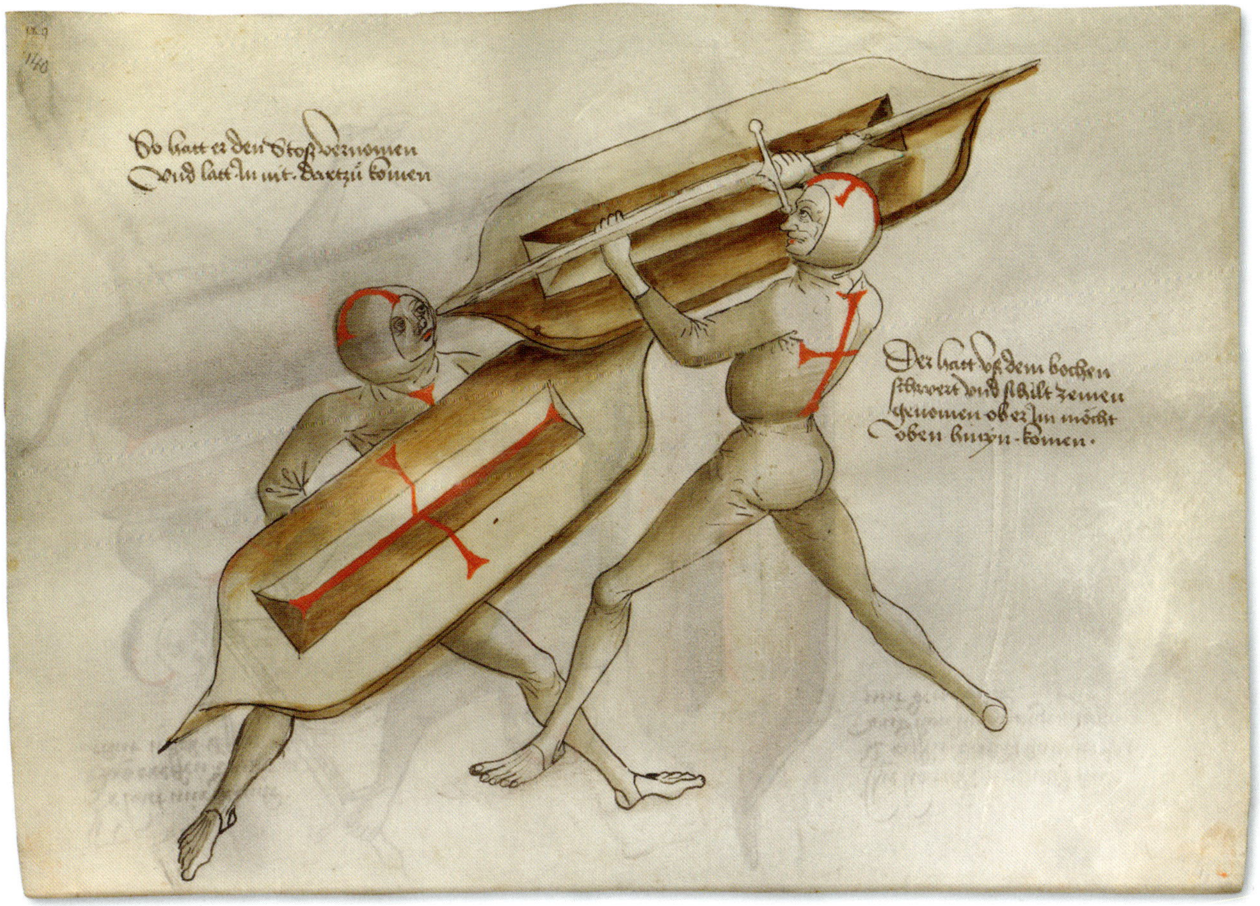

So he has perceived the blow and does not let the opponent come to this point.

From the beat, this one has taken sword and shield together in order to try to reach the other above.

Here, I have pressed between shield and sword so that I could come to your head.

When I have struck crossways, I have failed.

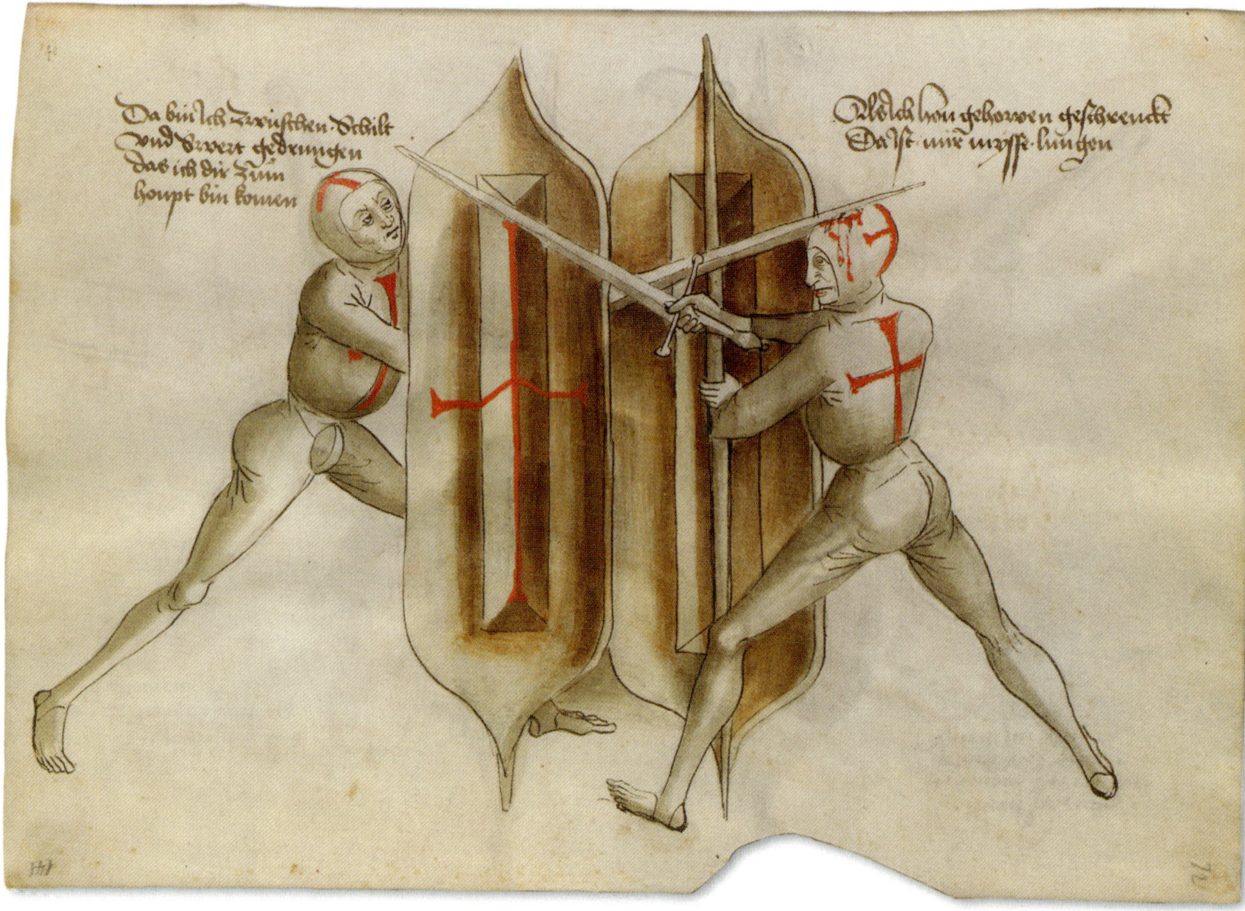

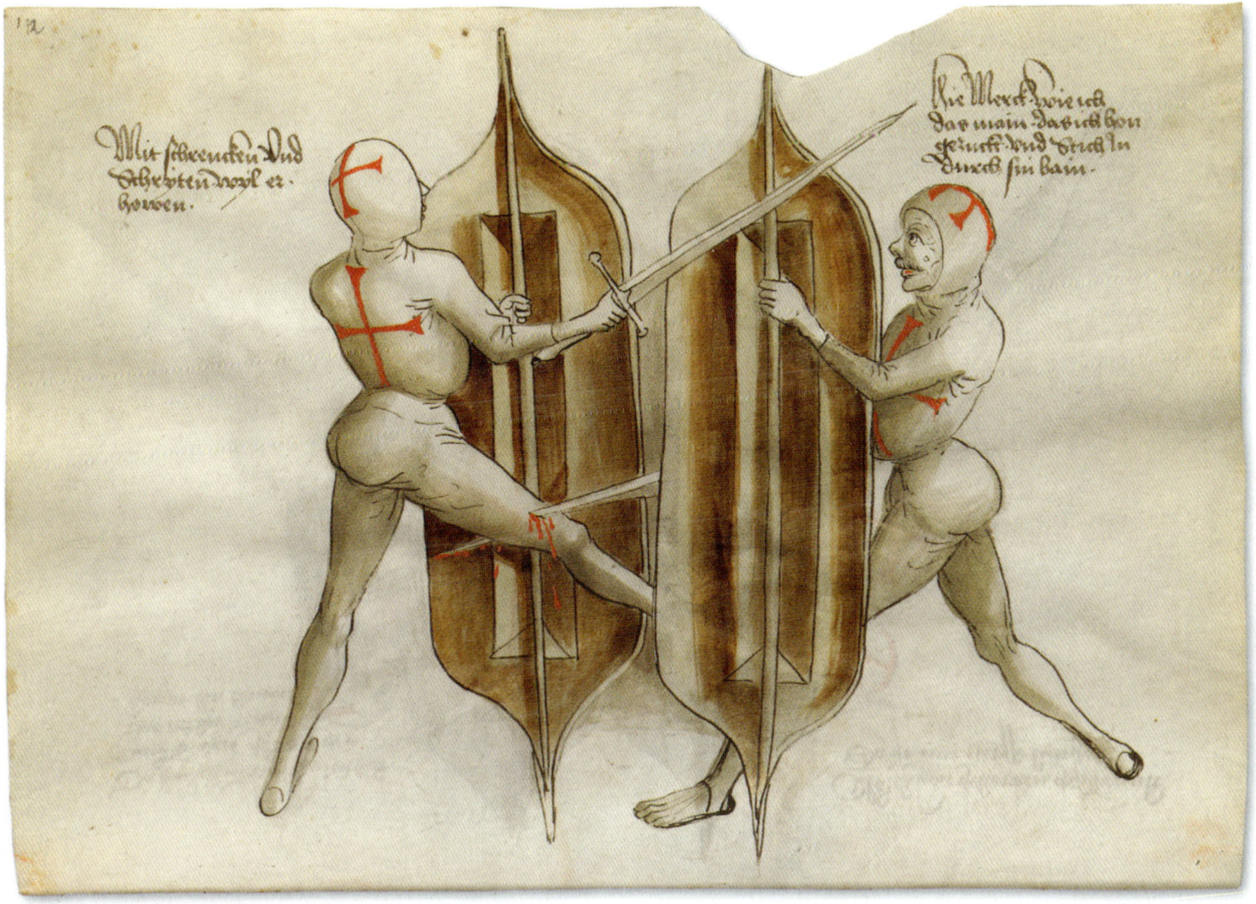

Crossways and stepping, he strives to strike.

~

Note what I have in mind: I have twitched away and thrust through his leg.

This one stands free in a displacement.

In this stance, he secretly changes the sword into his left hand and stabs the other to death over the shield.

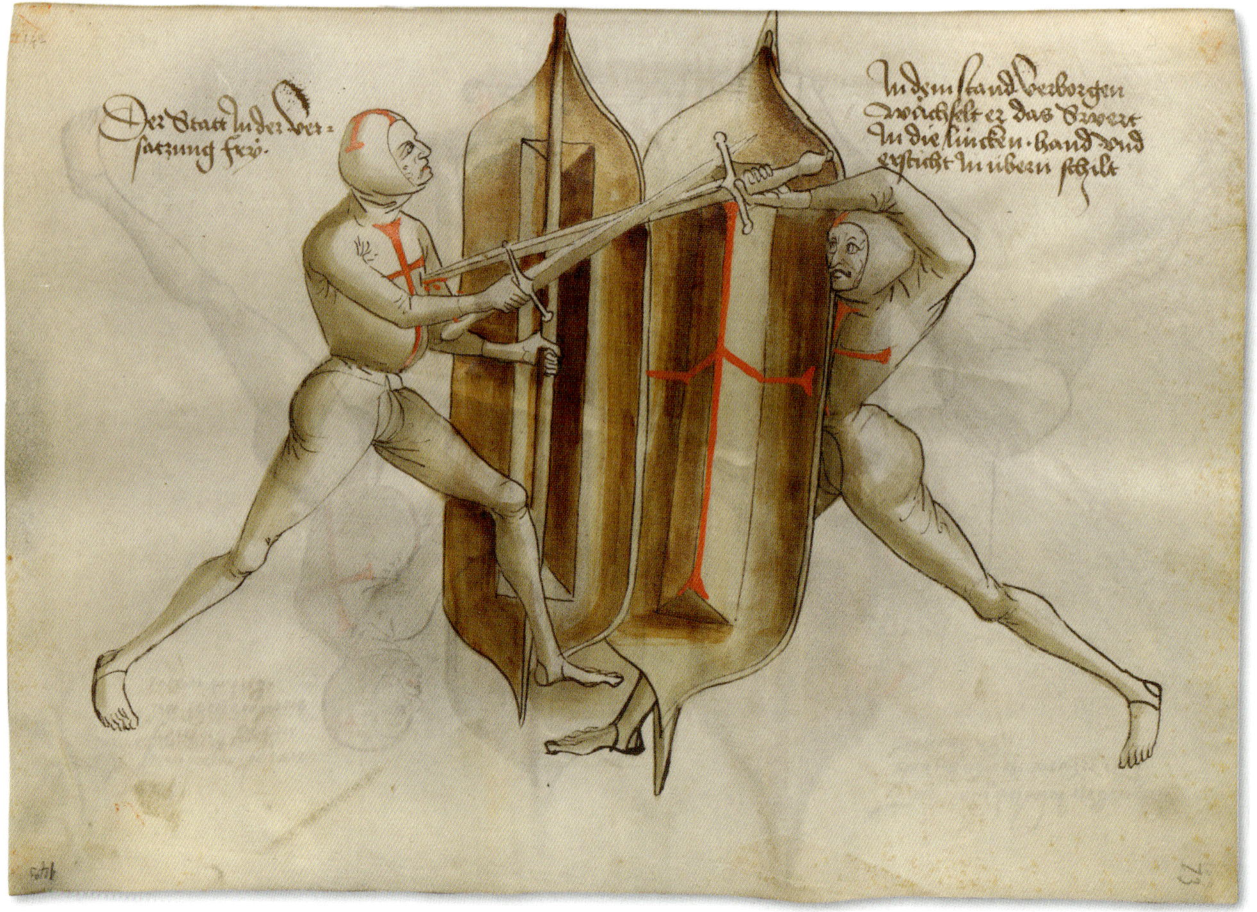

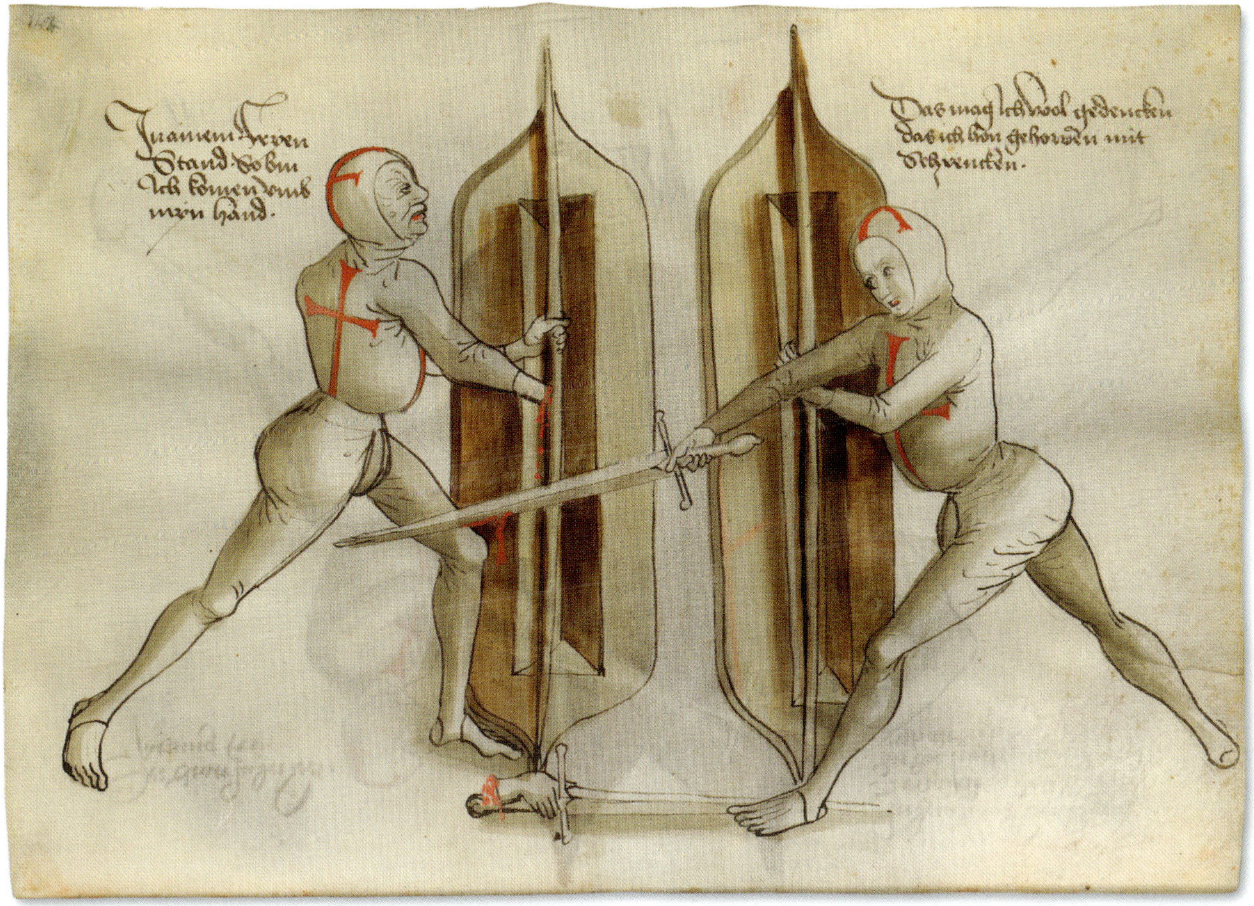

In a free stance, I have lost my hand.

I should remember well that I have struck with with arms crossed.

This one is positioned in the stance in which he lost his hand.

~

This one strives to wind over the shield in order to see whether he can find the other uncovered.

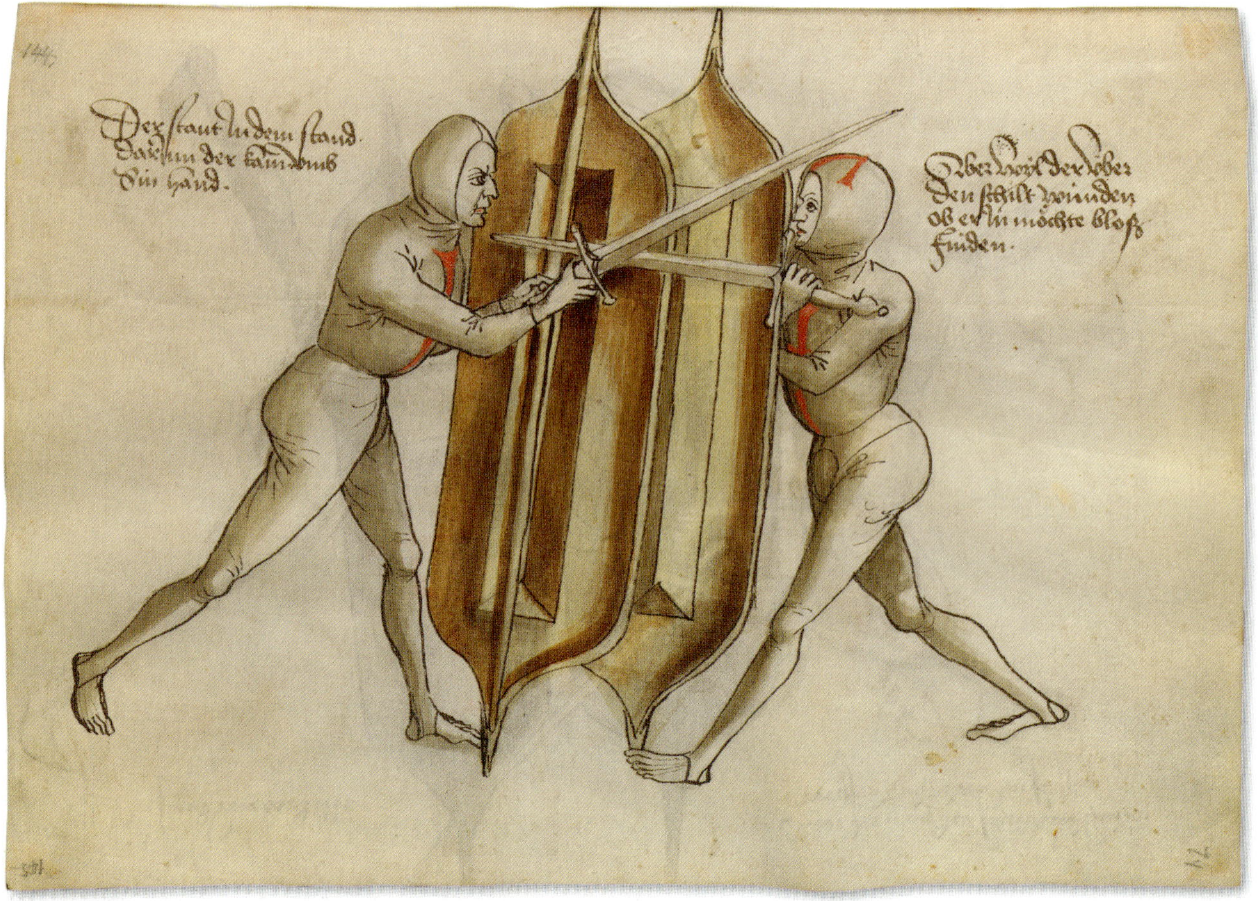

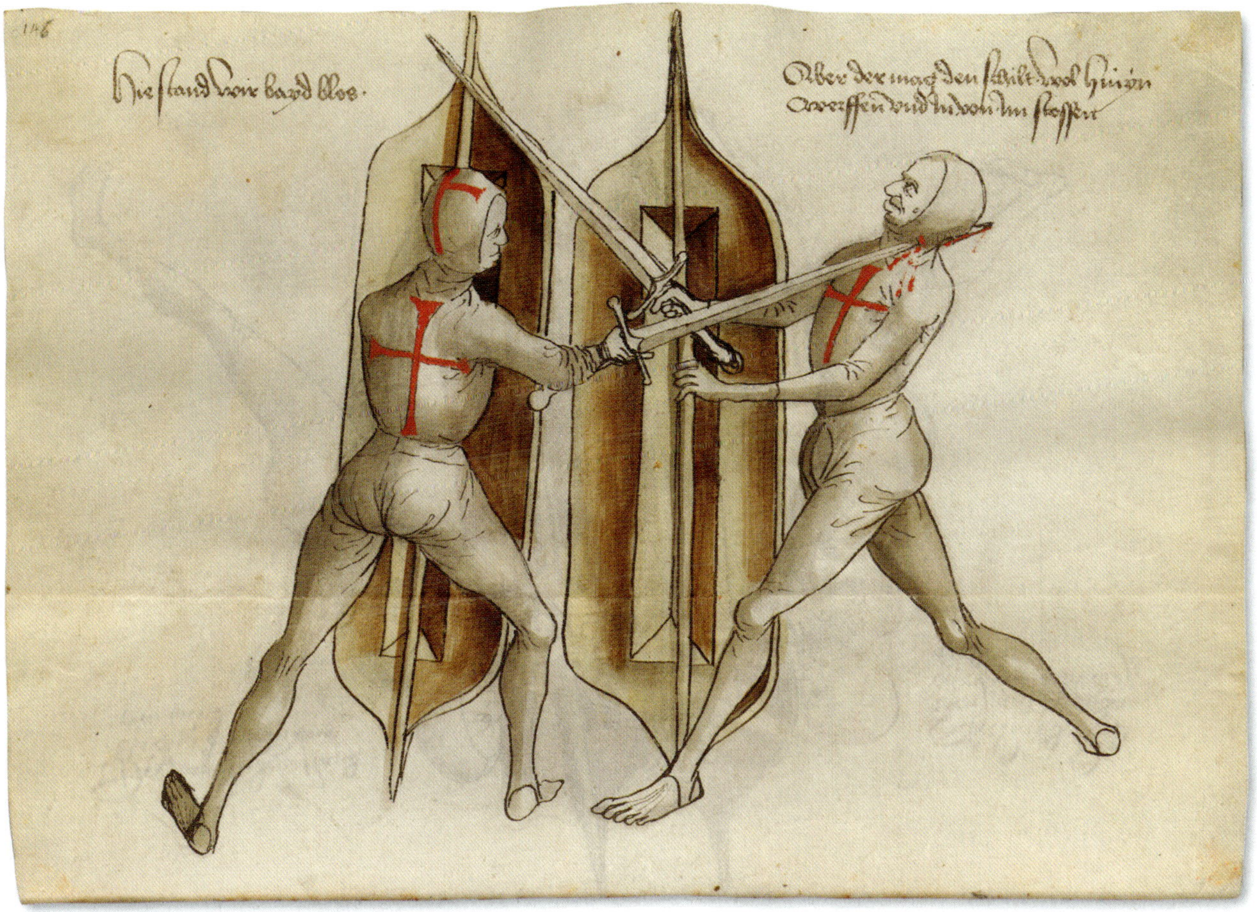

Here, both are standing uncovered.

~

But this one can throw in his shield and push the other away from him.

Here, one has pushed the other away. Because of the throw and thrust he must bend, and so I strike his head and back.

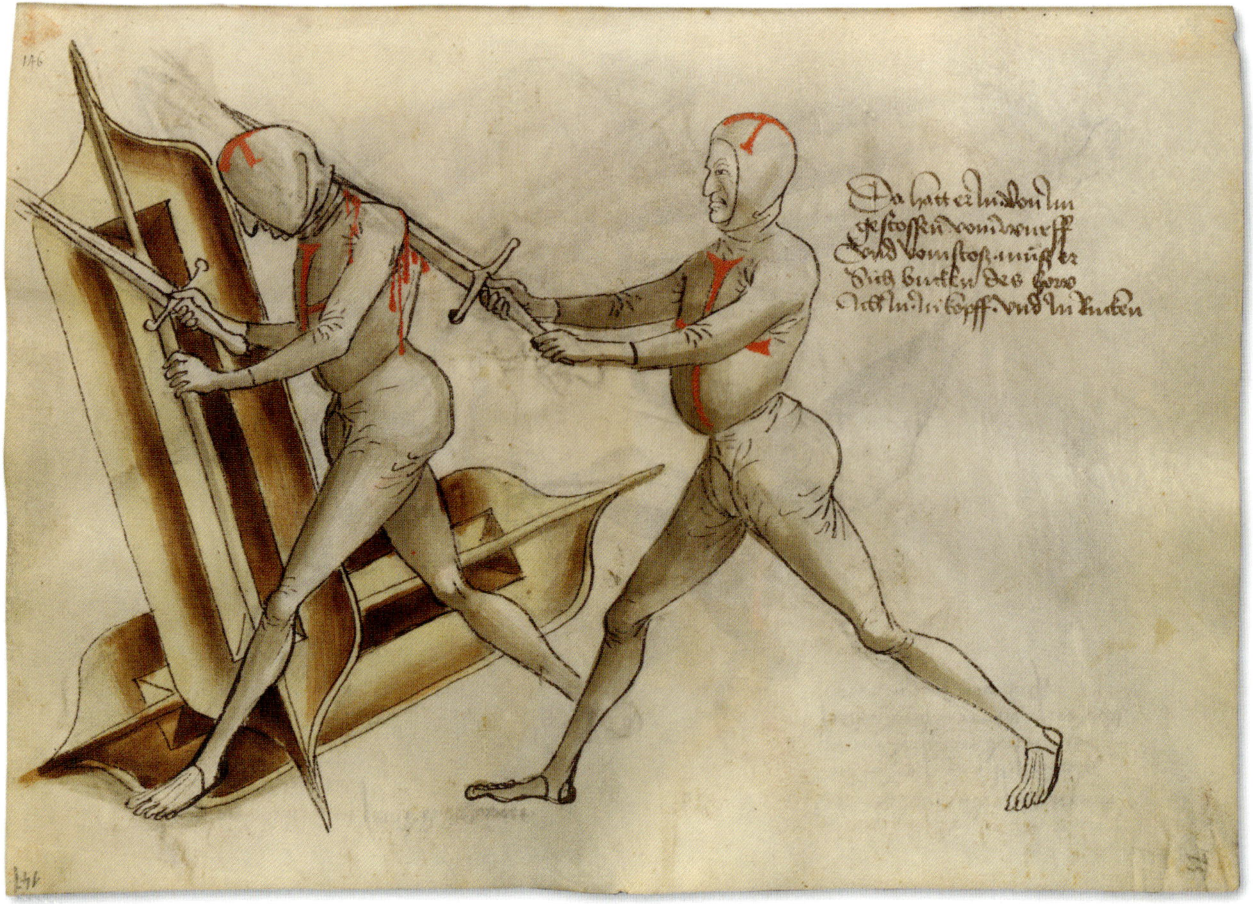

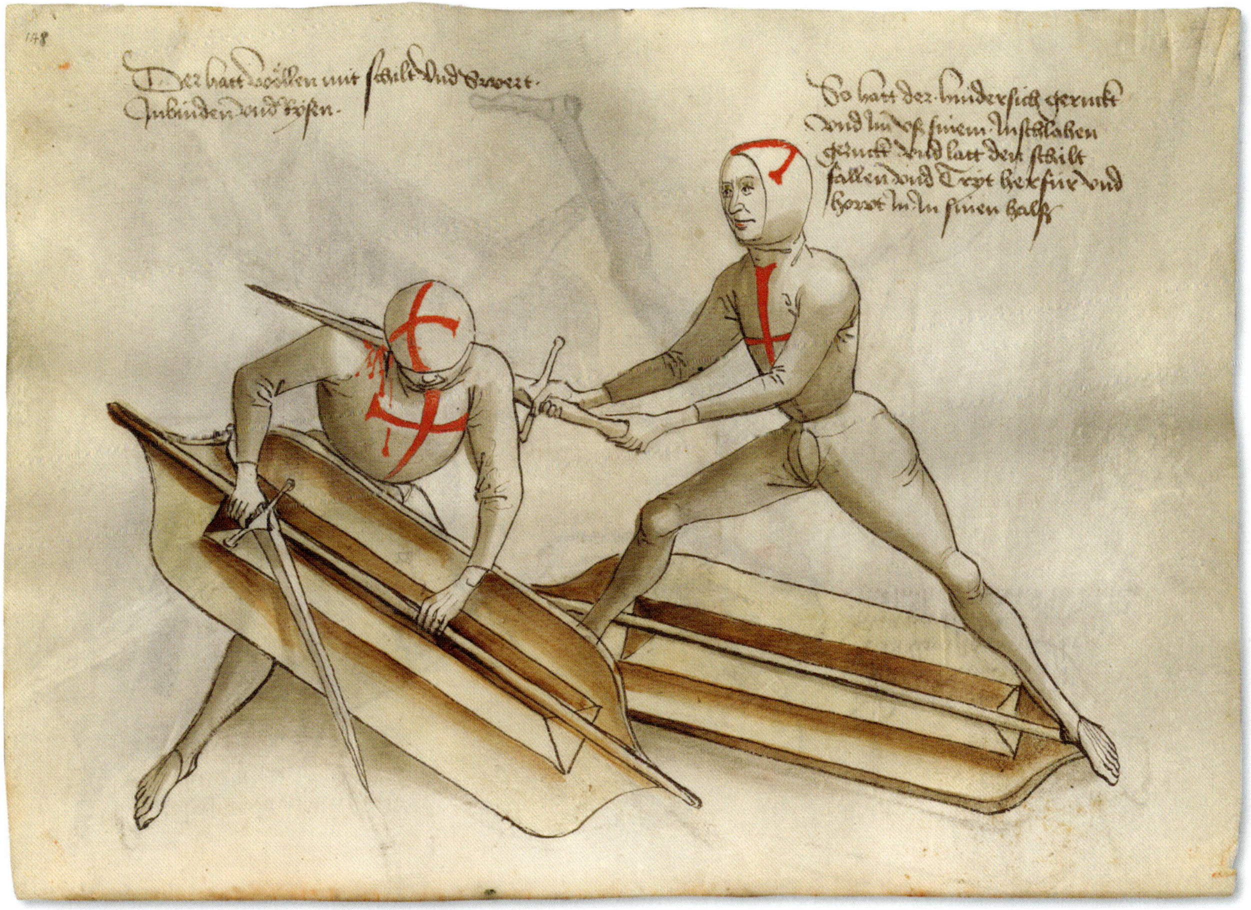

This one strives to bind with shield and sword and wrench at his opponent.

~

That one has wrenched backwards and twitched with his striking-in. He drops the shield, advances and strikes his opponent in the neck.

He tried to wind
the shield inwards
above.

~

One shall see
the shield from
the inside. The
crossways thrust
shall happen to him.

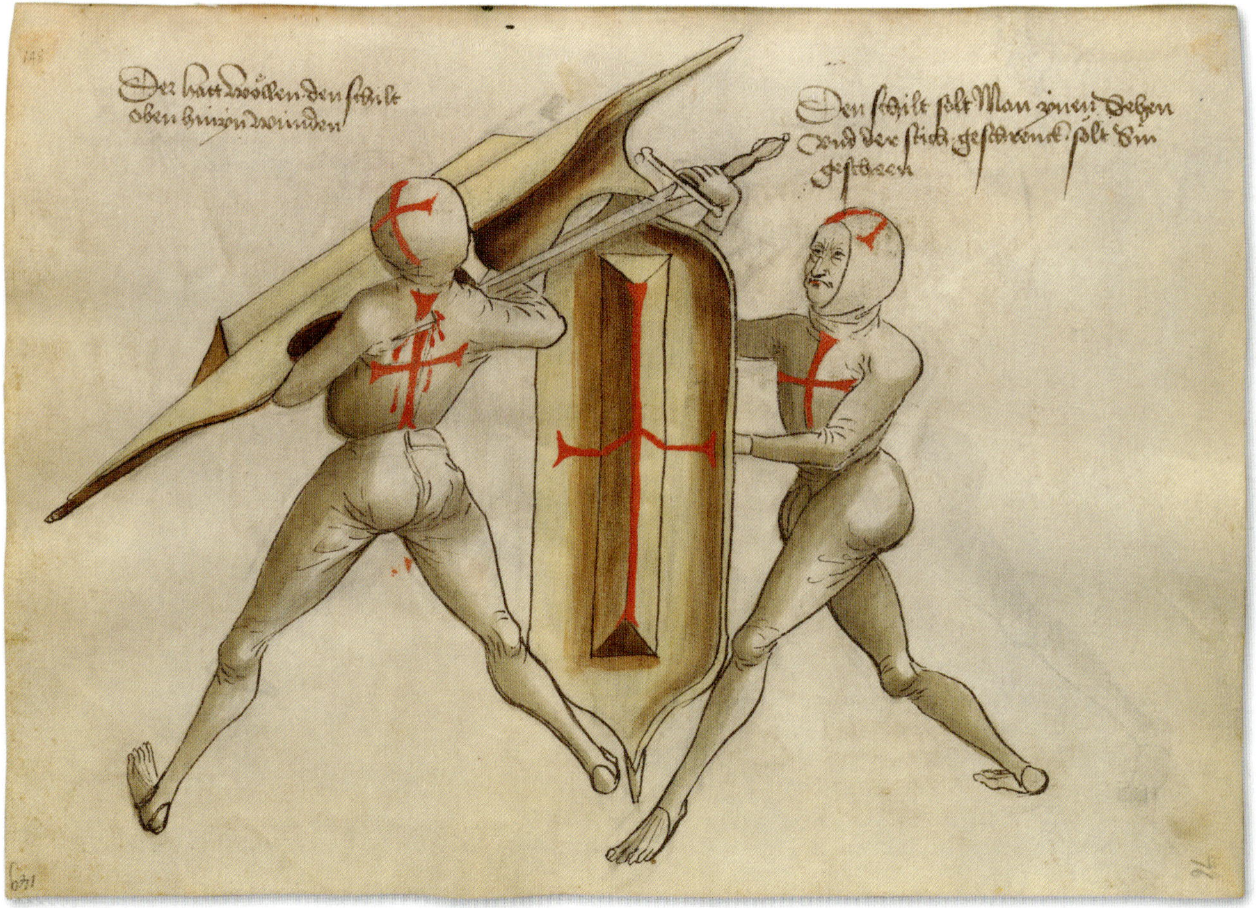

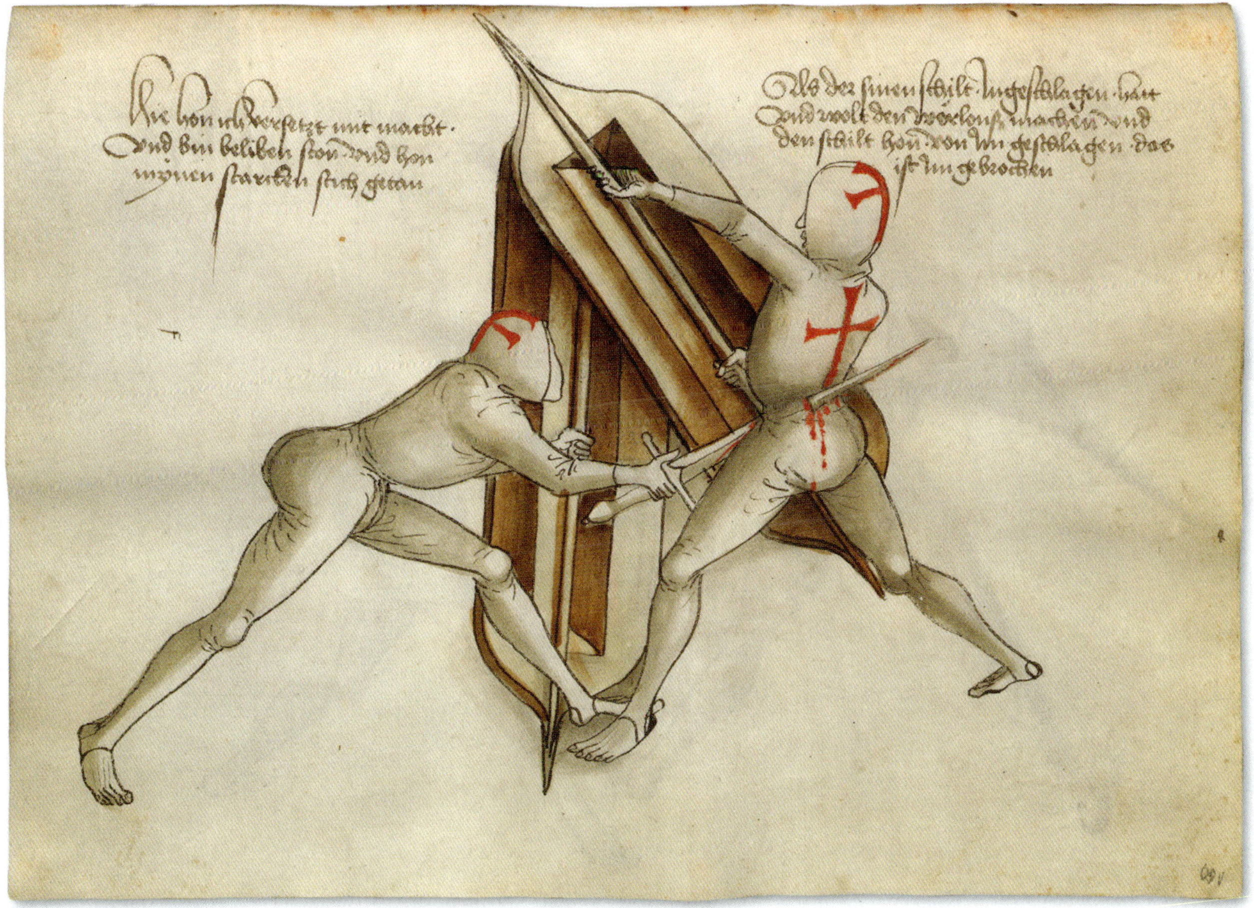

Here, I have displaced my opponent forcefully and stood still. I have made my powerful thrust.

~

When he has struck with his shield in order to render the other defenceless by striking the shield away from him, his piece was broken.

One has stepped forward, takes the other by the neck and throws him with the shield …

… after the other has struck the shield.

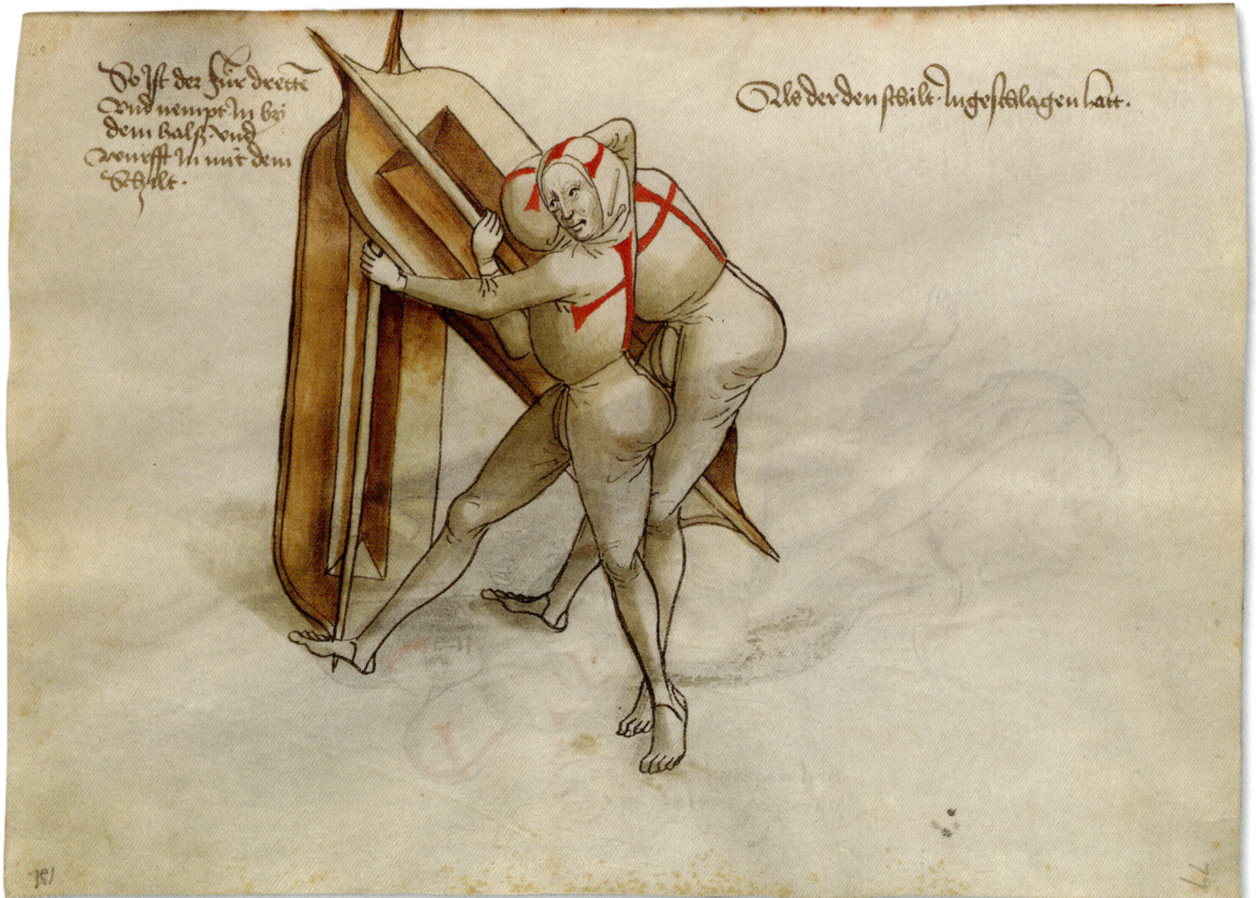

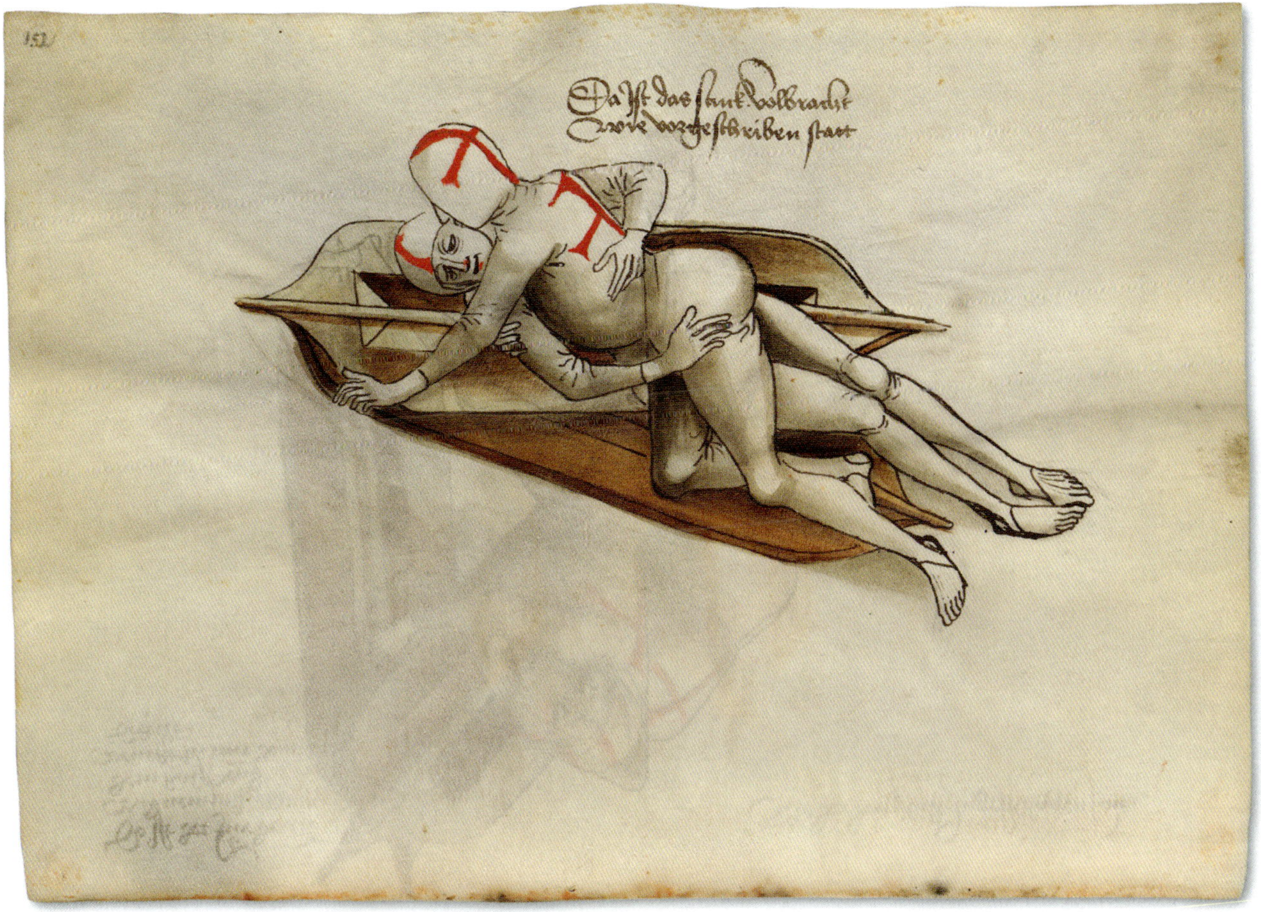

Here, the piece is completed as described above.

He has bound in between the shield and the opponent, struck the hook around his foot and strives to wrench at him.

~

I want to repay him appropriately for coming between me and the shield.

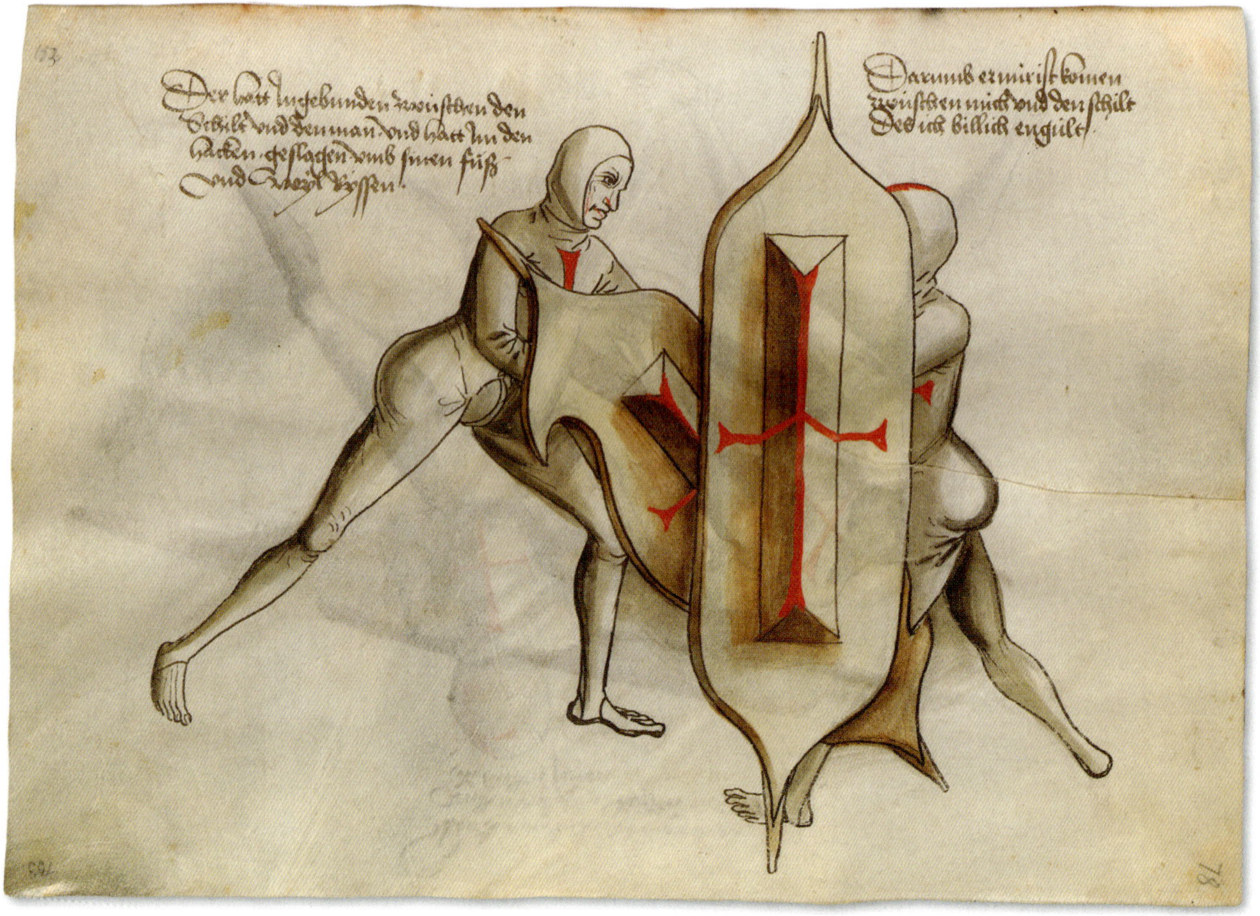

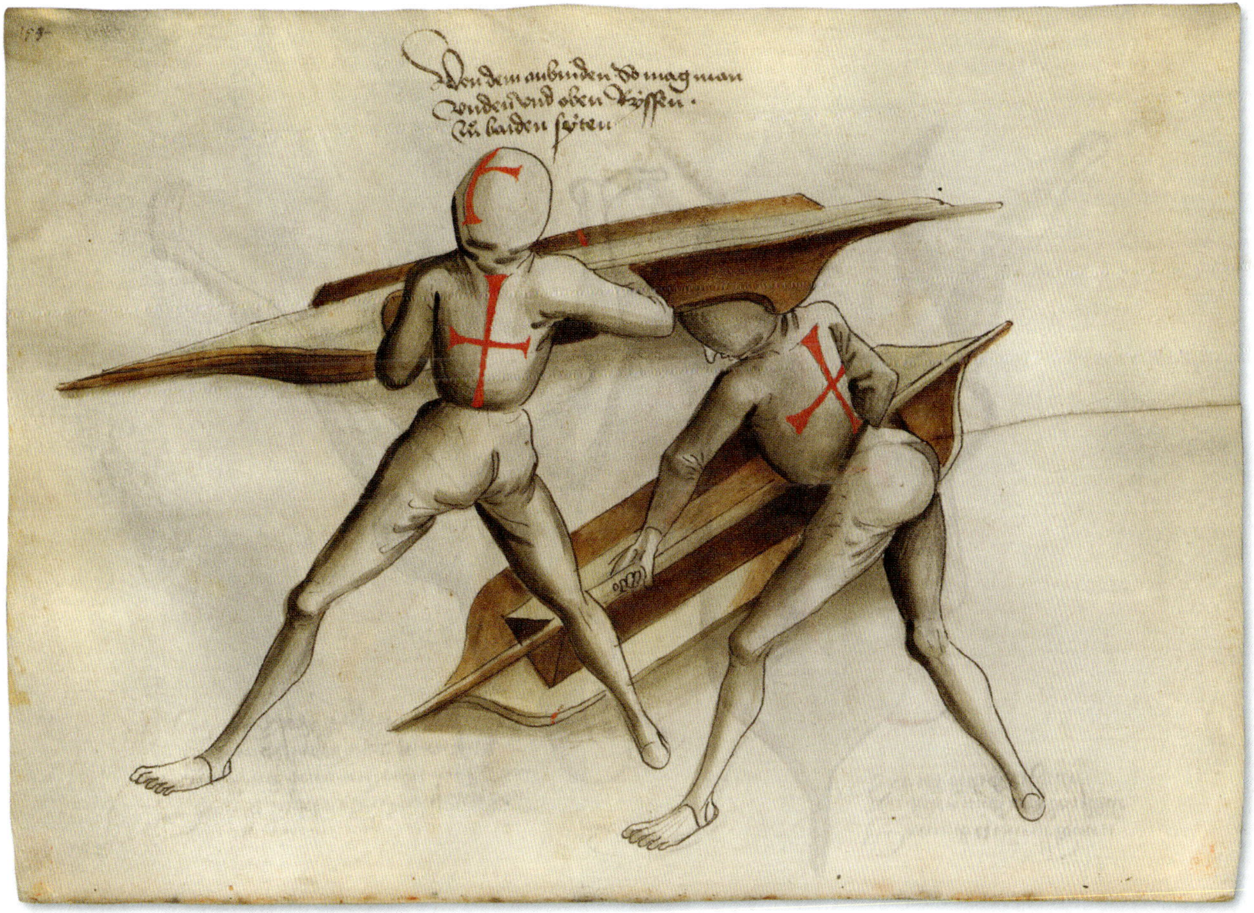

From the binding: this way one can wrench at both sides at the bottom and the top.

Both, one and the other, can do it below or above as it happens.

~

Again from the upper bind or beat: In this way either one can grab the other's shield and tear it from him if it so happens.

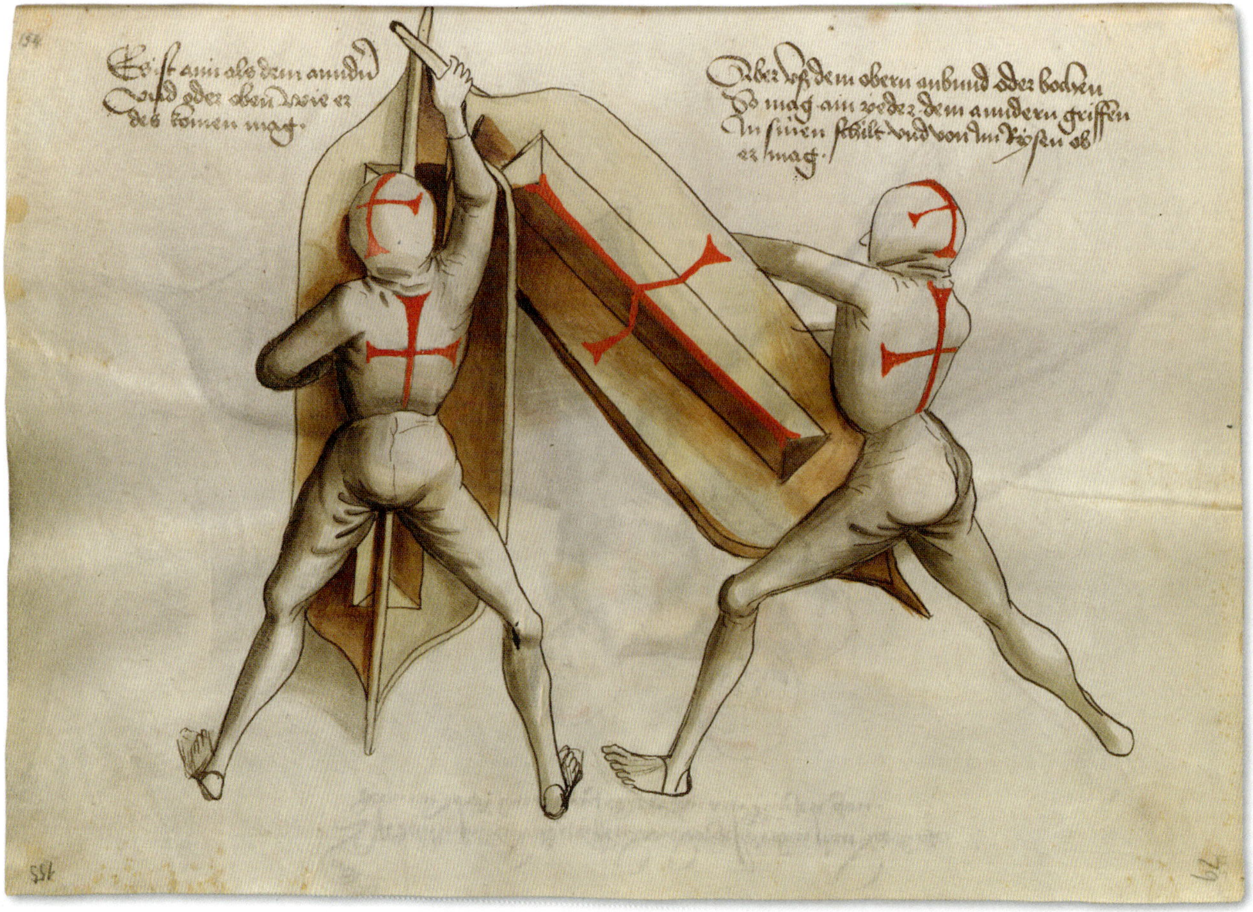

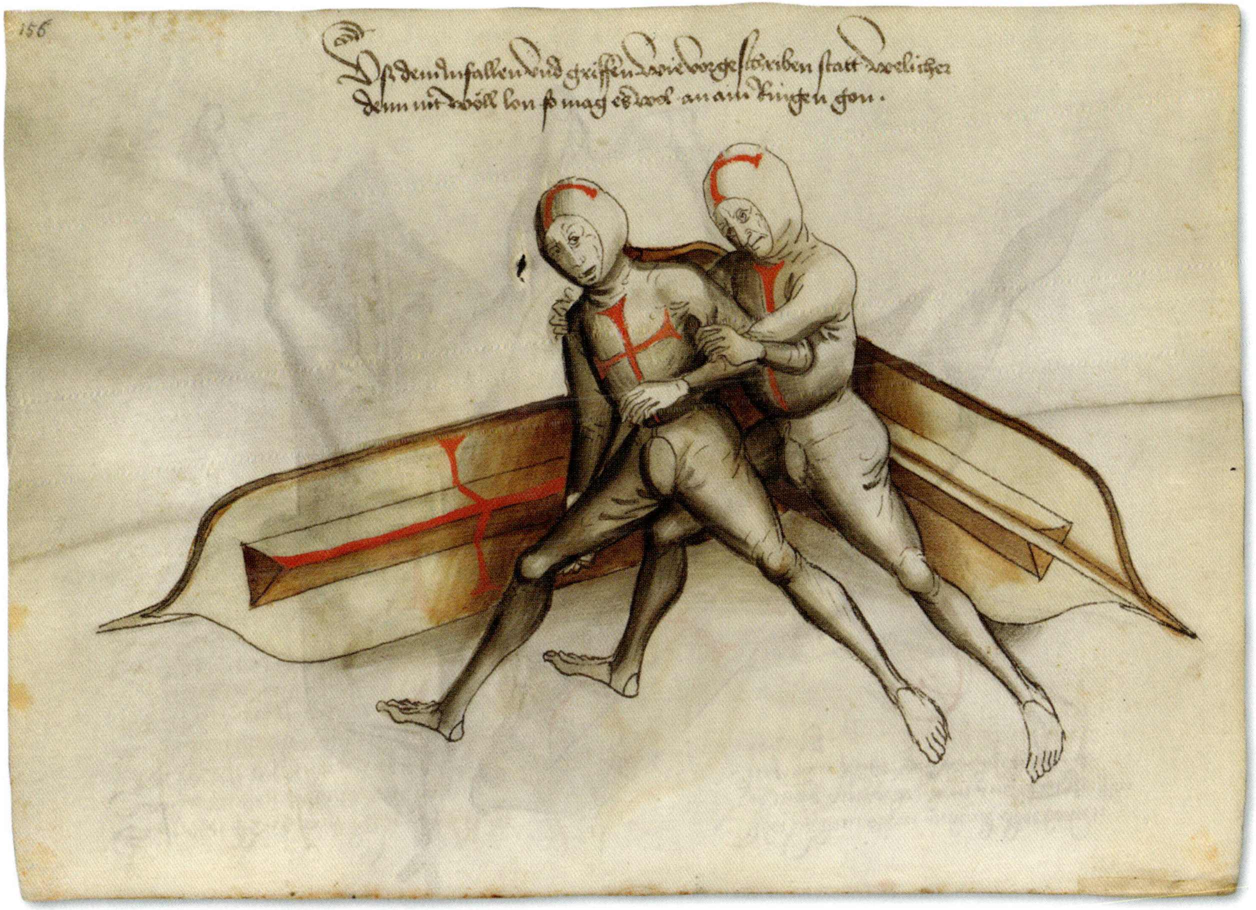

From the falling-in and grabbing, as described above: if one refuses to let go, it may come to wrestling.

Note this piece well: after one intended to strike and beat, he remained standing still. He releases the right hand from the shield, grabs the other's elbow and pushes him away so that he has to turn.

Here, he has struck into the other's shield and bound behind.

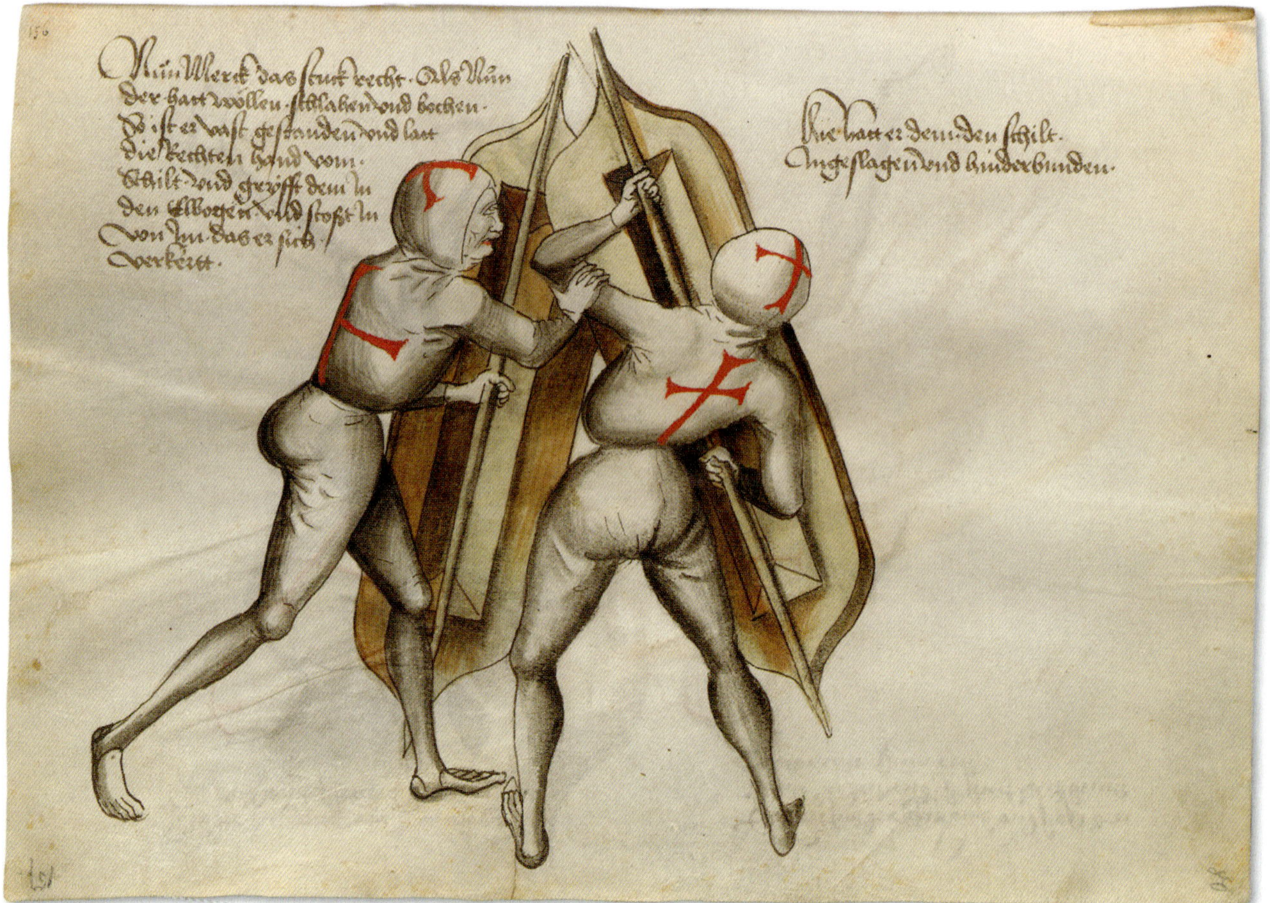

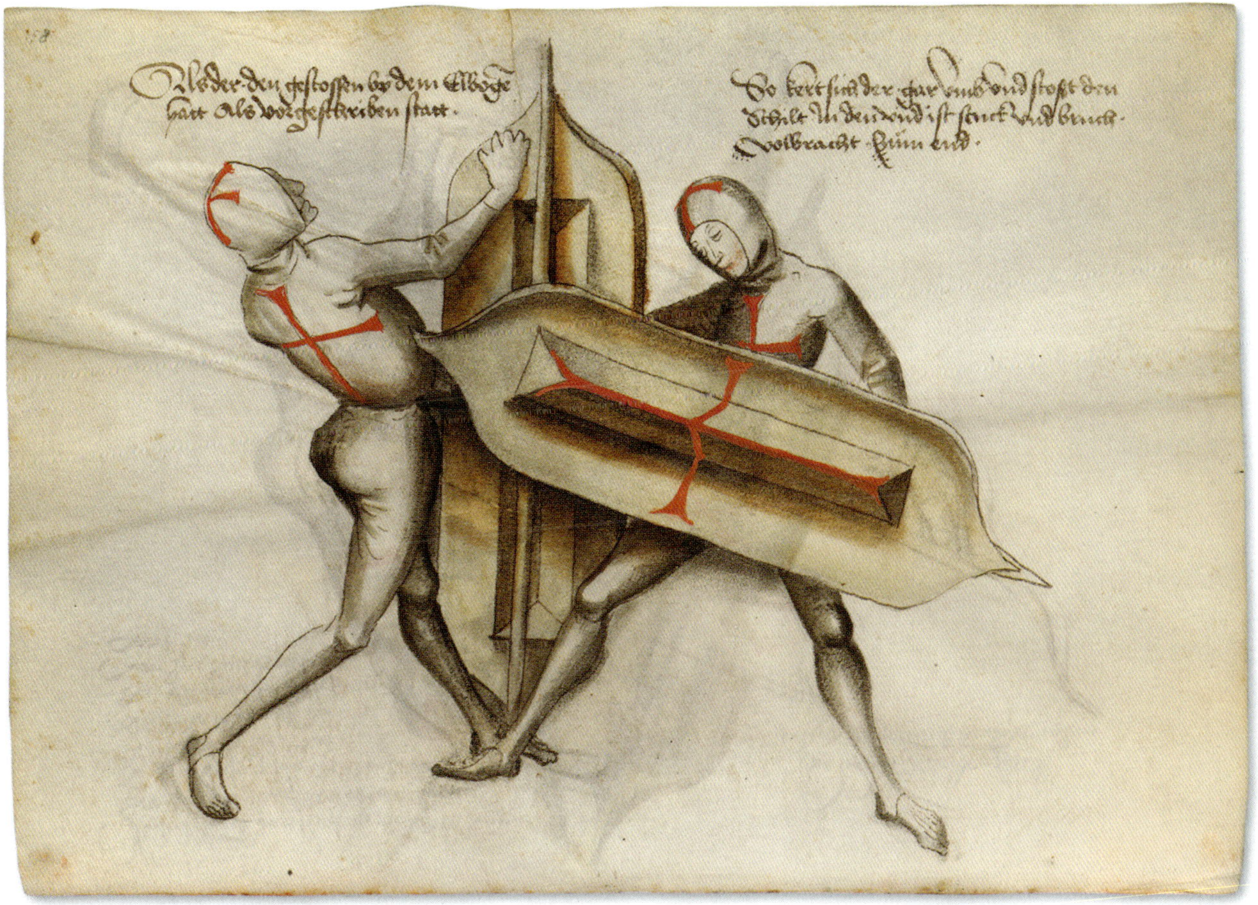

After one has pushed the other at the elbow, as described above, …

… the other turns around and beats the shield into him. Thus the piece and break have come to an end.

While running to him, one has stepped behind the other and brought him upwards. He fears he will be thrown.

This is another strong running in from which one has bound behind the other. He will throw him, or else he will become defenceless.

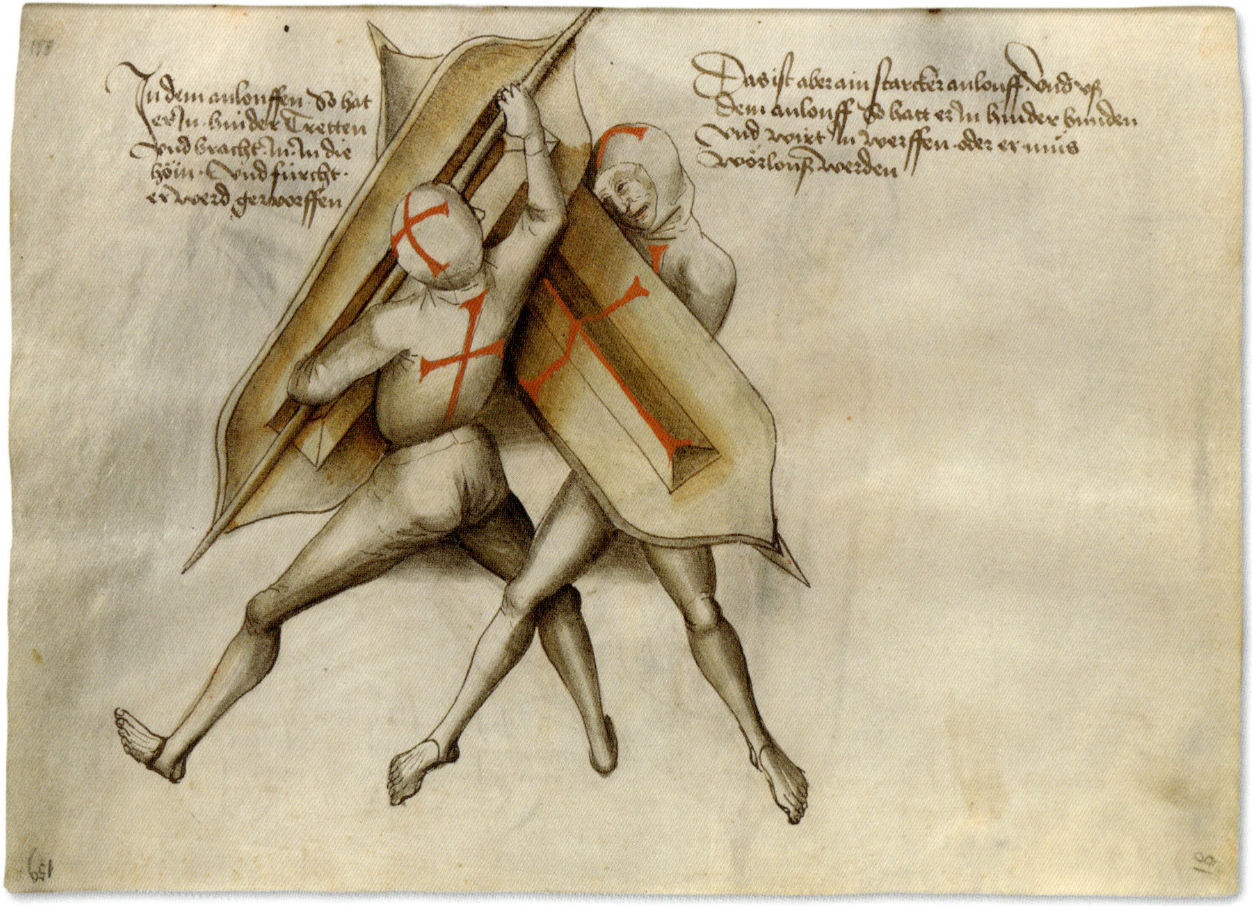

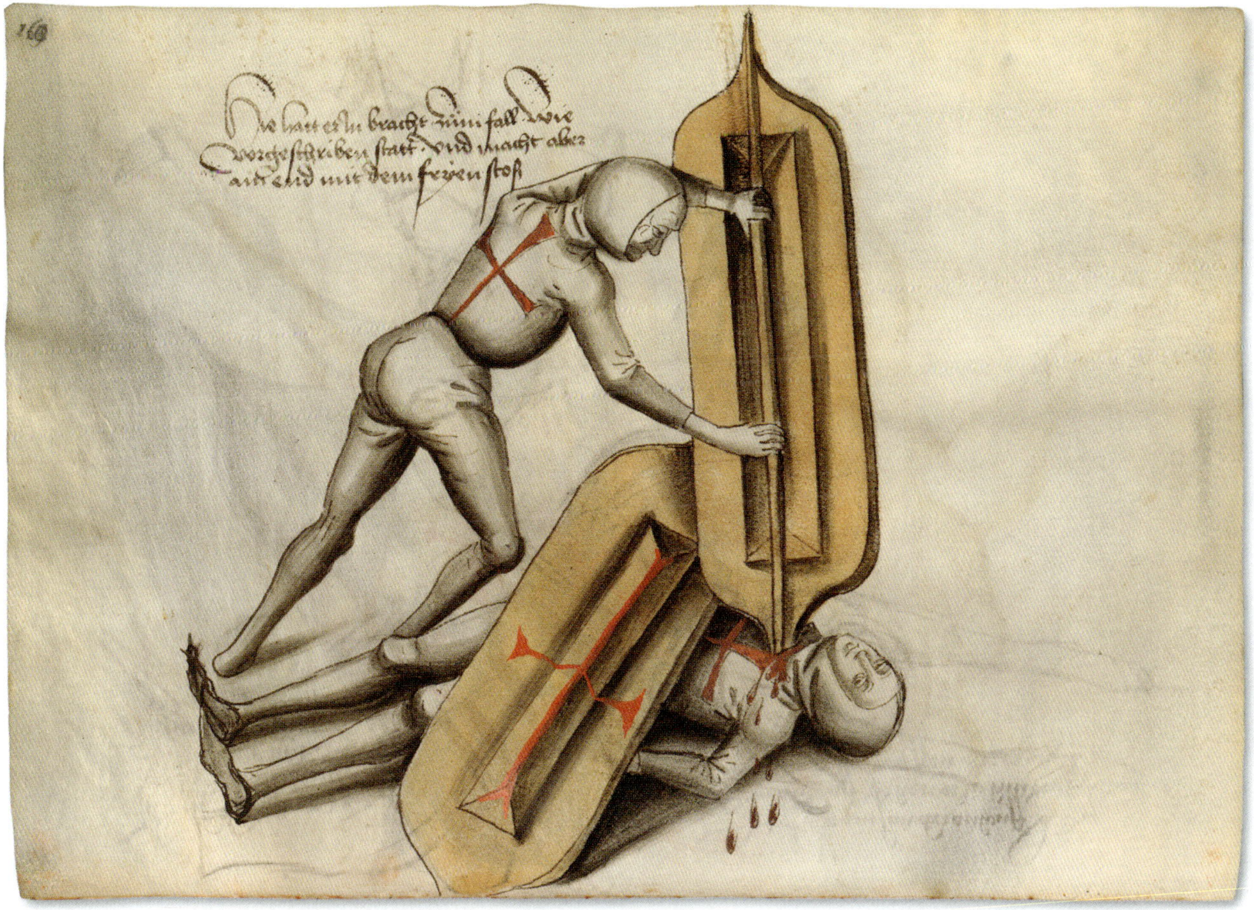

Here, one has brought the other down as described above. He also makes an end with a free thrust.

Here, both stand in a twisted binding while striking, and they oppress each other.

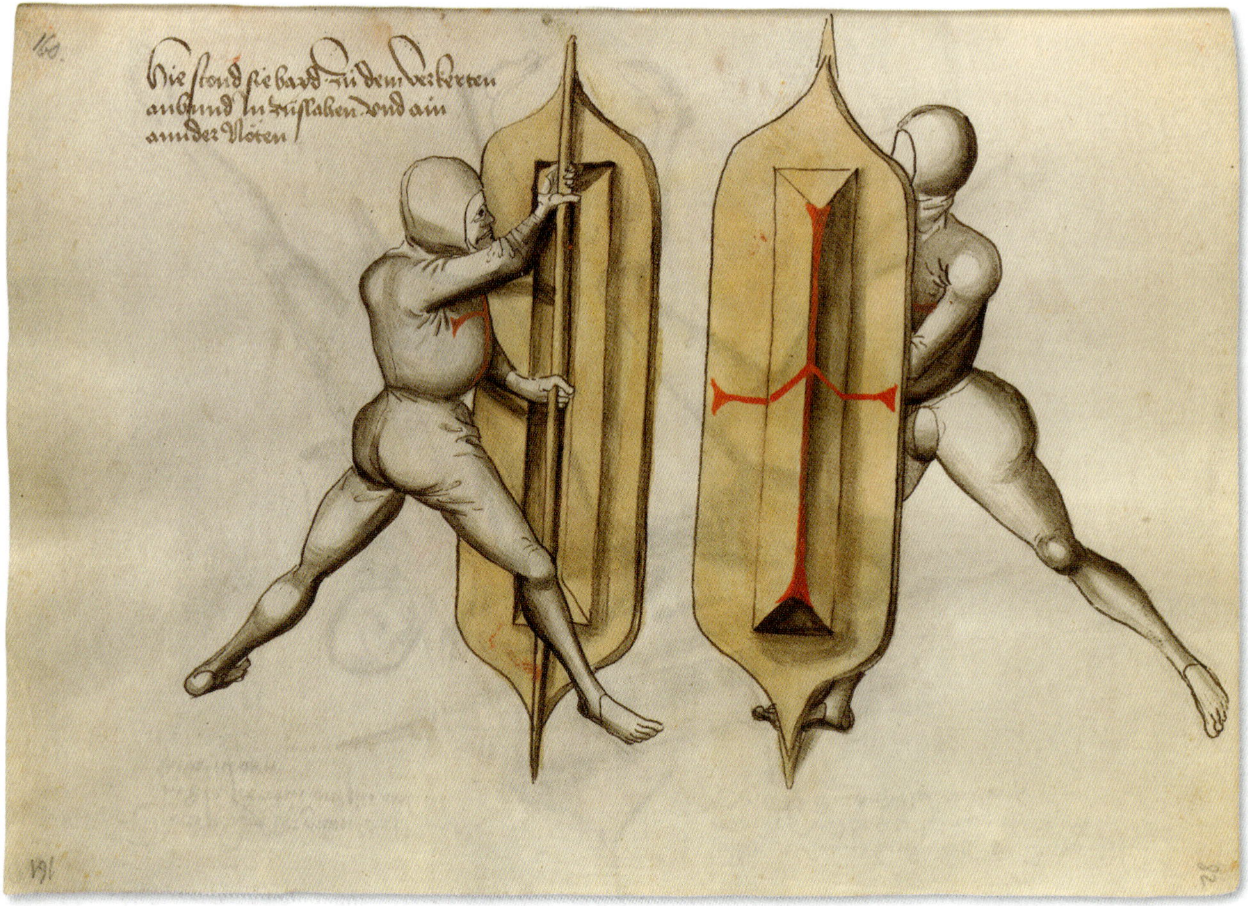

But this one strives to give the weakness into the strength in this way. Also note his work with this.

~

With the beat, each one would like to bind out his opponent so that he can come behind his shield to make him defenceless.

Here, he has succeeded in bringing the weak to the strong, and he can either push or wrench.

Here, the other has lost the strong and is being wrenched down.

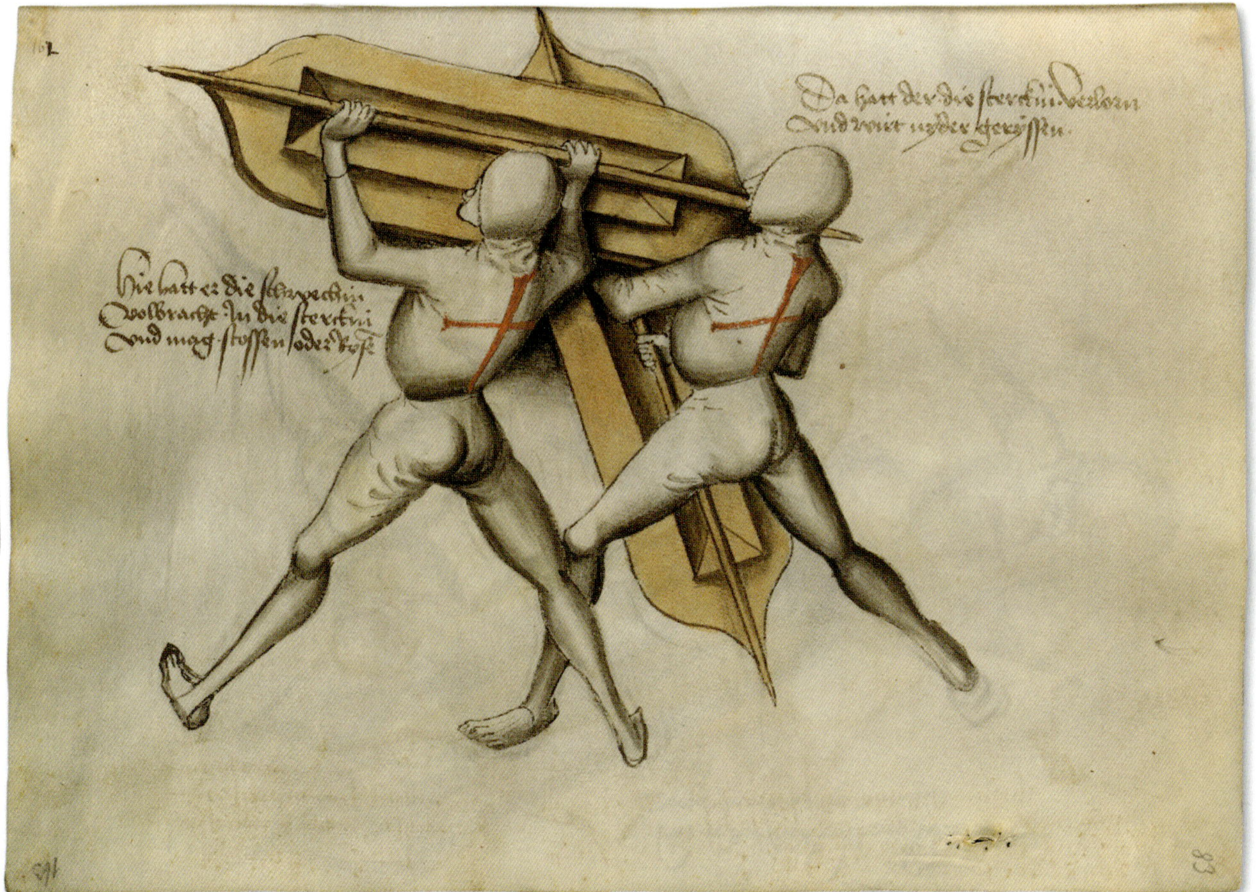

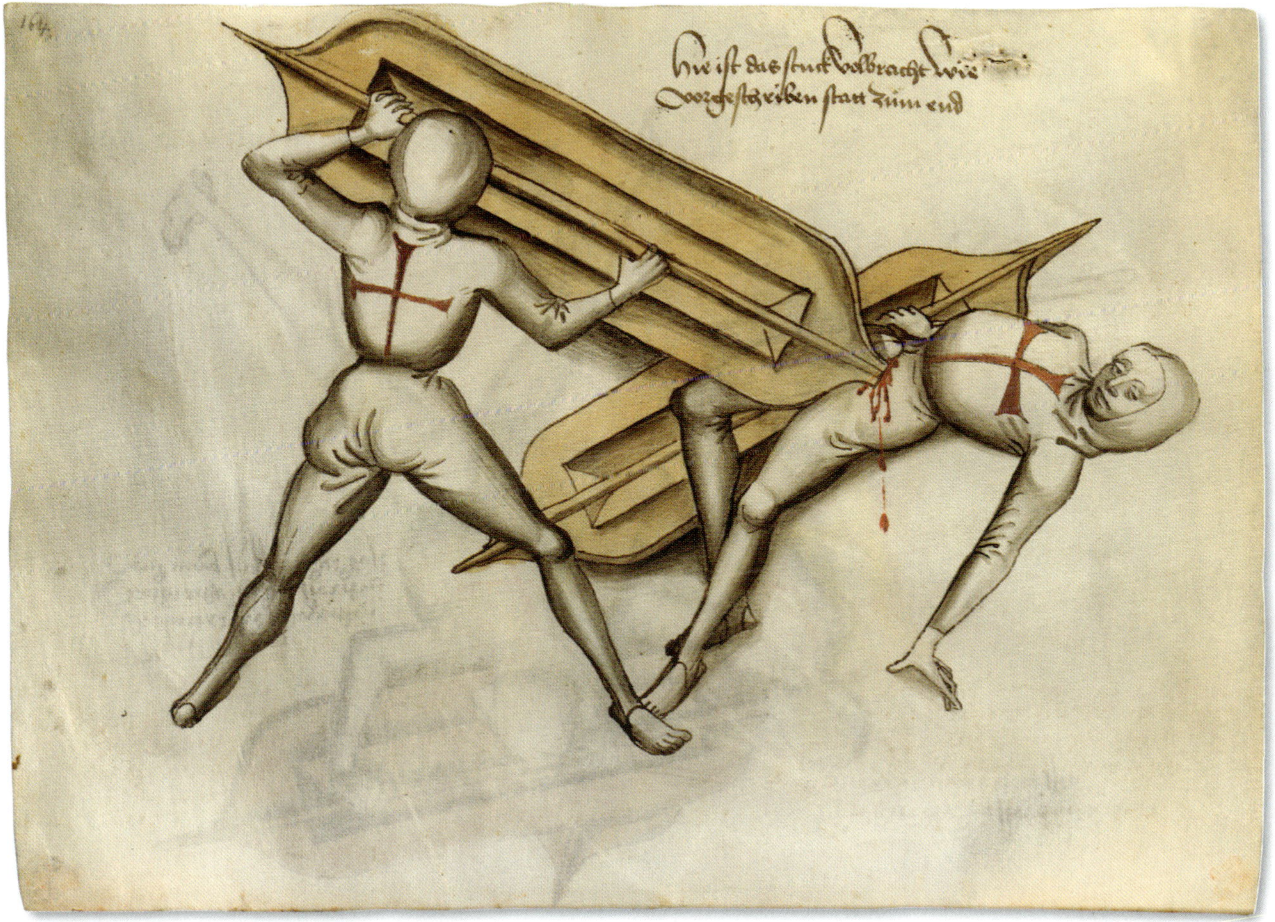

Here, the piece has come successfully to an end, as described above.

Now this one lets himself be oppressed and goes below into the weak. He takes the foot of the other one and wrenches.

Also from the binding; another oppression: The other has overpowered the first one in this way.

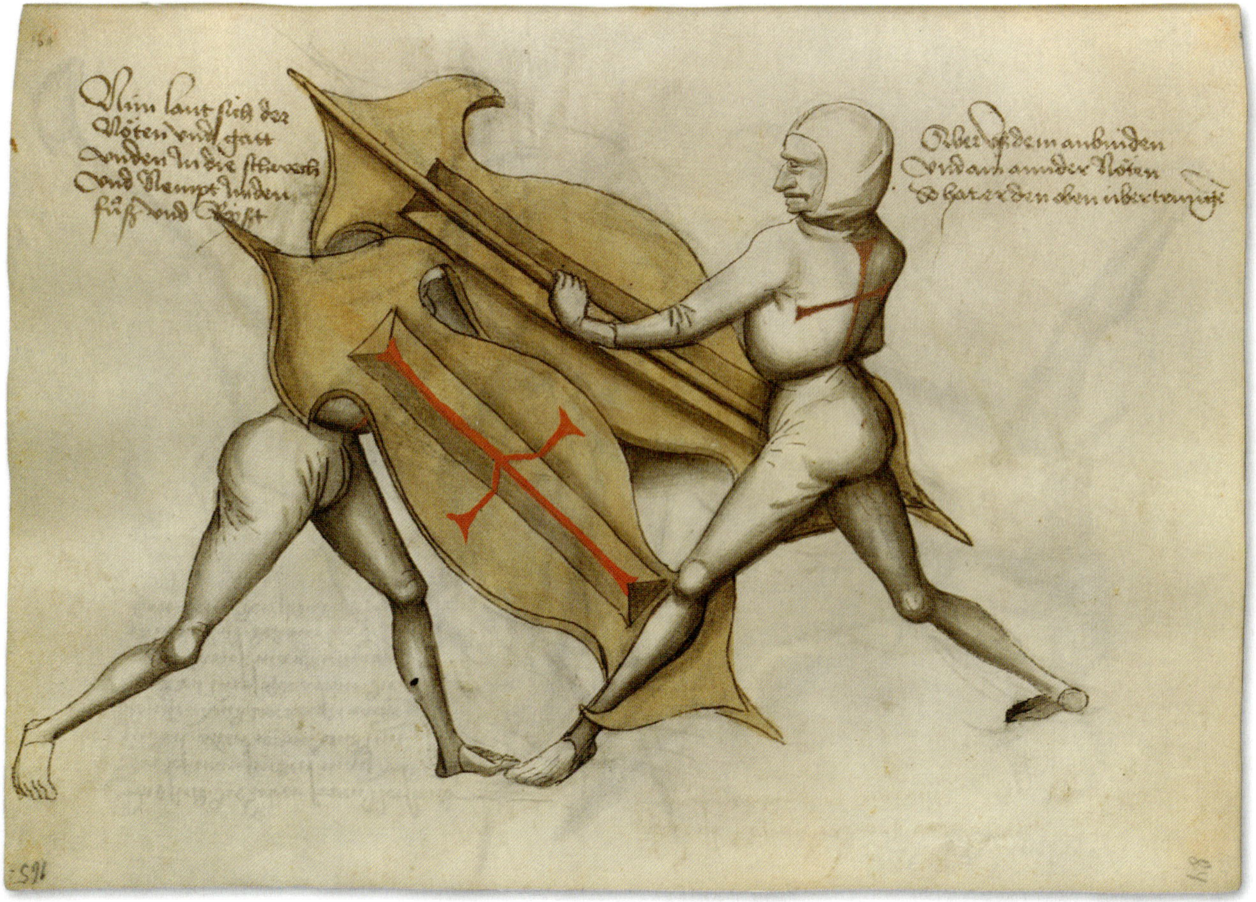

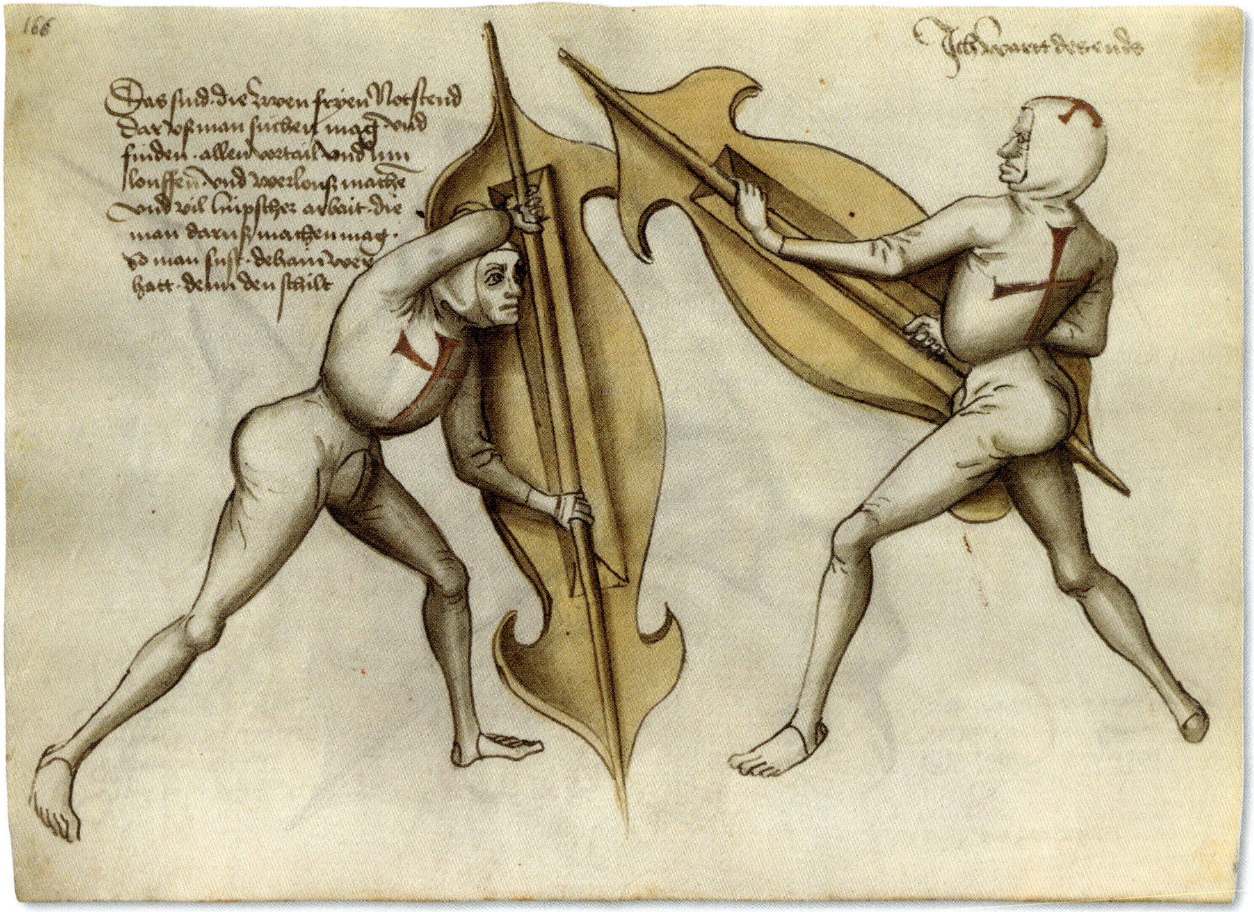

These are the two free distress stances in which one can look for and find every advantage: running in, making defenceless, and other excellent work, that comes from them when one has no other weapon than the shield.

I'm waiting for the end.

I managed to strike his neck.

Turn around, and you will see that the piece is done.

This one has twitched away from the hit and thrust above into his shield.

Here, the other intended to bind and beat.

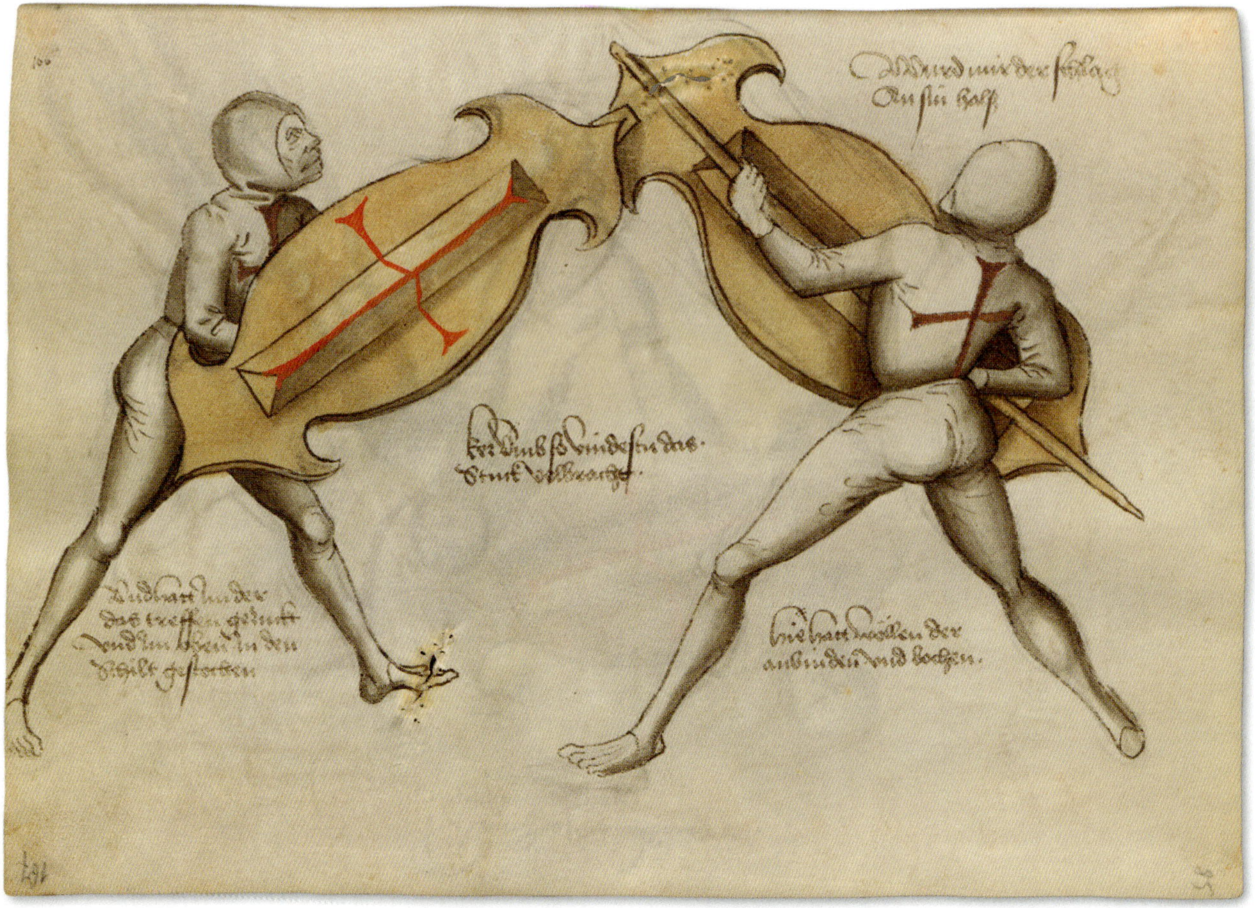

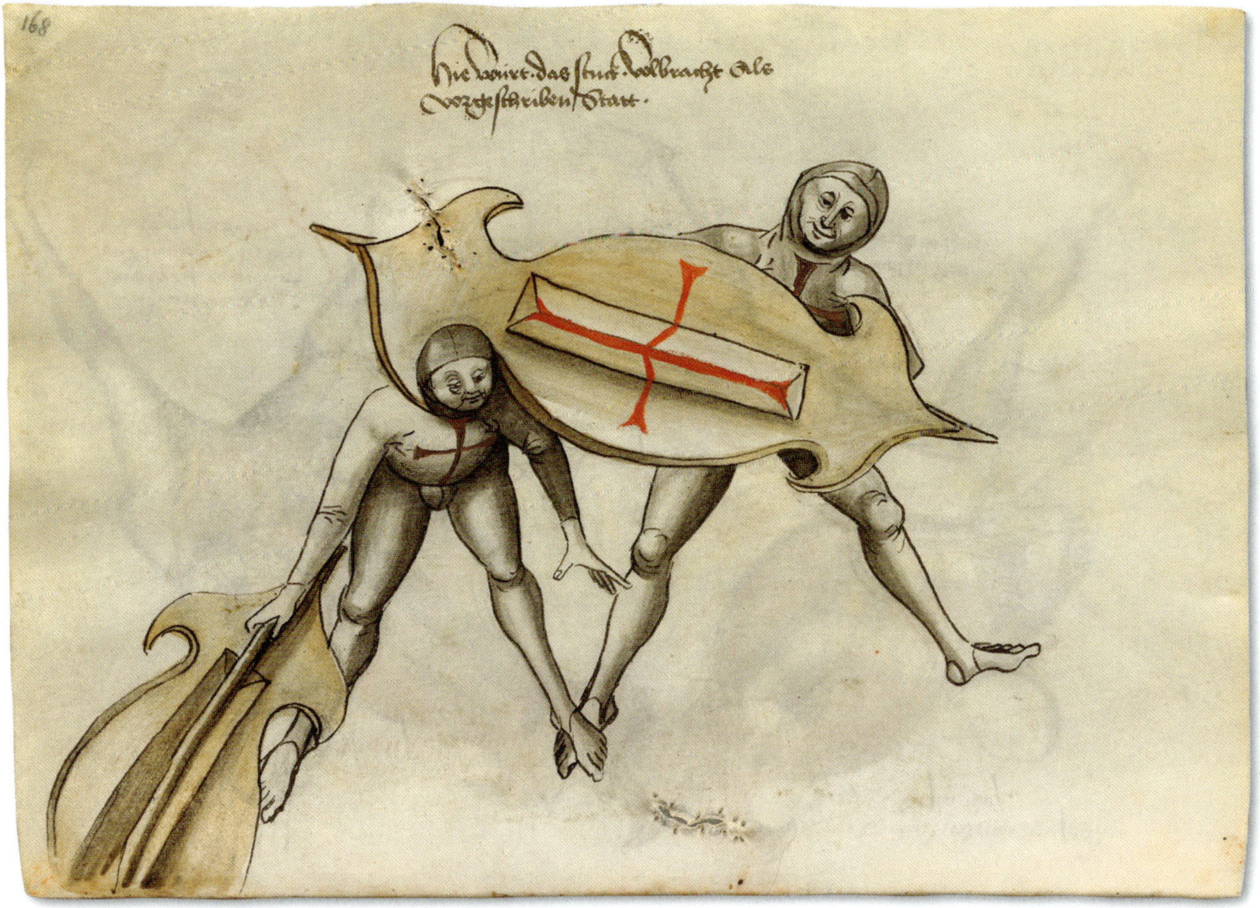

Here, the previously described piece has come to an end.

Here, the shield fencing ends. May God avert all our sorrow.

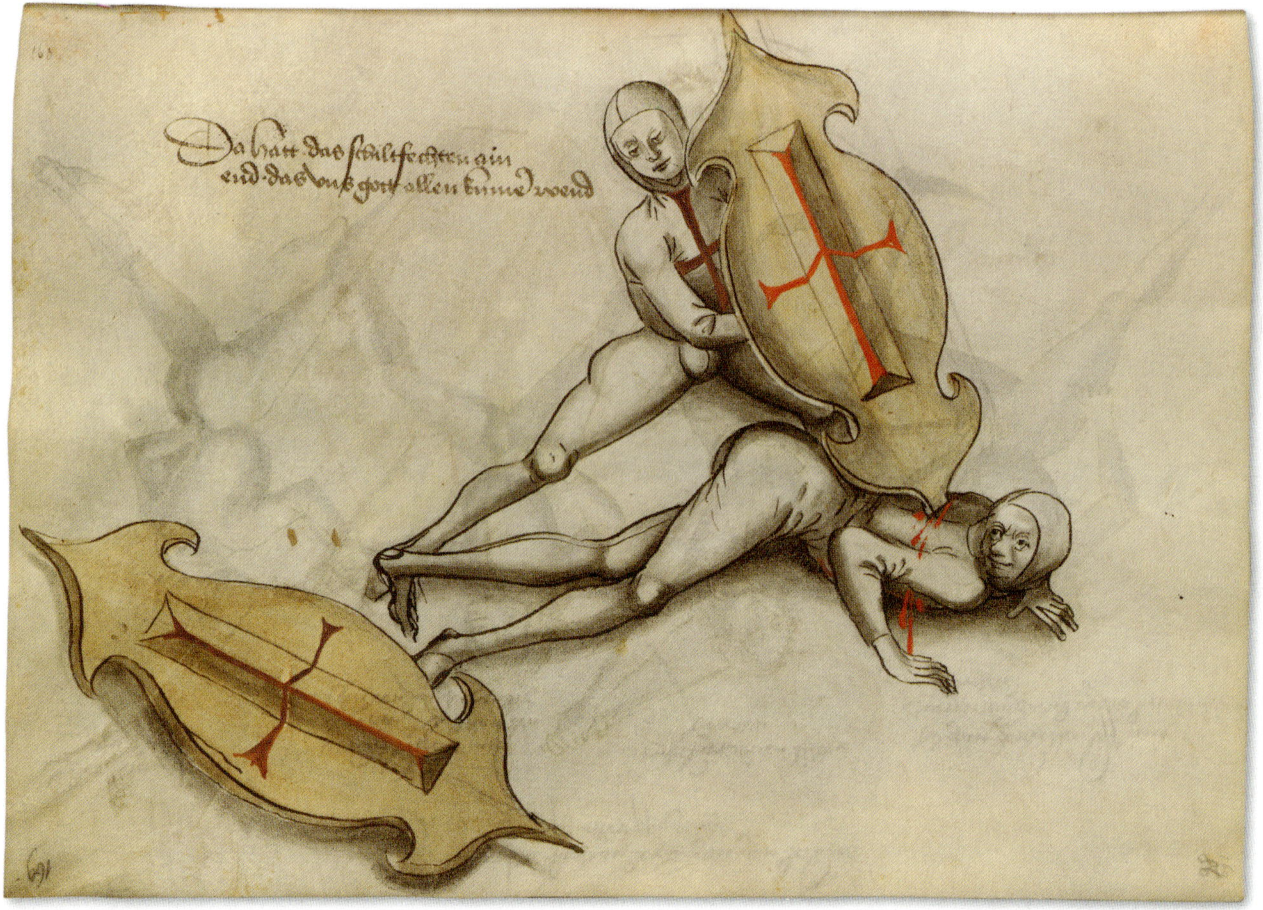

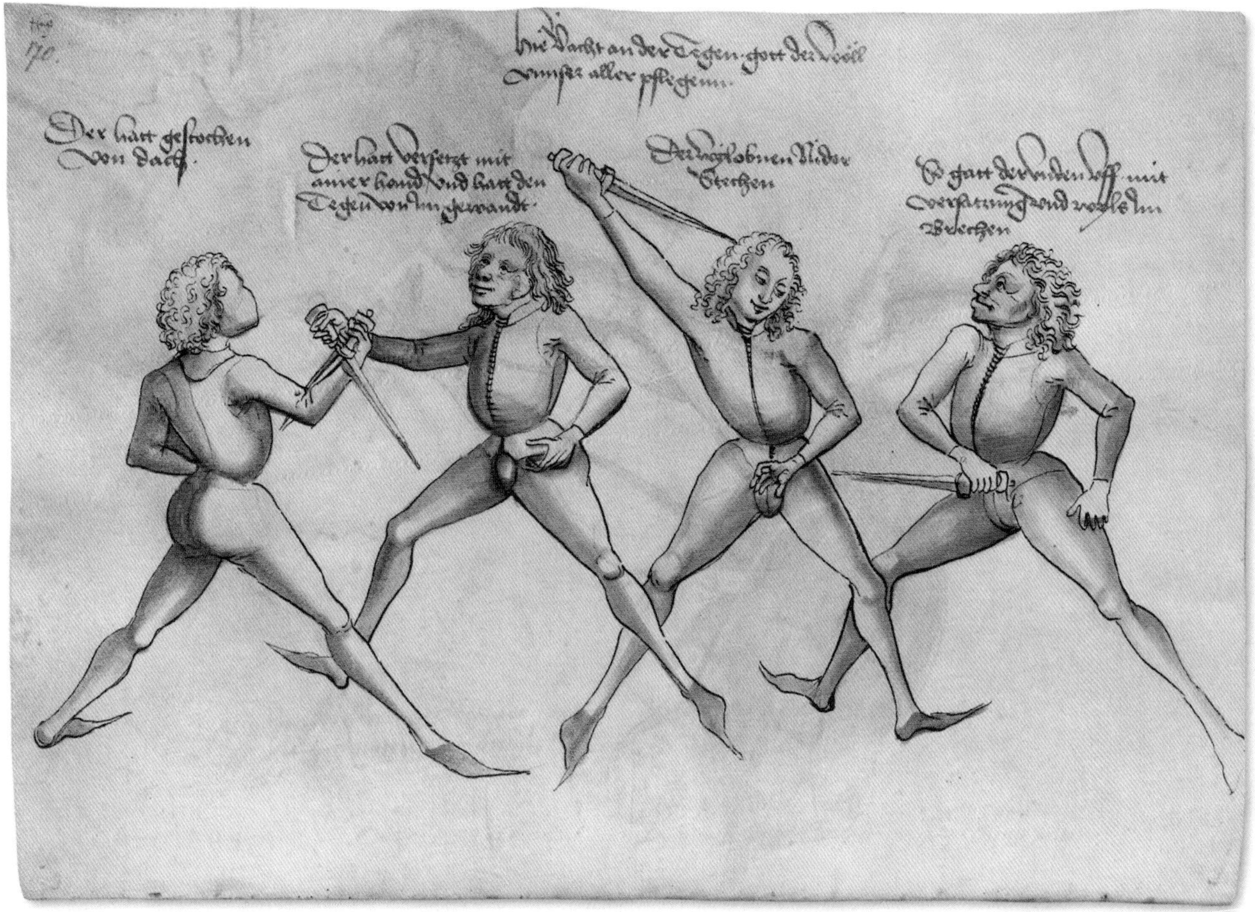

This is where the dagger begins. May God protect us all.

This one has thrust *vom Tag*.

This one has displaced his opponent with one hand, then turned the dagger away.

This one strives to thrust downwards from above.

This one comes up with a displacement from below and strives to break it.

The upper shield against the thrust.

This one thrusts freely *vom Tag*.

So the other makes the lower shield with a blow.

This one thrusts to the other below.

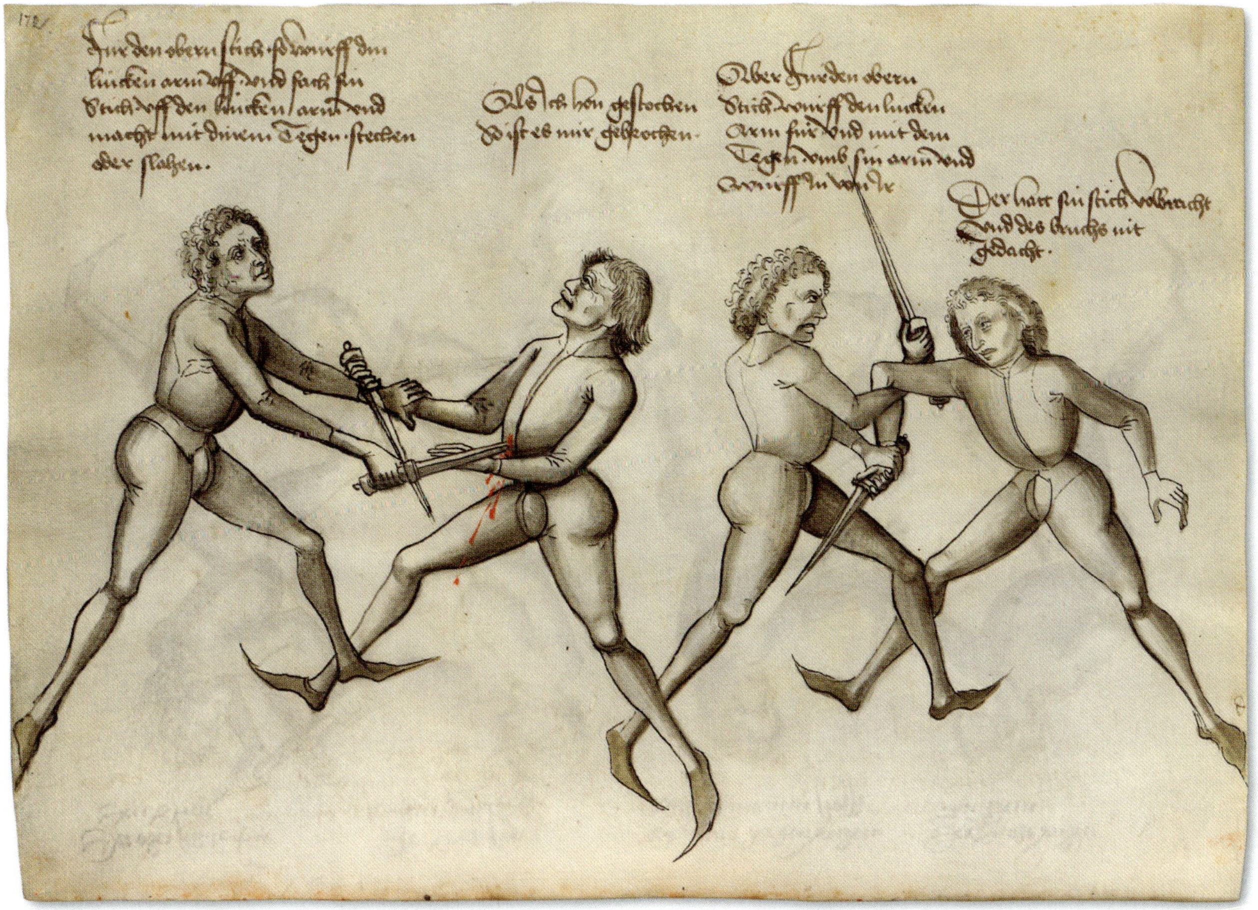

Against the *Oberstich*: Throw up your left arm and catch his thrust on it. Now you can thrust or hit with your dagger.

After I have thrust, it has been broken.

Also against the *Oberstich*: Throw your left arm forward and move your dagger around his arm. Throw him from you.

This one has accomplished his thrust, but did not think of the break.

After one thrusts the other above, he has broken it with the left arm and reaches between the other's legs with his dagger. He wrenches him upwards and throws him from him.

After one has thrust from above, the other displaced with the right arm and turns the dagger around his arm. With this, he steps forward and throws the other from himself.

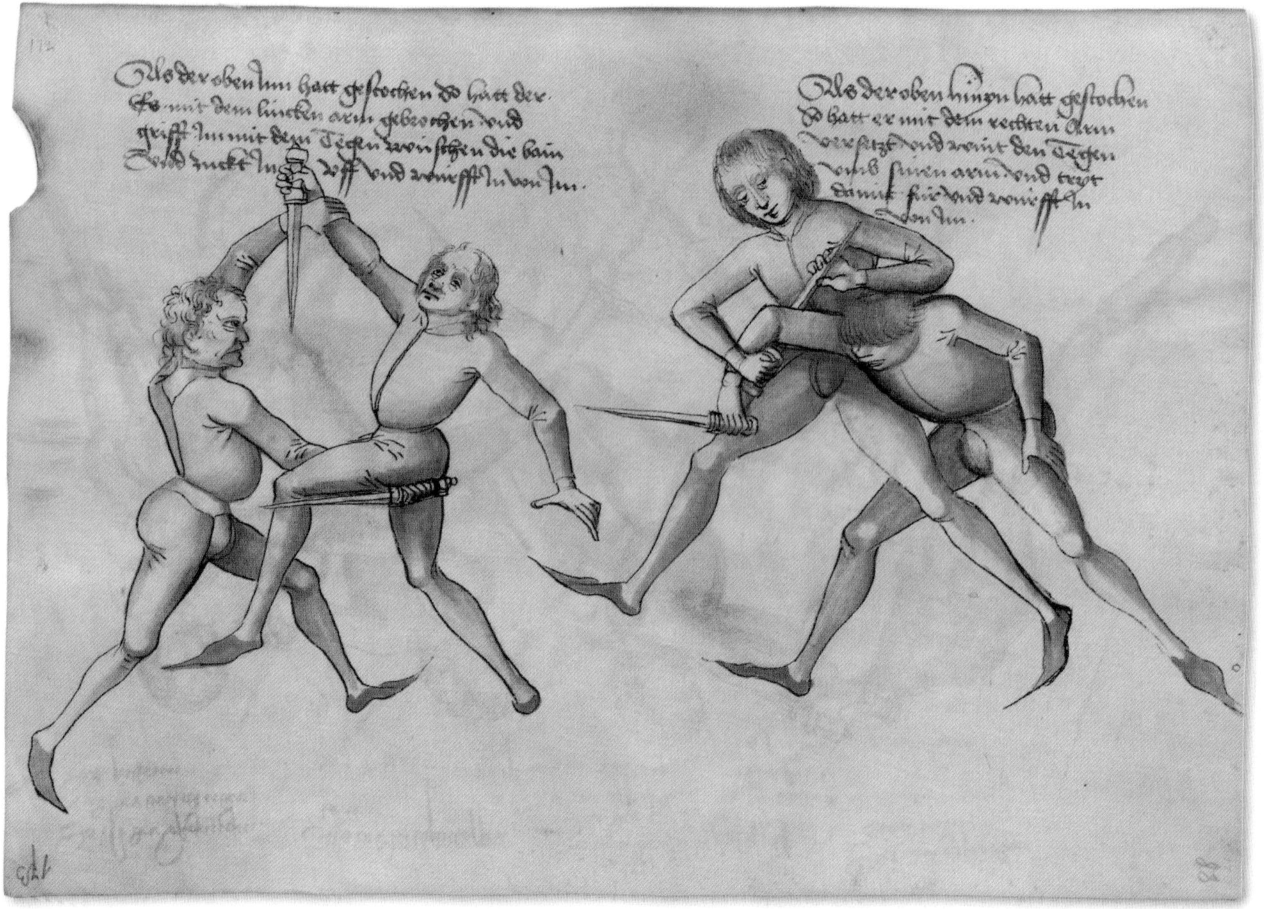

Here, this one has dodged, which is the break against catching …

… after the other has enclosed him.

In this way, one has enclosed the other with the scissors.

That one has thrust from above.

A powerful throw against the *Oberstich*: Go up with the right arm and fall with the left arm over his right arm. Throw him from you.

Here, the piece and the throw are finished.

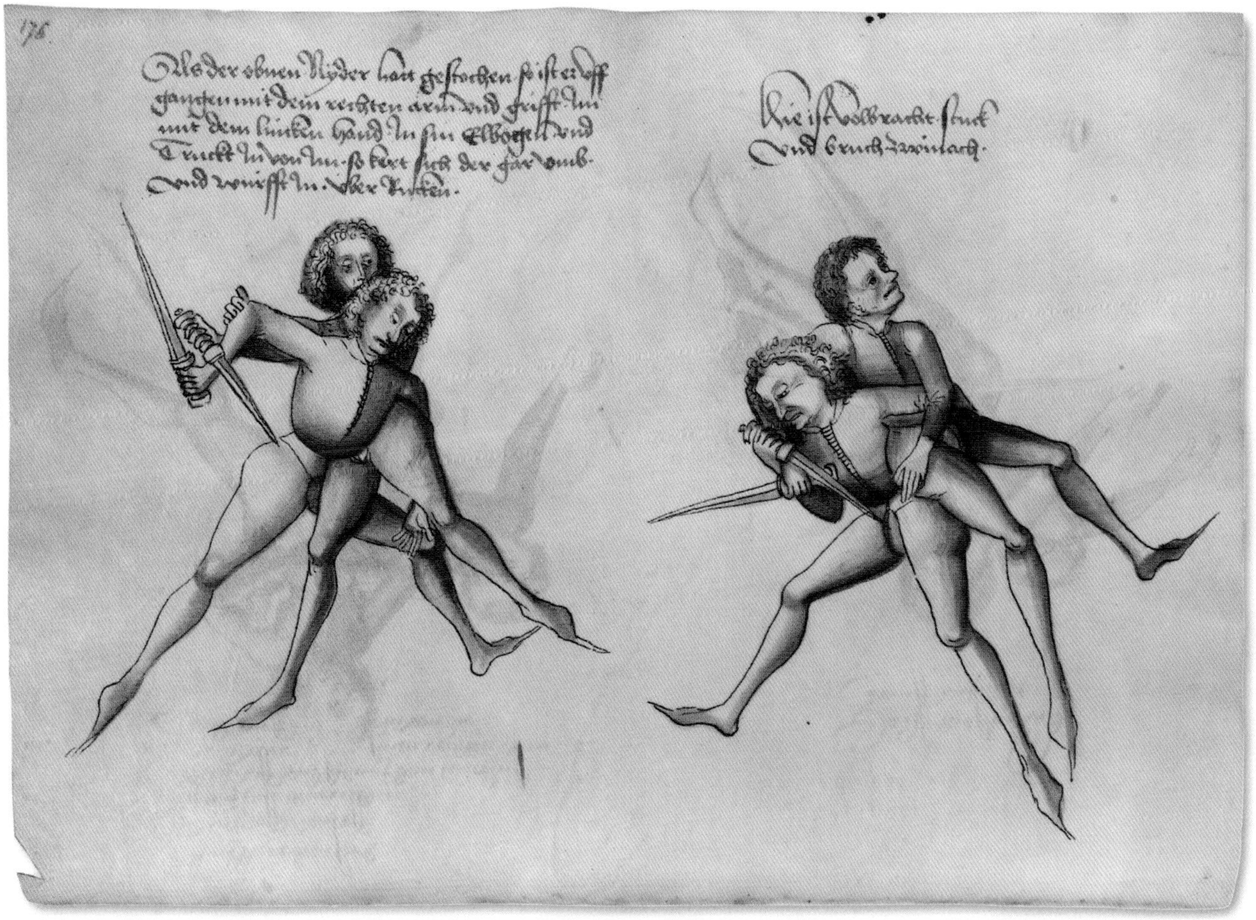

After one has thrust downwards from above, the other went up with his right arm and grabs his elbow with his left hand. He presses the first one away so that he turns around completely and is thrown backwards.

Here, the piece and break are completed a second time.

Against the *Oberstich*: One has thrown up his right arm, fending off the other's thrust, and stabbing him to death.

This one has thrust, and his thrust has been broken.

This is supposed to be a catching of the arm or the neck from the scissors.

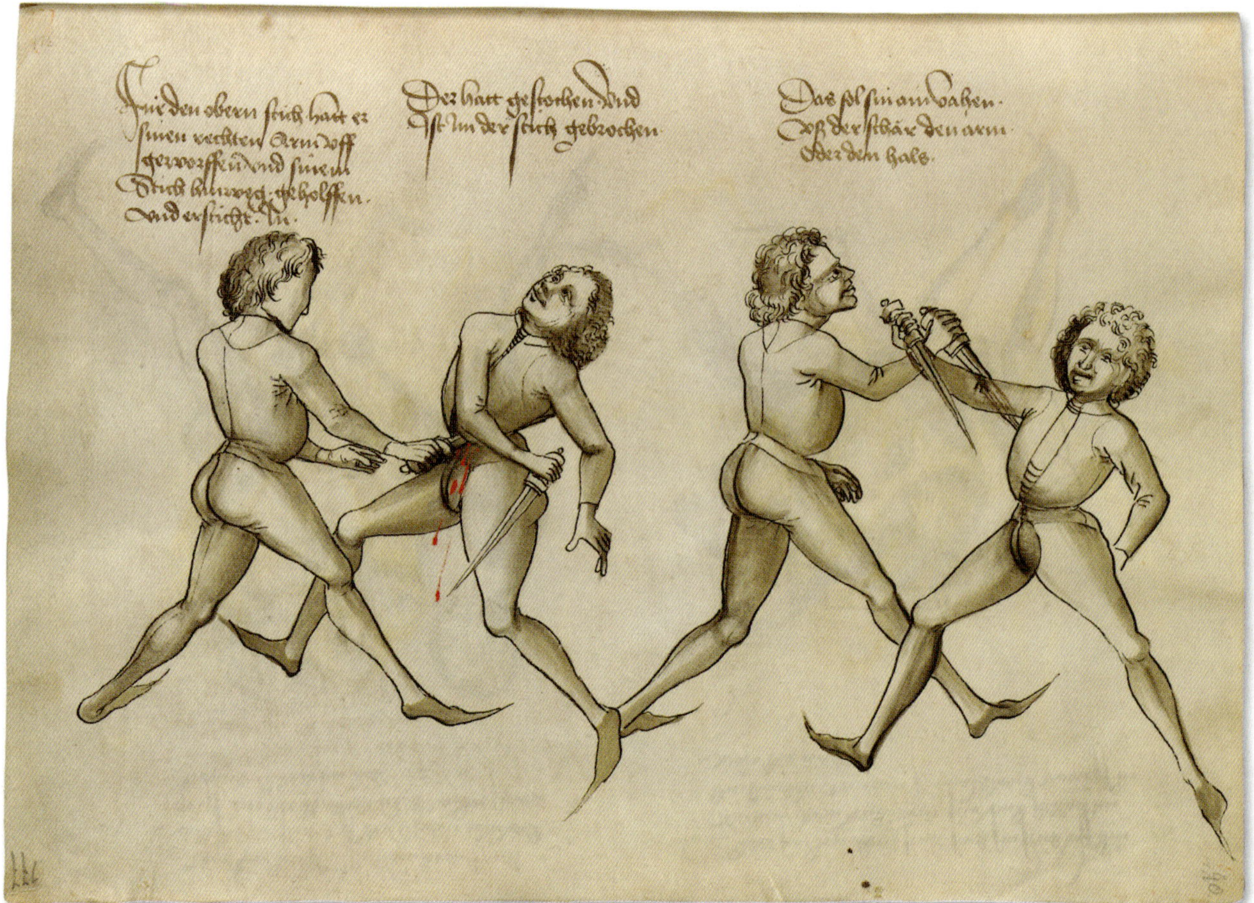

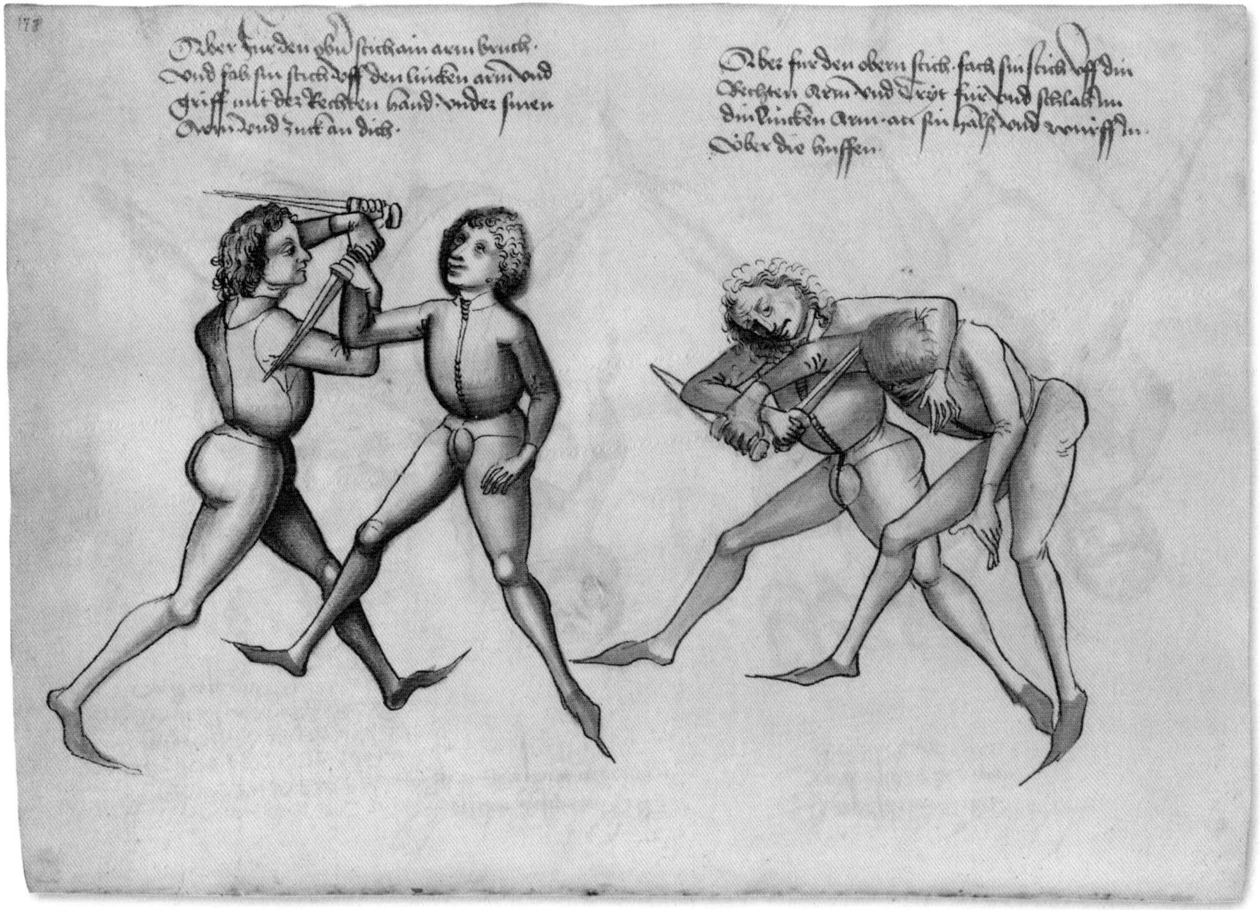

Also against the *Oberstich*: breaking the arm. Catch his thrust on the left arm, grab under his arm with the right hand and drag him towards you.

Also against the *Oberstich*: catch his thrust on your right arm and step forward. Hit him with your left arm on his neck and throw him over the hip.

Also against the *Oberstich*: throw up your right arm and catch his thrust on it. Grab his elbow with your left hand and push him to the ground in front of you.

The break: turn around completely and throw him over the hip.

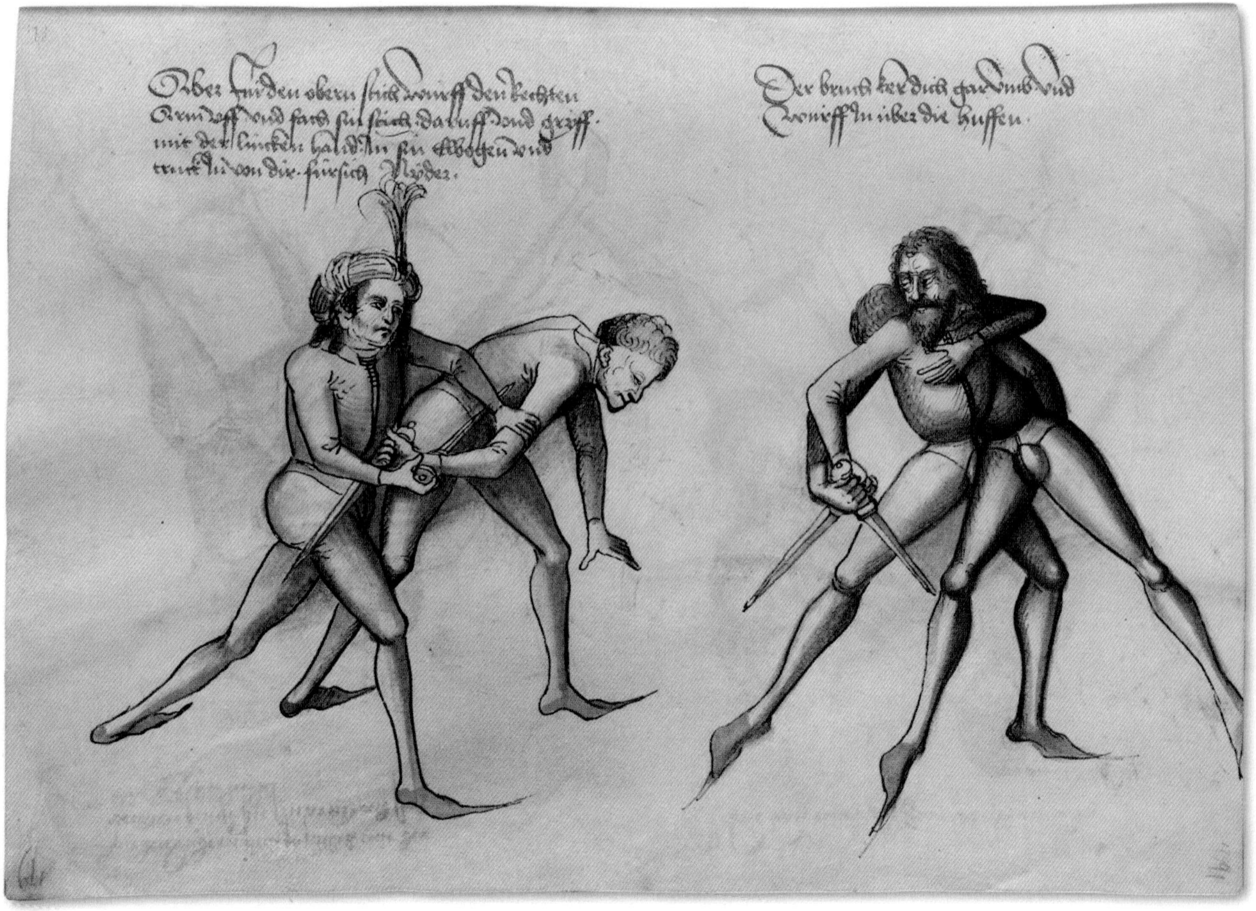

Against the *Unterstich*: hit him in the arm with your right fist and stab him with the dagger.

Here, one does as well as the other.

Also against the *Unterstich*: strike away his thrust with your right hand and hit his neck with your left hand. Step in front of him and throw him from you.

Against the *Unterstich*: catch his thrust in both your hands, turn his arm on his left shoulder and break it.

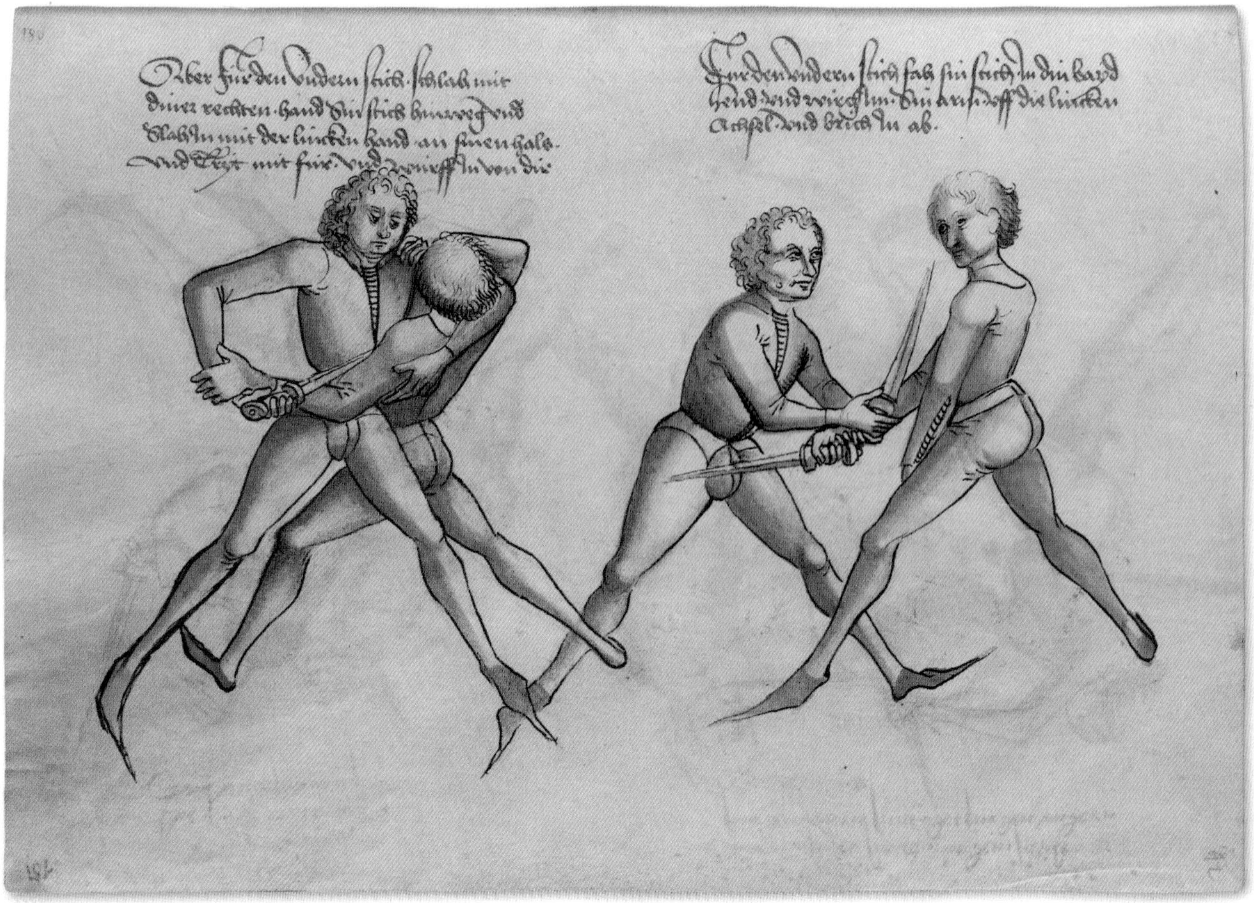

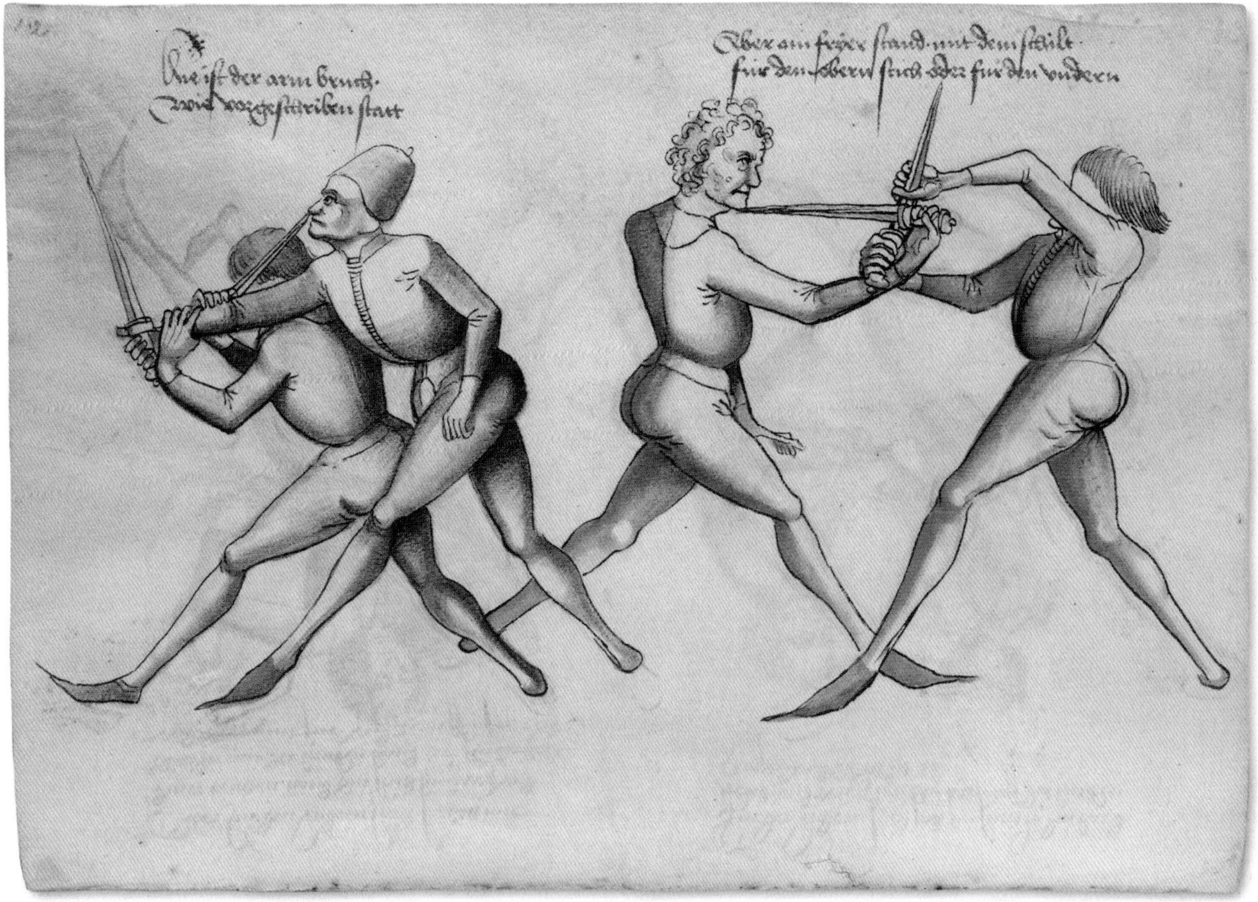

This is the break from the arm as described above.

Again, a free stance with the shield against *Oberstich* and *Unterstich*.

After one has made the movement of the shield …

… the other has turned around the *Oberstich,* thrusts the hilt of the dagger between his arms above the shield, and throws him from him.

Here, he performs the same piece.

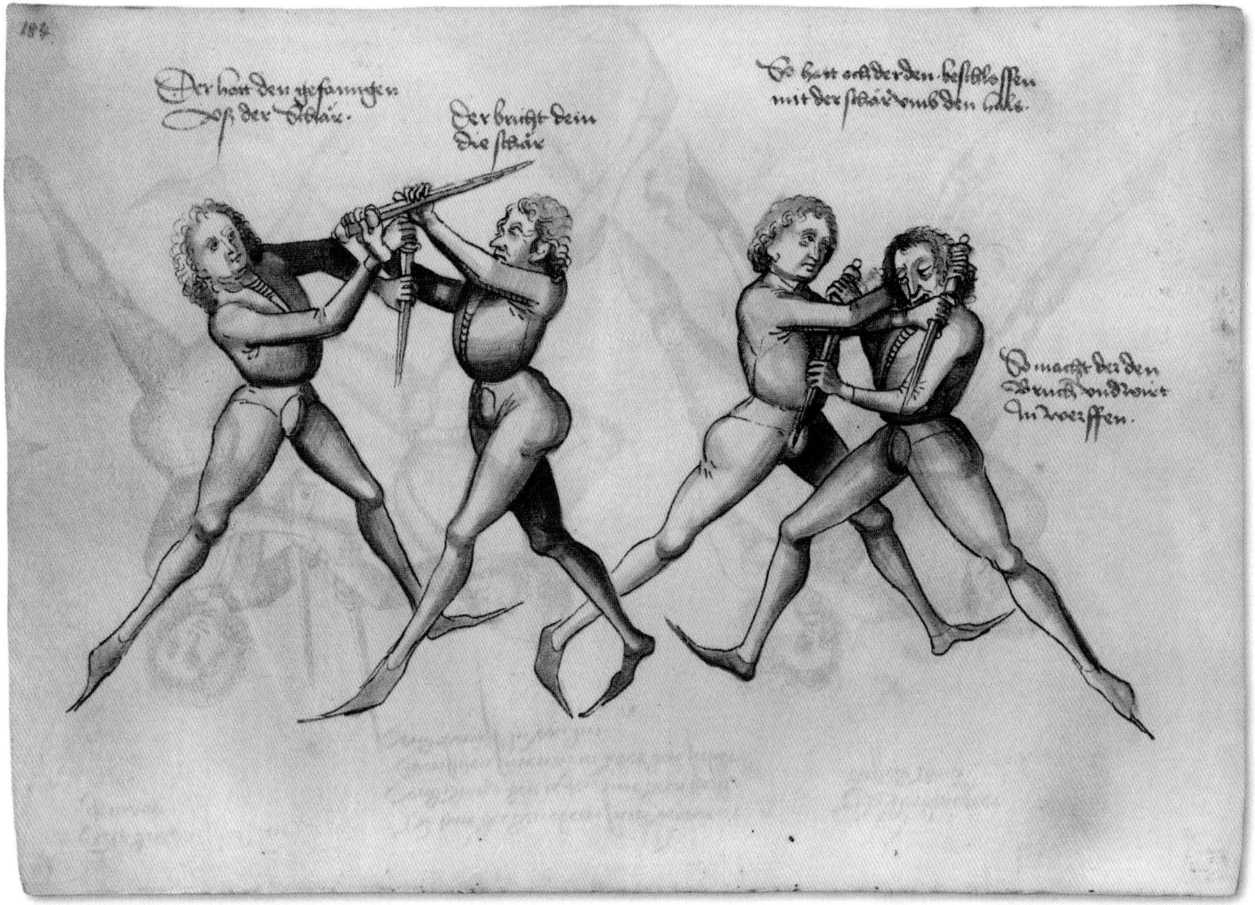

One has caught the other with the scissors.

The other breaks his scissors.

In this way, one has enclosed the other around the neck with the scissors.

Then the other makes the break and strives to throw him.

Here, he went up against the *Oberstich* with the shield.

So one has faked the thrust, strikes up from below and thrusts.

This one has also made the upper shield.

But the other has also faked the thrust; he grabs him over both arms and throws him.

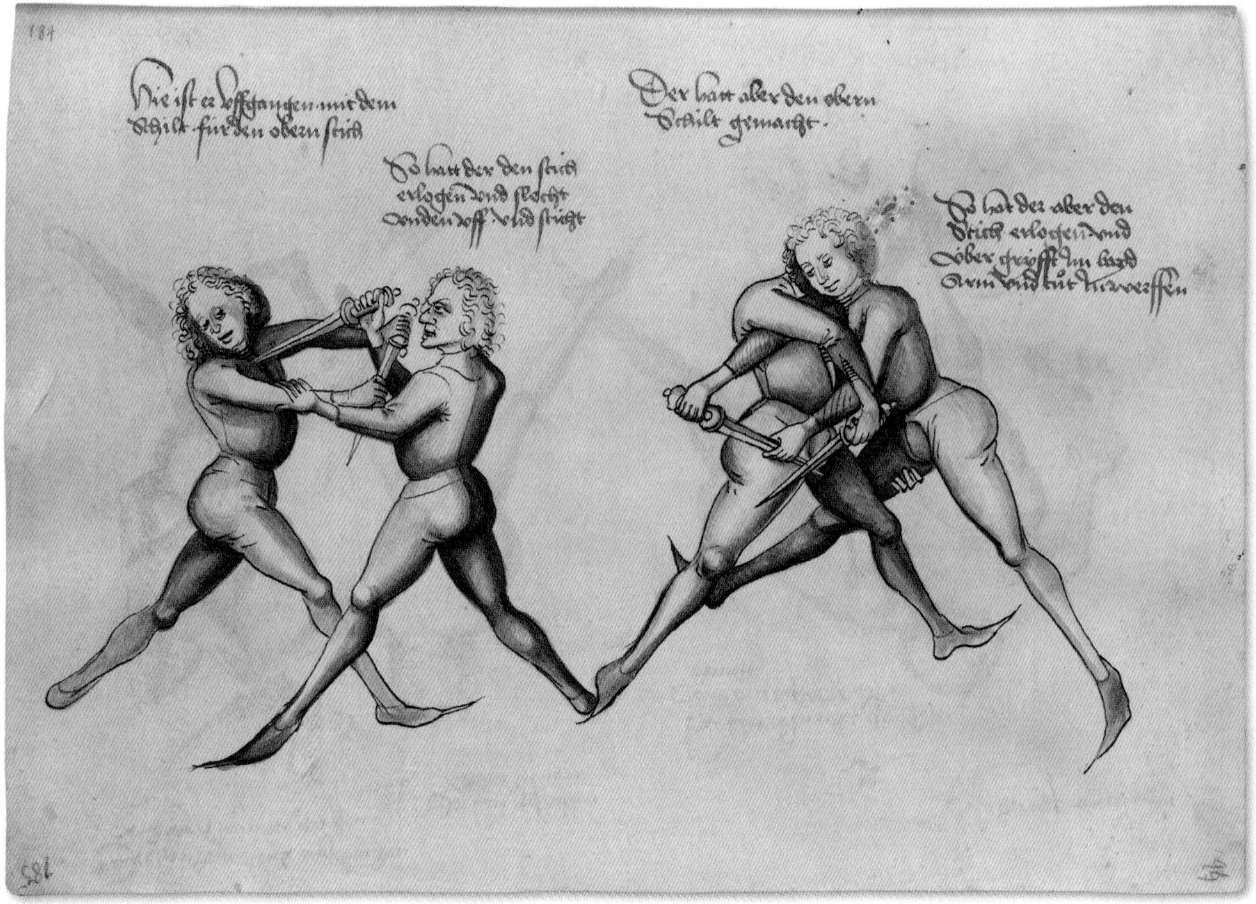

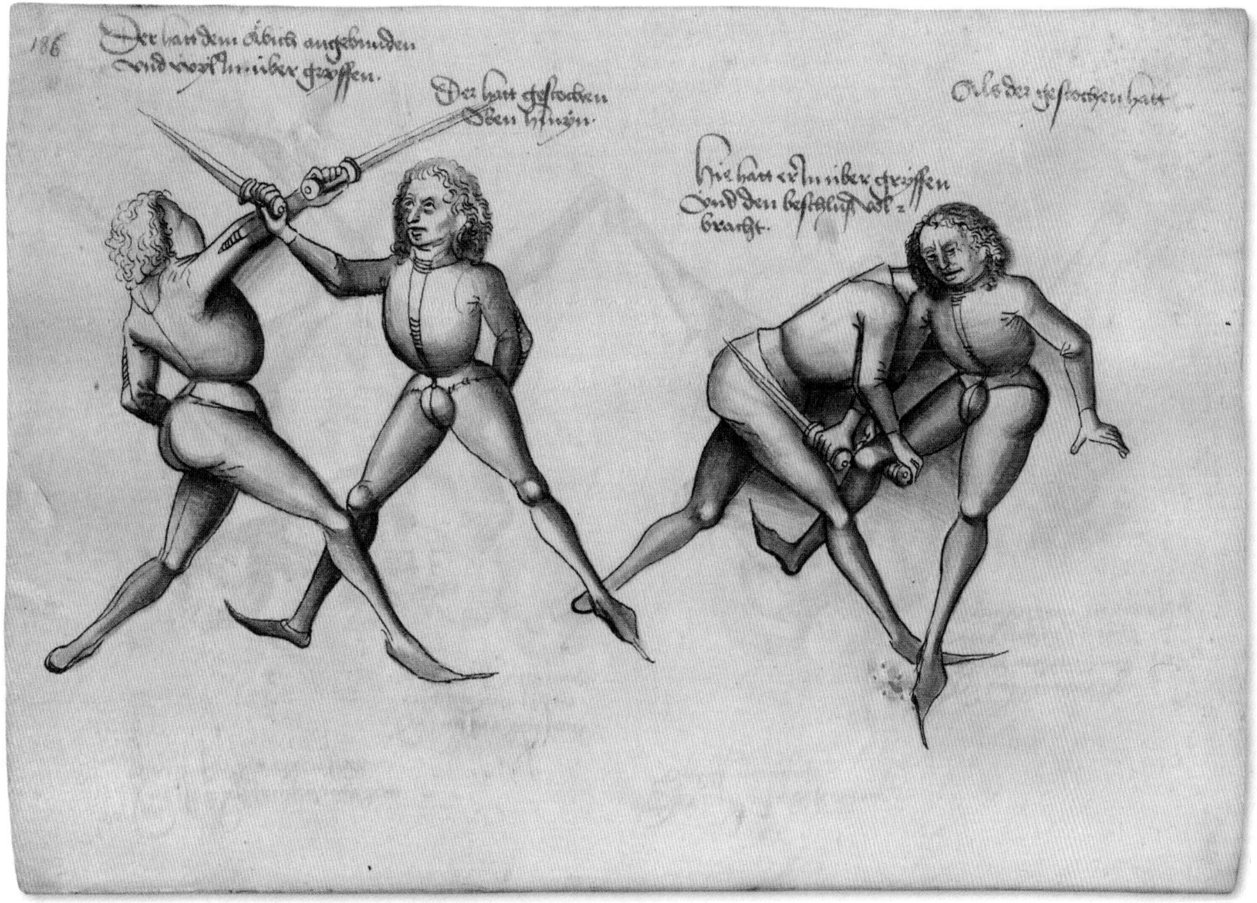

One has bound the other in a twisted way and strives to grab over him.

He has thrust from above.

Here, one has grabbed over the other and completed the enclosure …

… after the other has thrust.

Here, one has displaced with the scissors. He can push the other away with either hand.

That one has accomplished his thrust.

He stands in a free guard.

This is how he behaves with the thrust of doubt, having the dagger on his back. He can thrust with either hand.

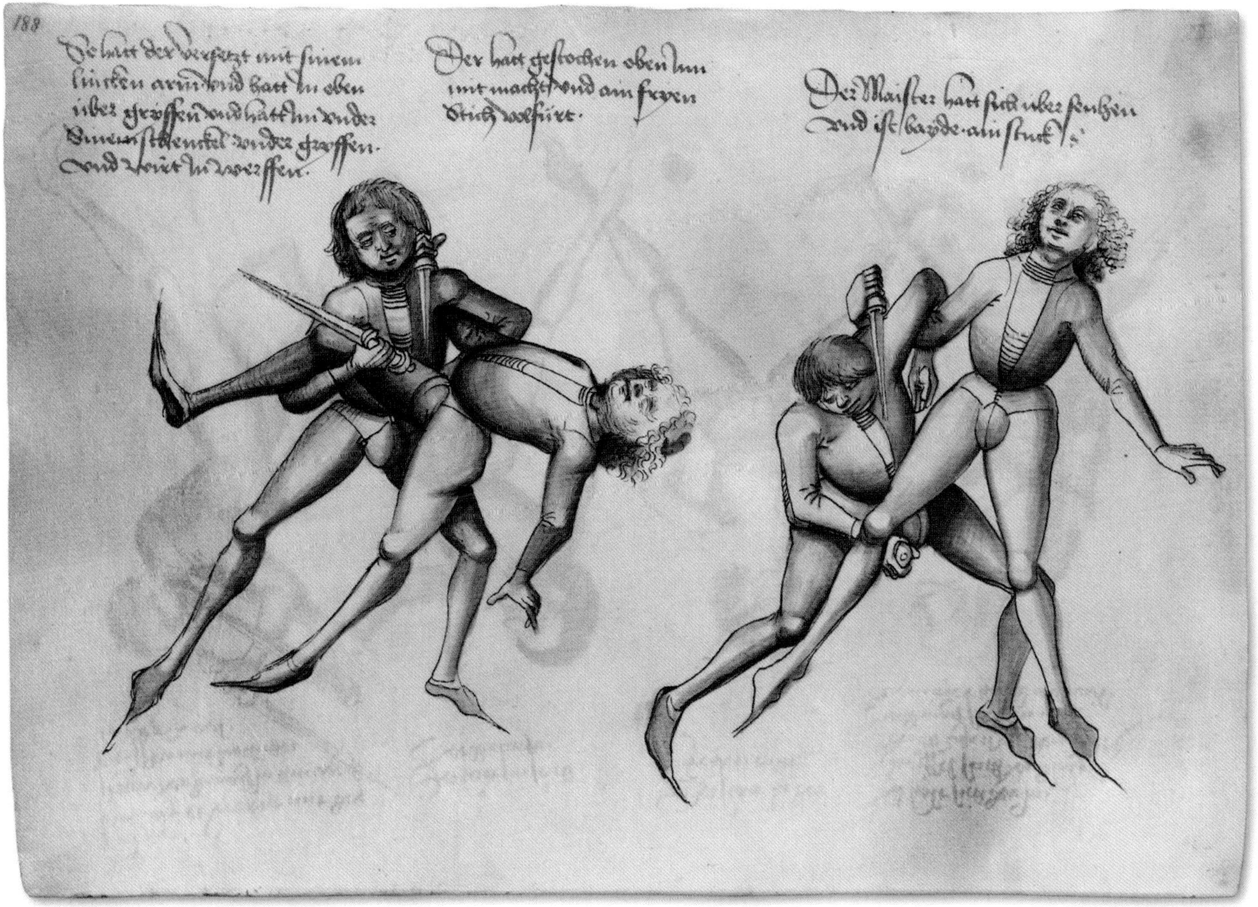

Thus one has displaced the other with his left arm and grabbed over the opponent above. He grabbed through under his leg and will throw him.

That one has thrust powerfully from above and performed a free thrust.

The master has overlooked it. Either is a piece.

After he has overrun me with a free thrust downwards from above, I threw up my left arm and displaced him with it. With my right arm, I grabbed him around the neck and now choke him down.

Here, one has choked the other down and puts an end to him.

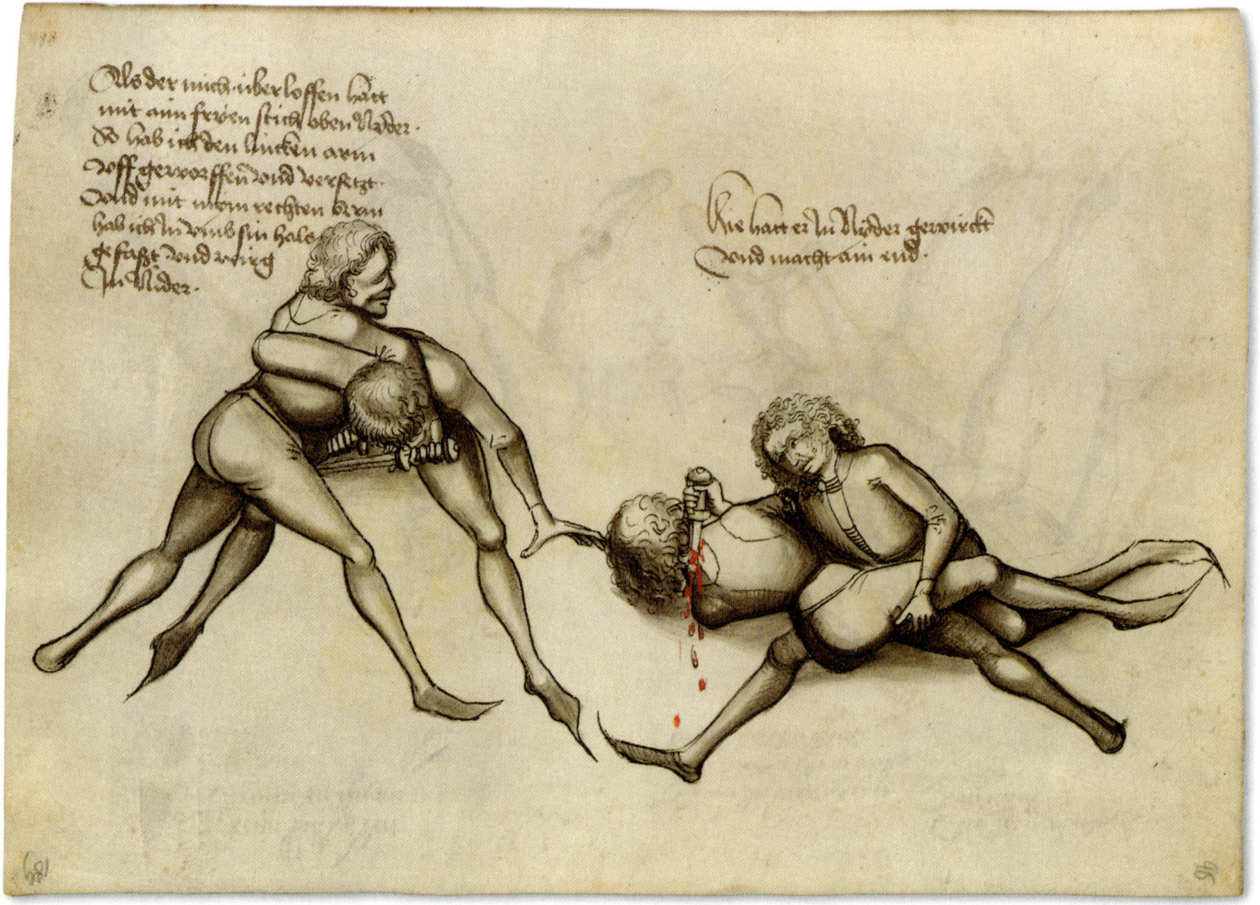

The beginning of the wrestling techniques from the arms.

~

Thus he has displaced the opponent with the left arm and thrusts *vom Tag* as well.

~

After one has thrust down *vom Tag* …

These have taken hold of each other in the manner of farmers: One arm down and the other up. So they lie in the struggle.

In the struggle, one slips his head through the arm of the other and pulls him up.

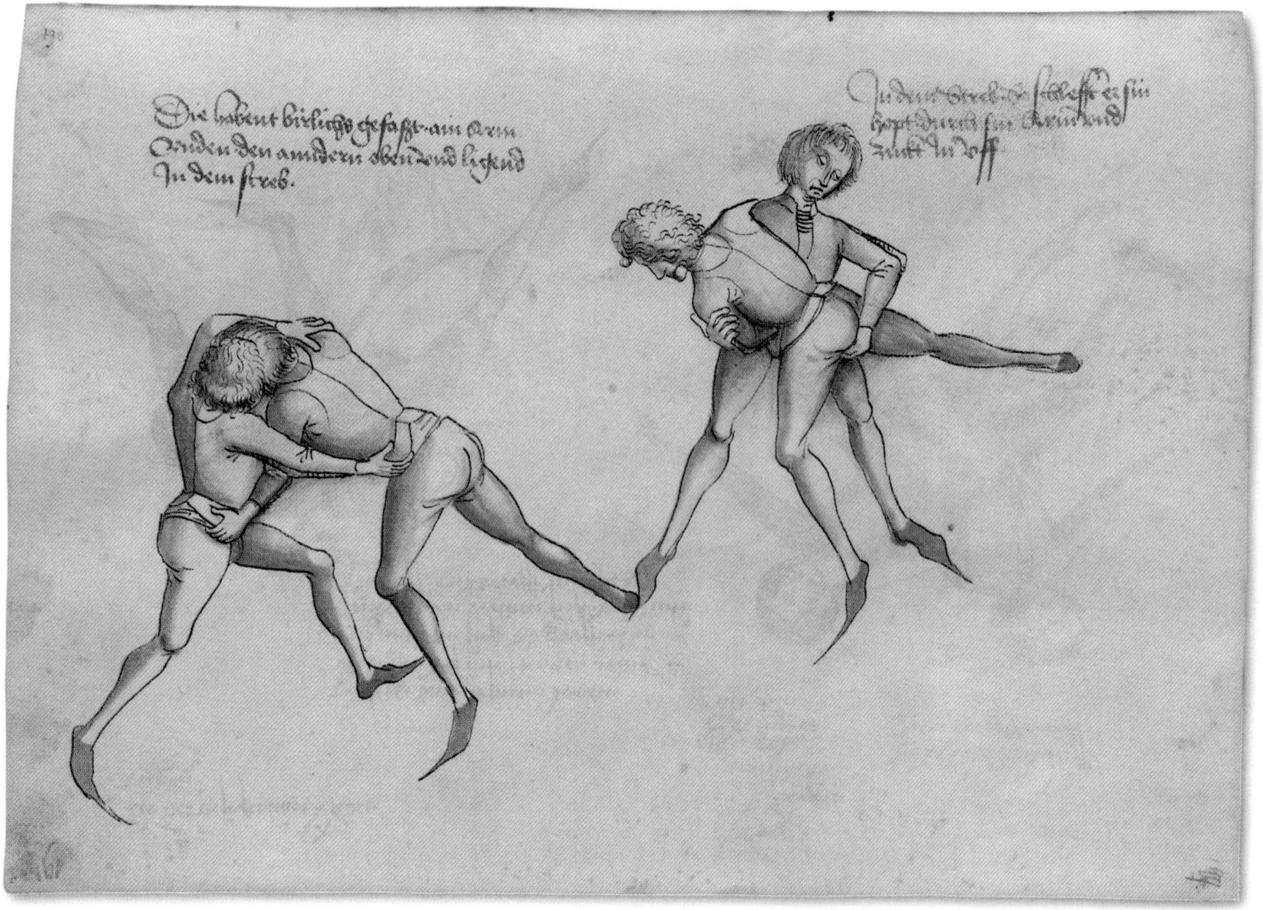

After one has grabbed with the arms …

… the other strikes his right arm with the right elbow. During the blow, he grabs him under his right foot, pulls him up and throws him up and away.

It goes badly for this one.

He strives to throw the other one over the leg.

Here, he strives to bring him into the *Trappe* and tries to break his arm.

Here, he has put his opponents's arm onto his shoulder and strives to break it.

From the hold: one throws the other over his leg.

Here, one has stepped behind the other and pushes him over his leg.

Another hold in the manner of farmers.

—

From the farmer's hold: one wrenches his lower arm away, grabs his opponents's head, and hurls him away.

During the hold, this one strives to go through.

Here, he has gone through and throws the other over his back.

While going through, this is the break: one chokes the other by his neck.

Another break against going through: meanwhile, twitch your foot backwards and grab his neck.

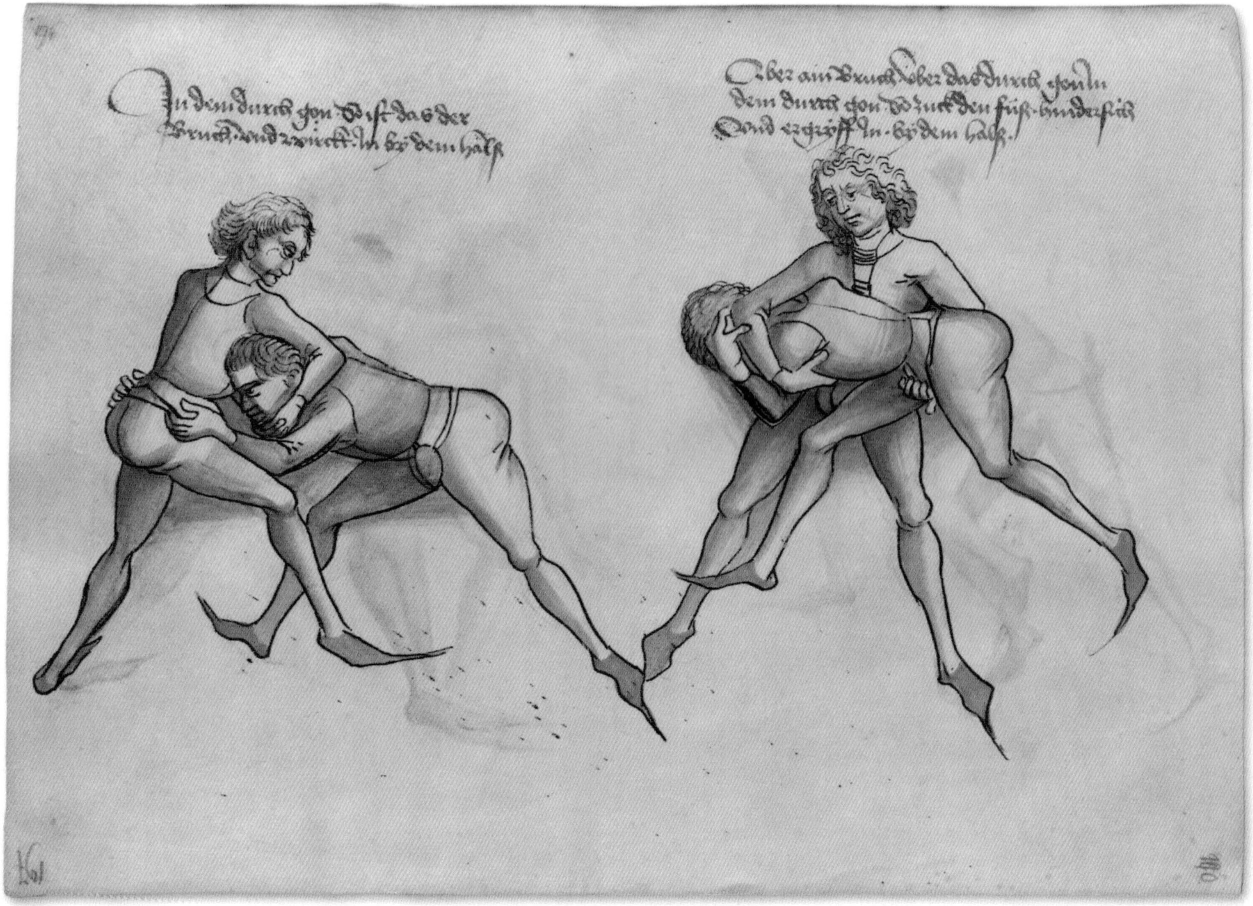

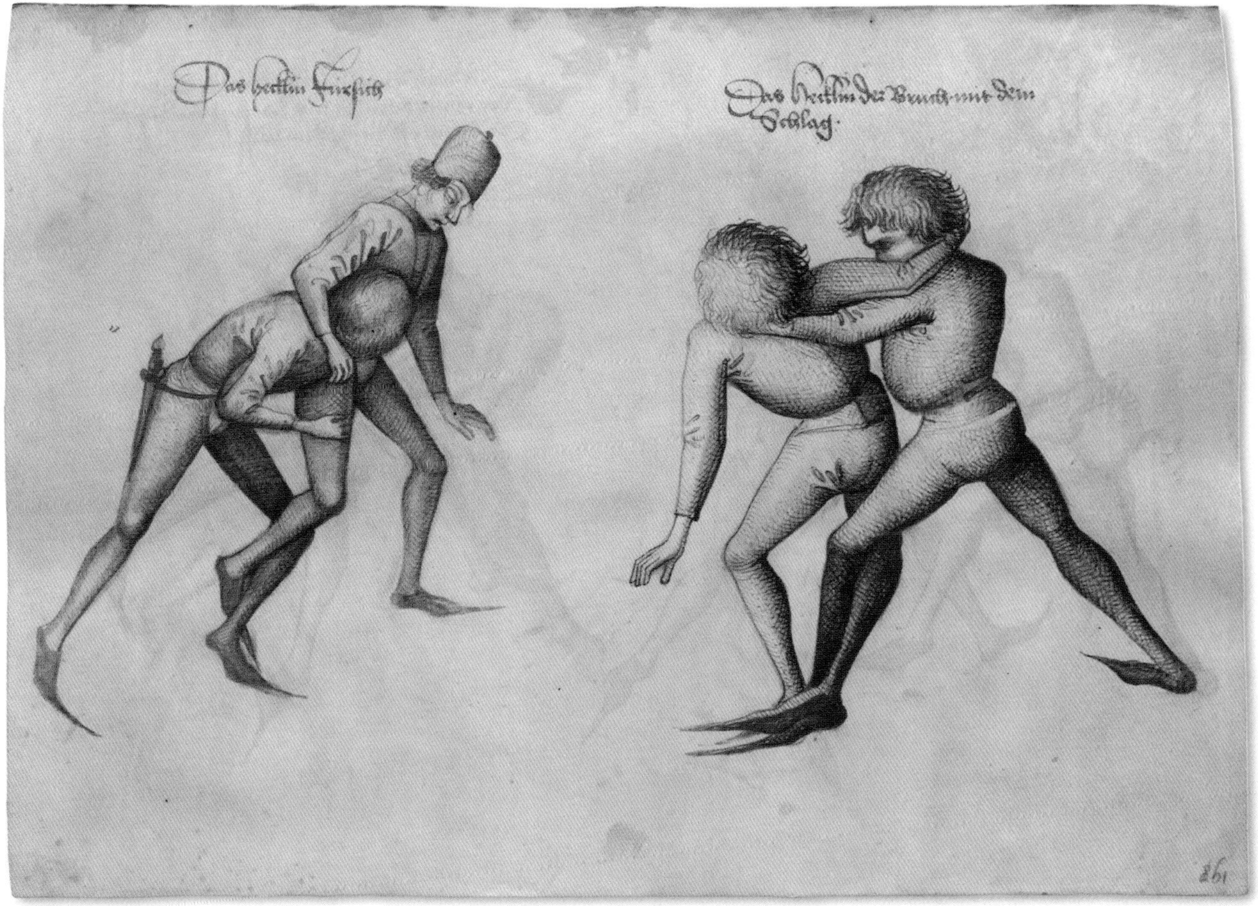

The hook to the front.

The hook: the break against the strike.

When one grabs the other from behind, note if you can reach his leg. Pull it through and stand up.

When again one grabs the other from behind, step backwards with your foot between his legs and grab his leg.

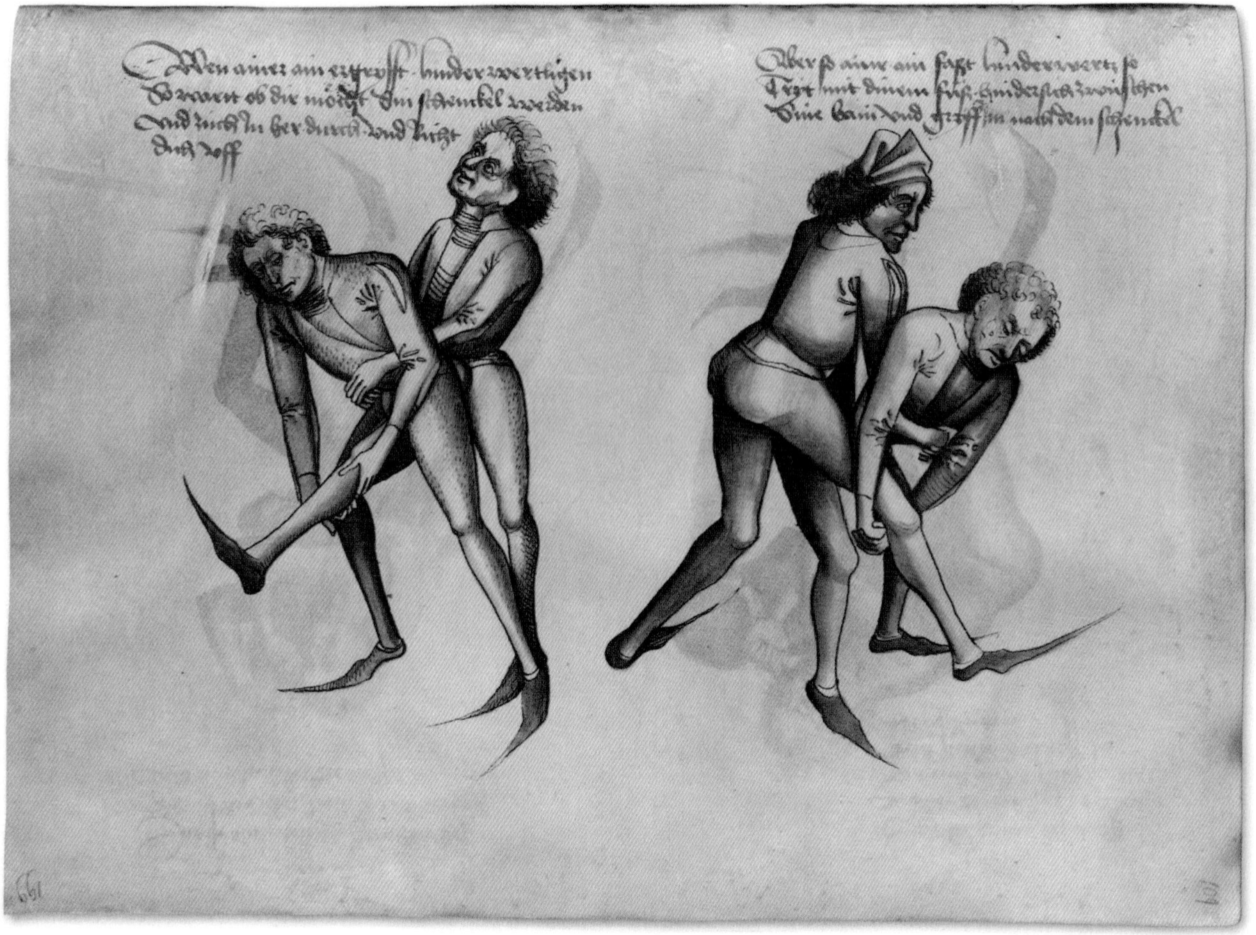

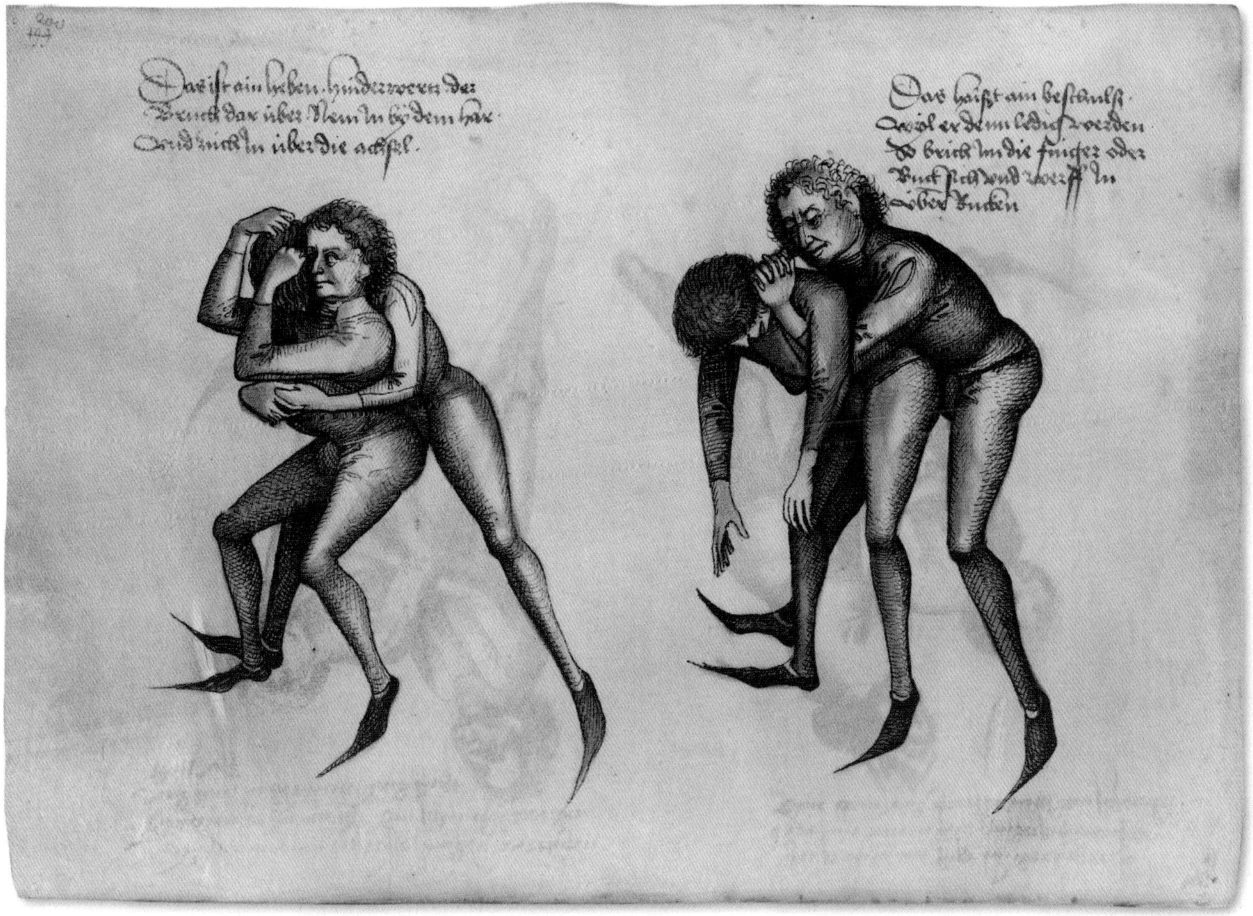

This is a lift from behind. The break against it: grab his hair and pull him over your shoulder.

~

This is called enclosing. If he strives to free himself, break his fingers or bend over and throw him over your back.

When one grabs the other, has both arms down and strives to drag you up, grab his face with both hands and push him away from you; so he lets you go.

This one lets himself be caught voluntarily and yet the other cannot throw him off. Instead, this one throws the other.

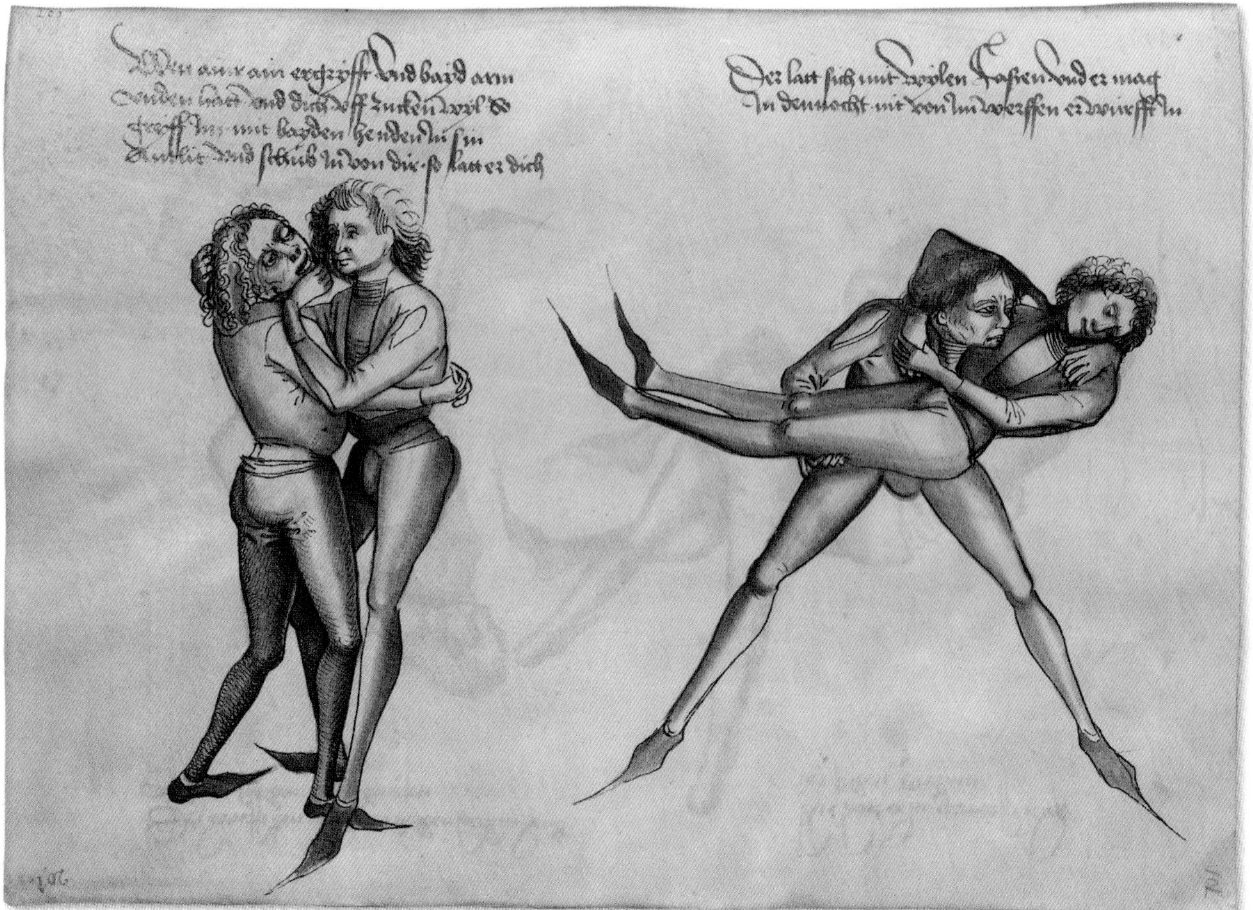

One throws the other one with the half hook and from the half hip wrestling move.

Here, one has thrown the other with the same work.

An upper hip wrestling move.

—

Here, he attempts a break against the upper hip wrestling move.

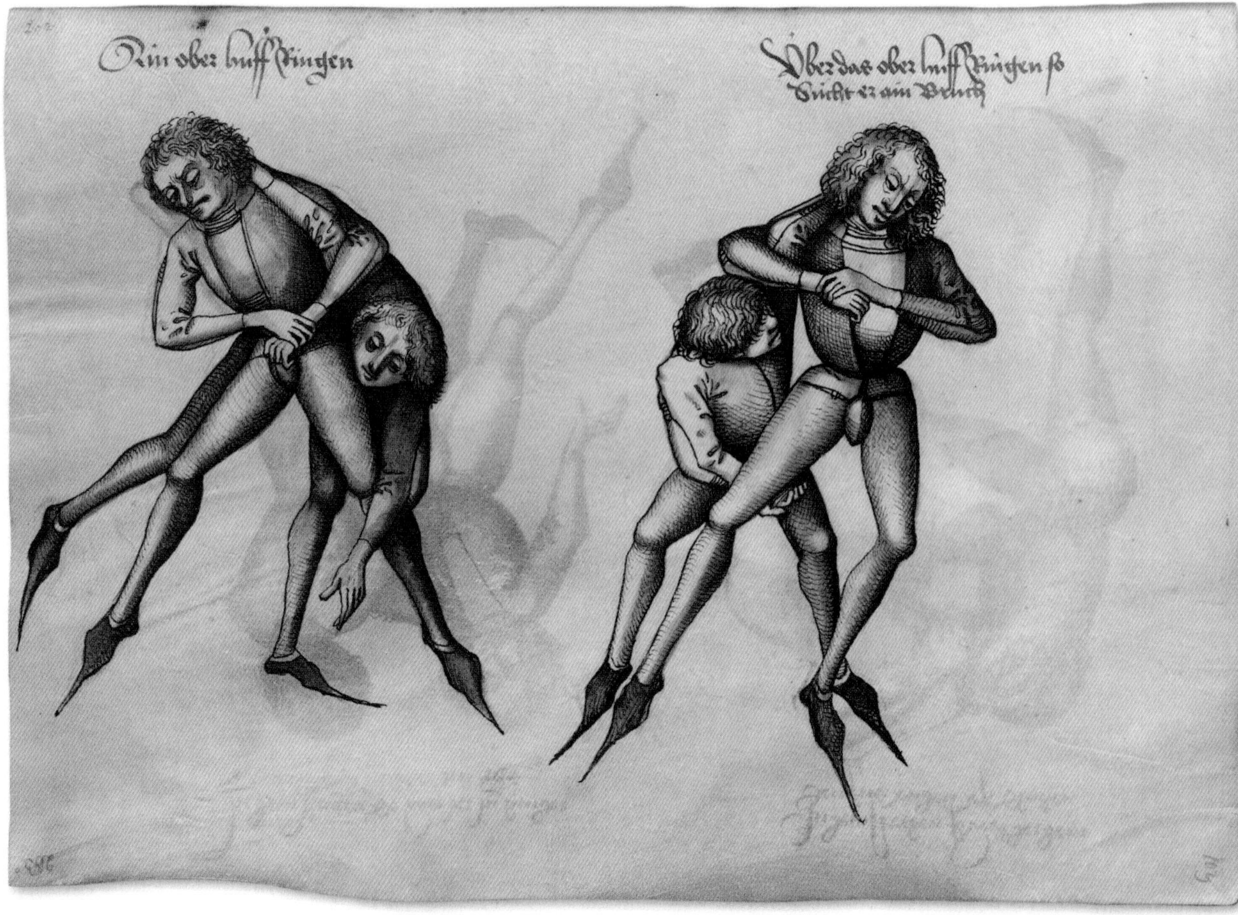

While wrestling, one has stepped behind the other and [throws] him over his foot.

In the struggle, this one strives to knock the other's feet from under him.

While wrestling, one has pulled the other one up and walks around with him. This is called wrestling to dizziness.

—

One has grabbed the other by the neck with both hands and shakes him. That is how he pushes him away.

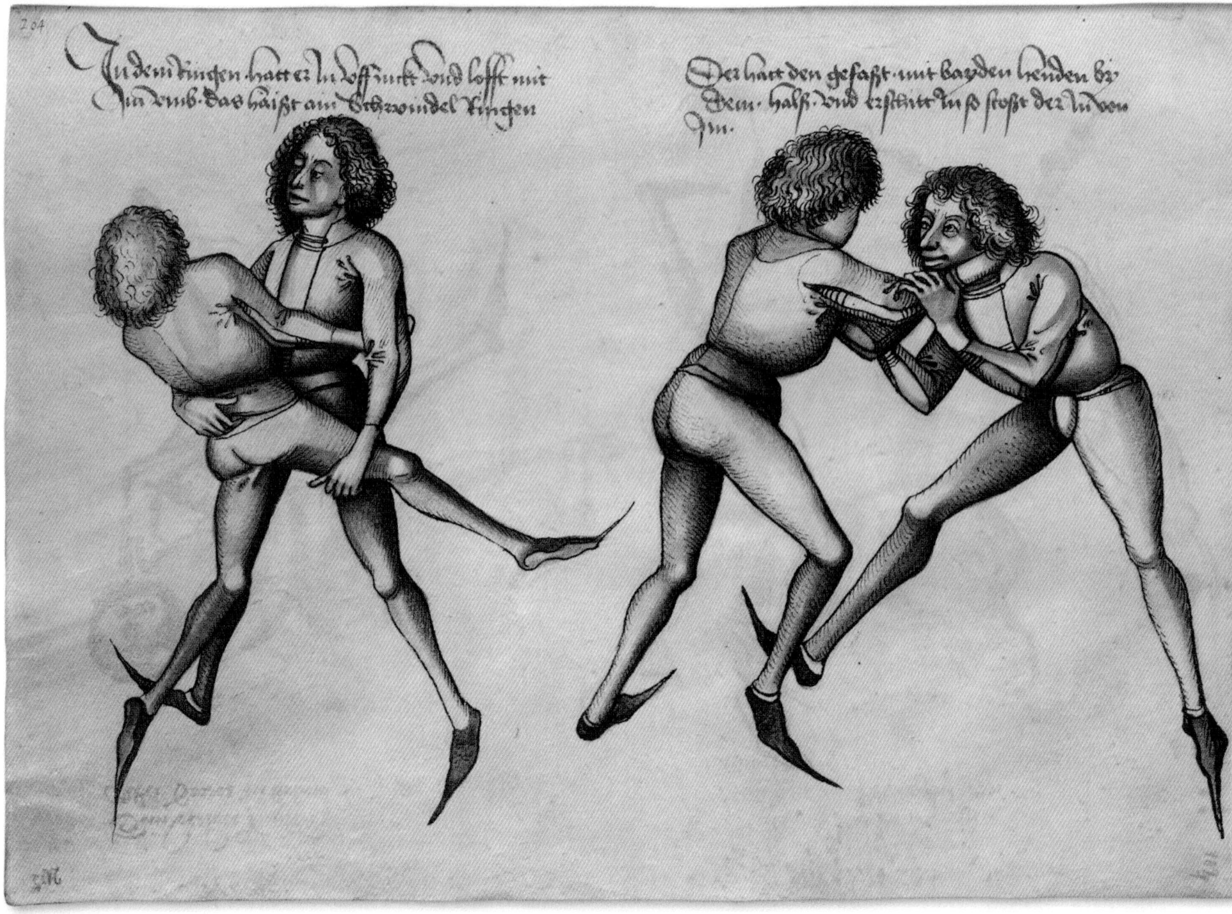

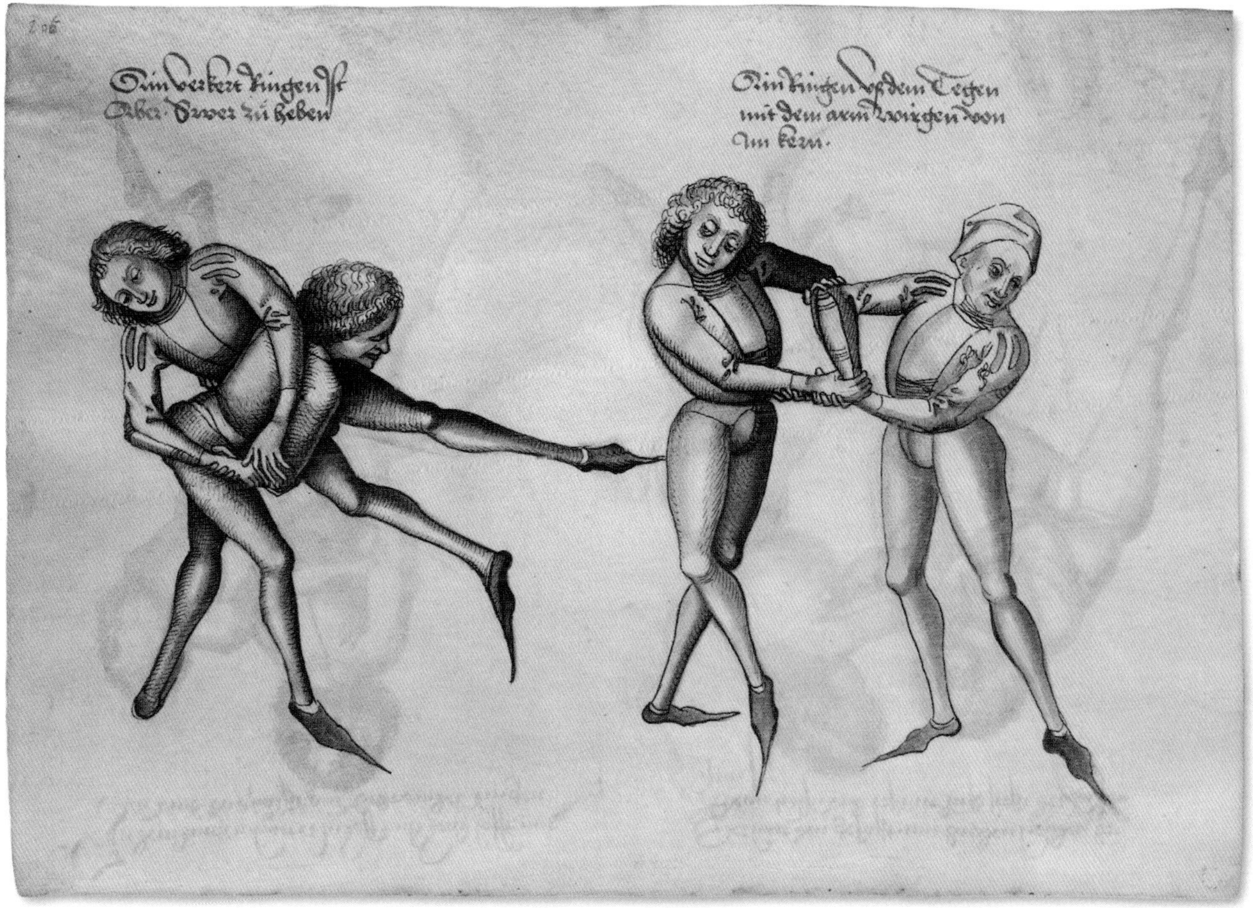

A twisted wrestling move that is difficult to lift.

A wrestling move from the dagger fight: press with your arms and turn from you.

This is called *Bubenringen.*

The break against it: fall and drive your knee into his manhood.

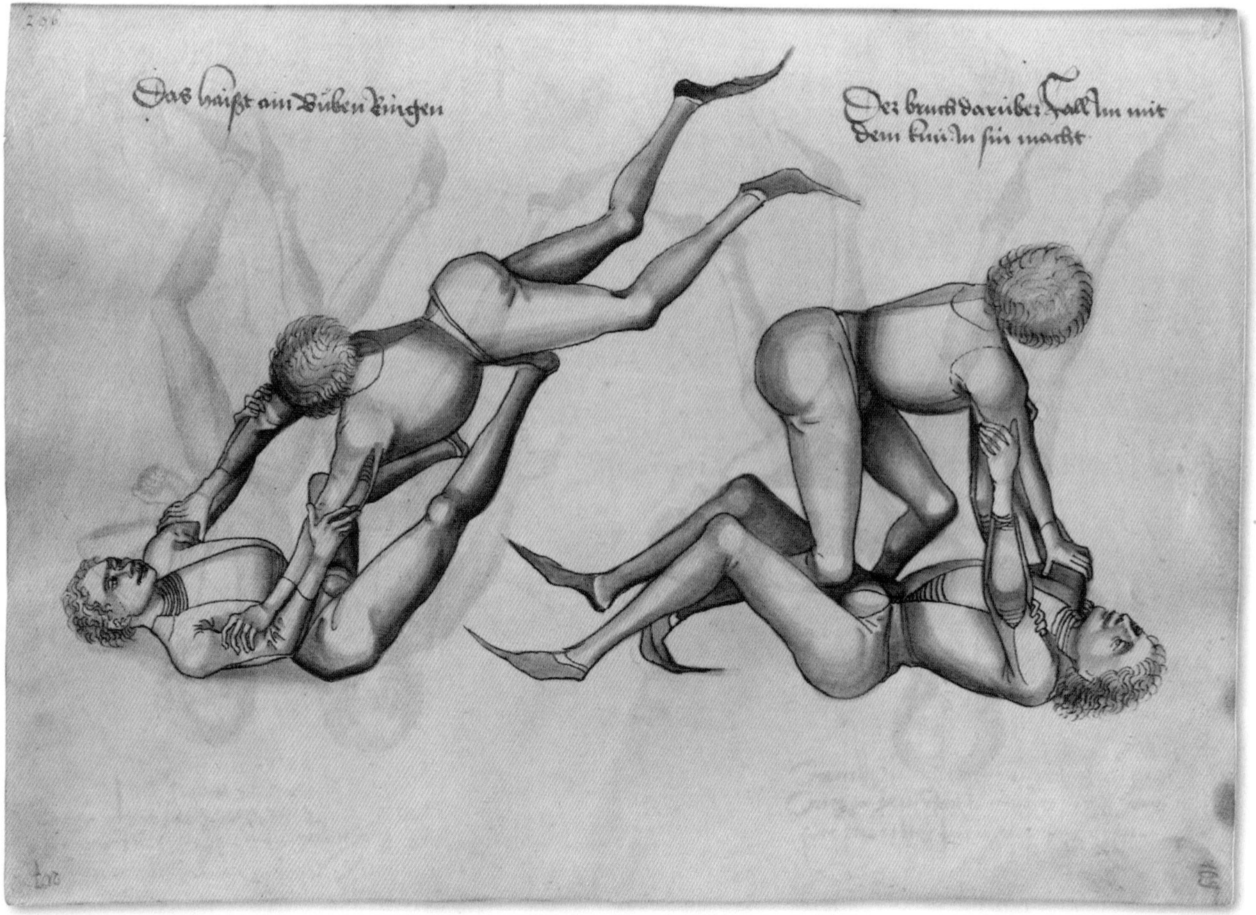

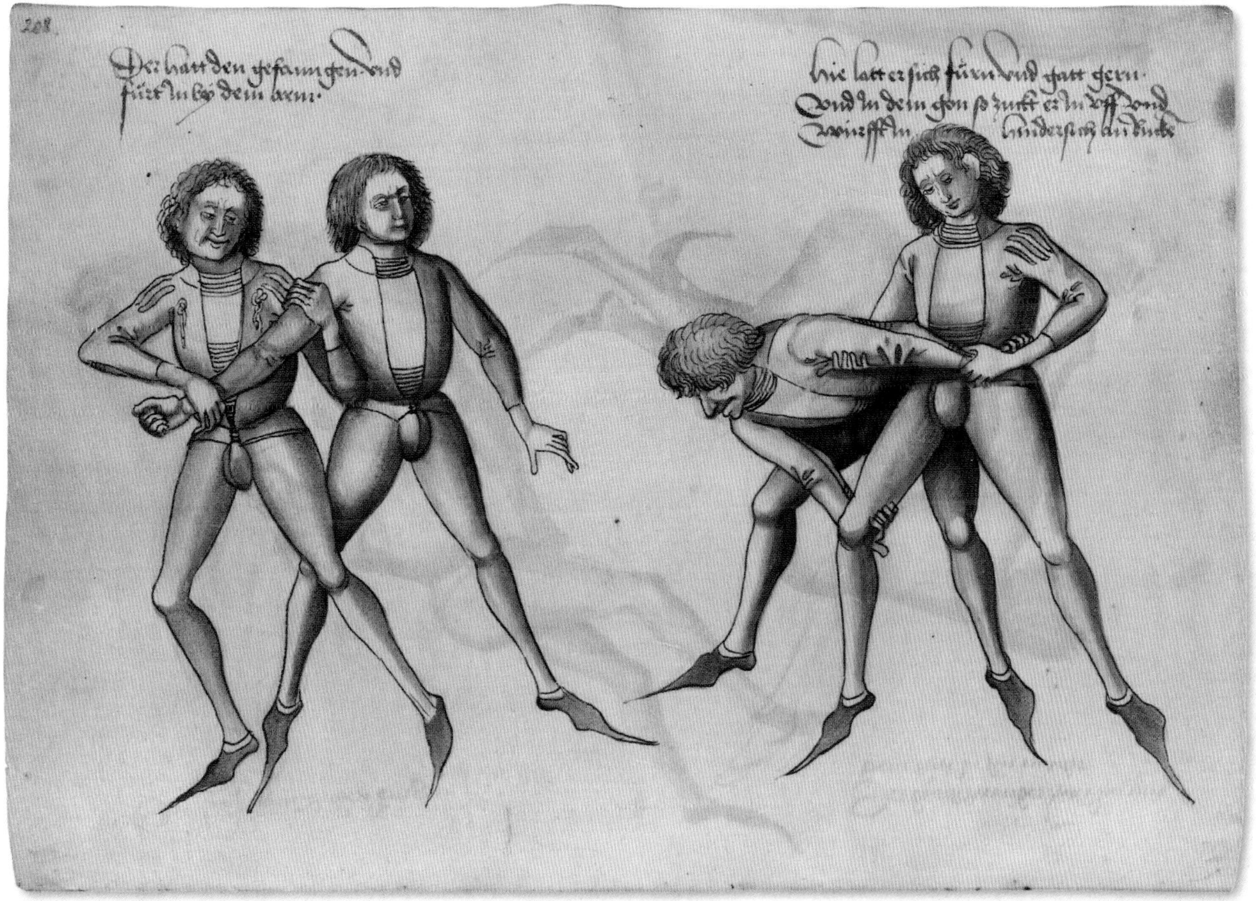

One has caught the other and leads him by the arm.

Here, he lets himself be lead and goes voluntarily. Meanwhile he pulls the other up and throws him backwards on his back.

When one tried to go through it, the other pulled him. He grabs him and pushes him down.

This is called shoulder wrestling and comes from grabbing the neck.

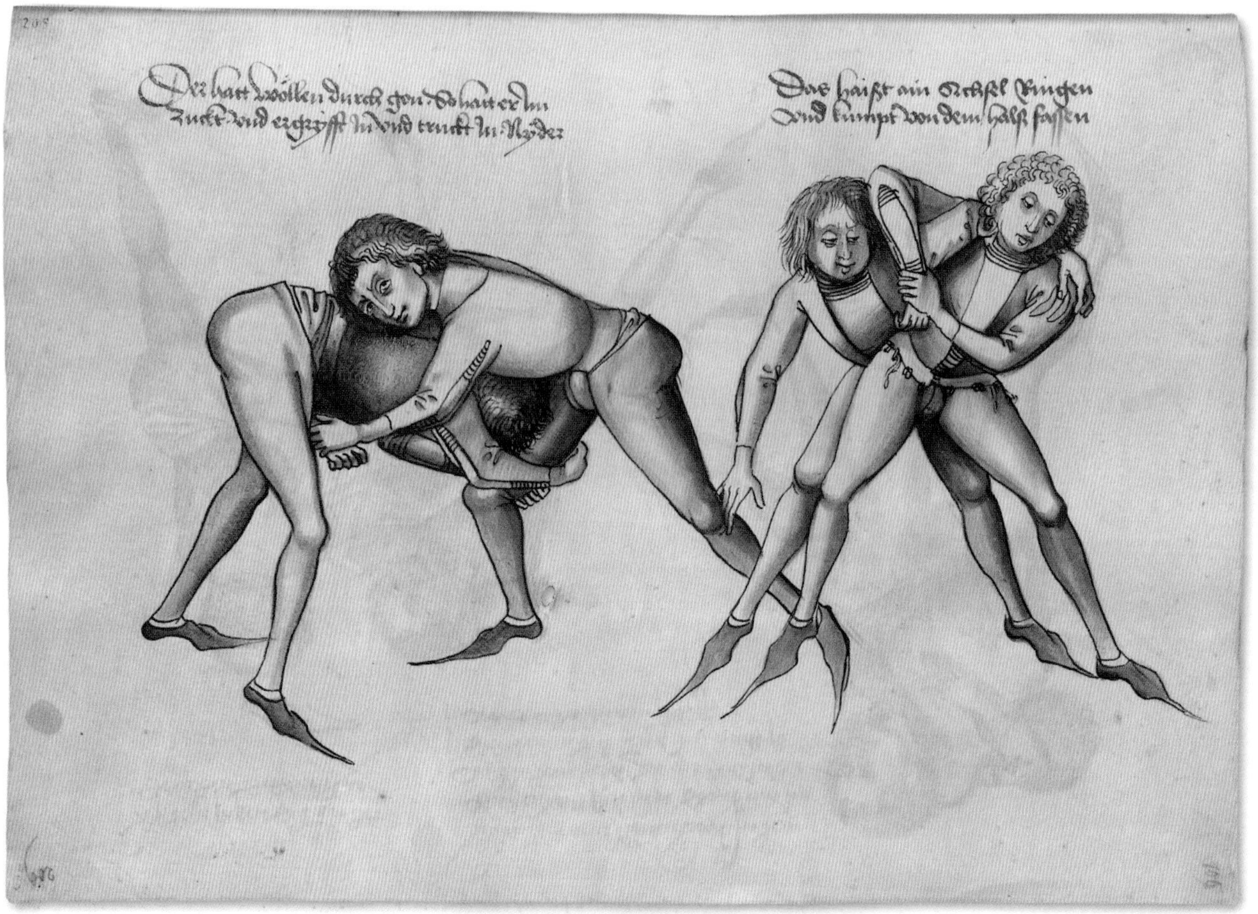

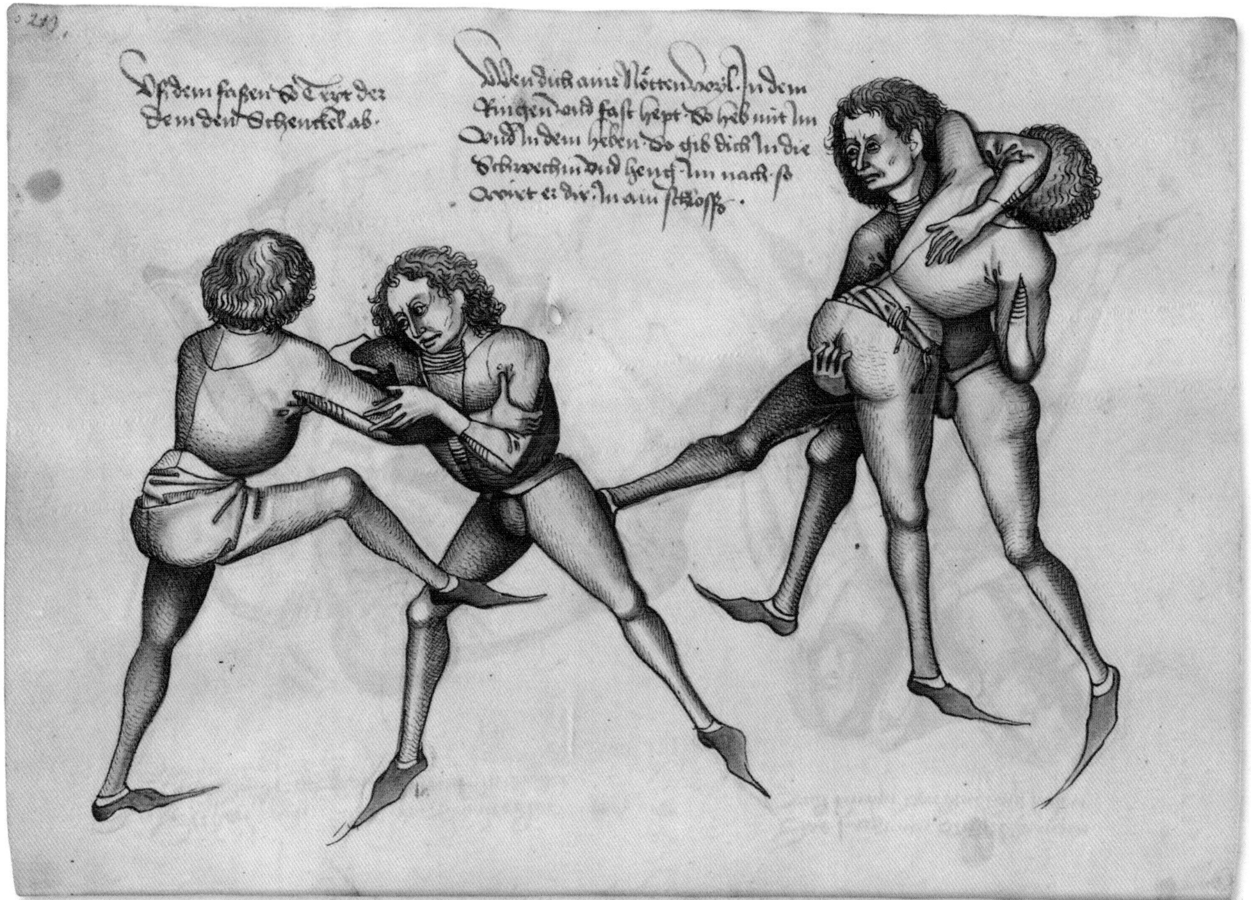

From the hold: this one crushes the other's leg.

When someone strives to oppress you with wrestling and lifts you vigorously, lift him at the same time. Meanwhile, go into the weakness and cling to him, thus you will succeed in enclosing him.

One has grabbed the other one by the collar and strikes his foot away below.

From the previous work: in this way, one throws the other over the lower hip.

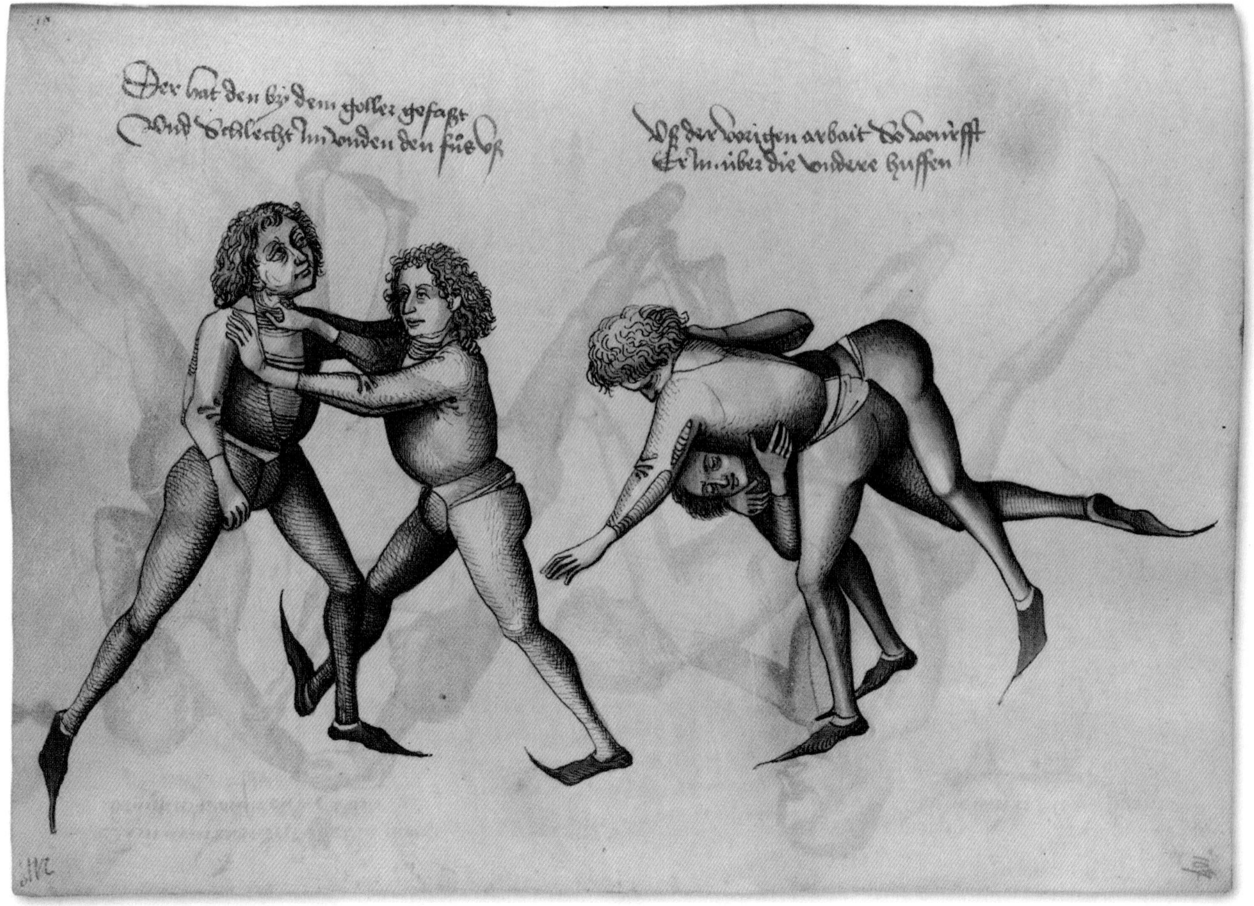

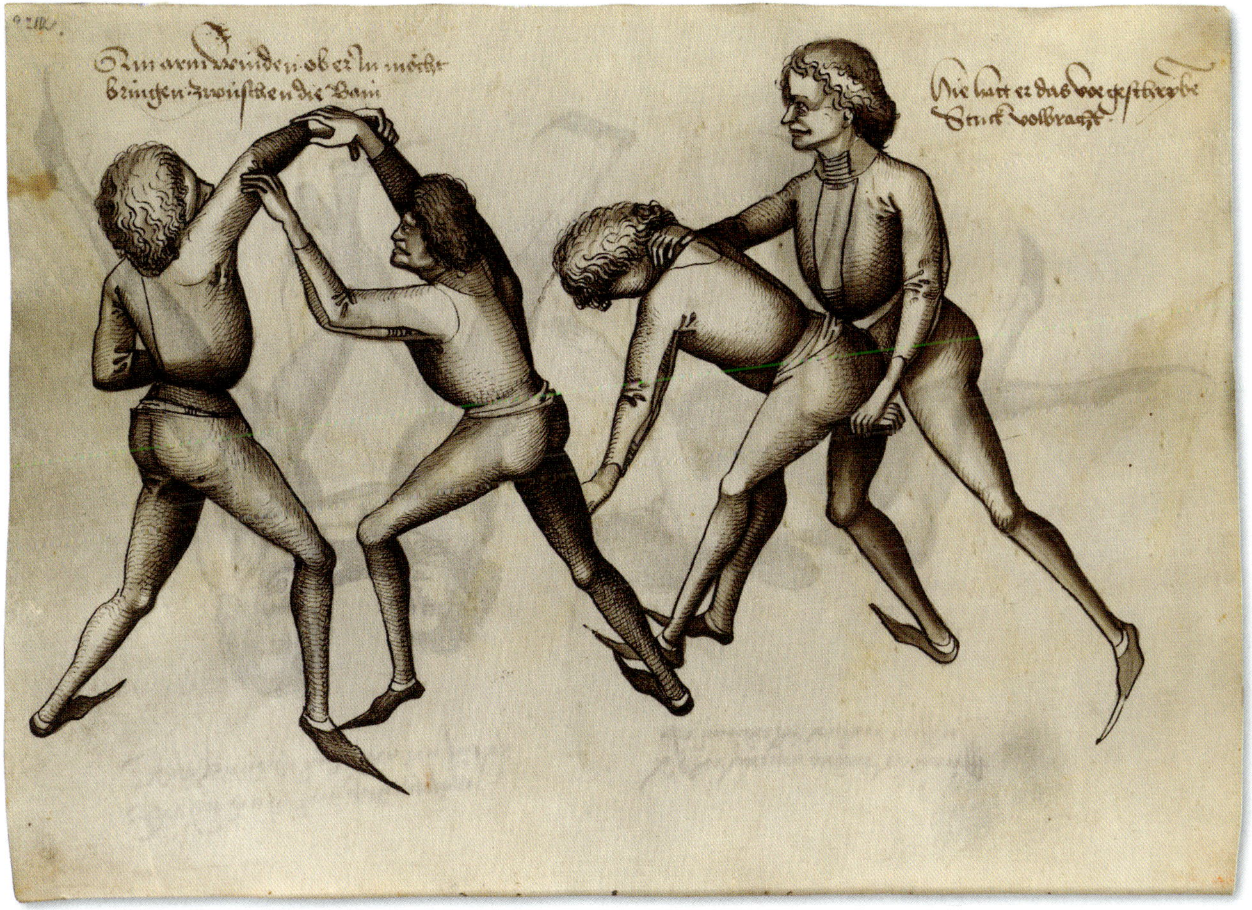

A turning of the arm when one strives to bring the other between his legs.

Here, he has accomplished the previous piece.

This one has brought his opponent into the *Trappe*.

In the *Trappe*, he pulls the other towards himself. Meanwhile, he hits his neck and throws him over the hip.

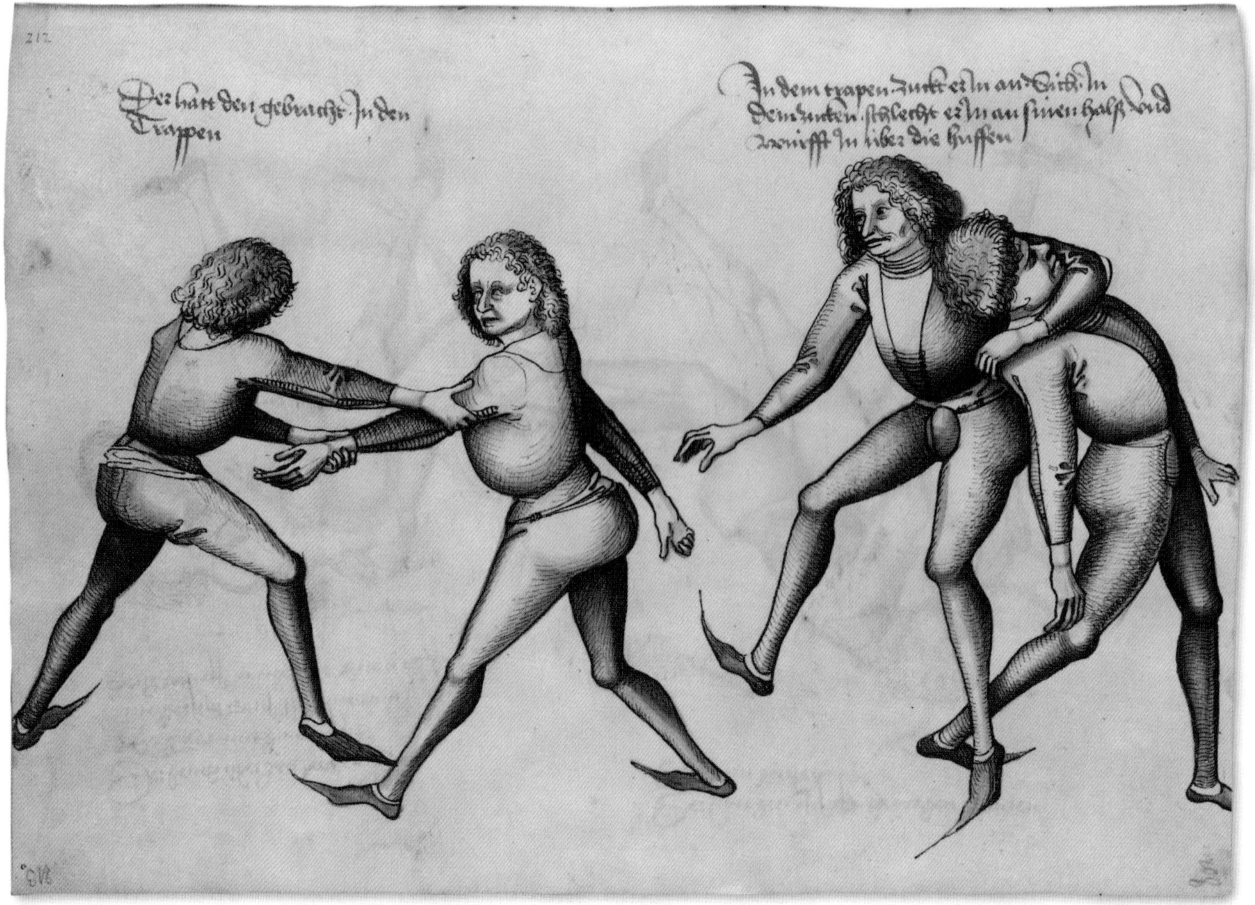

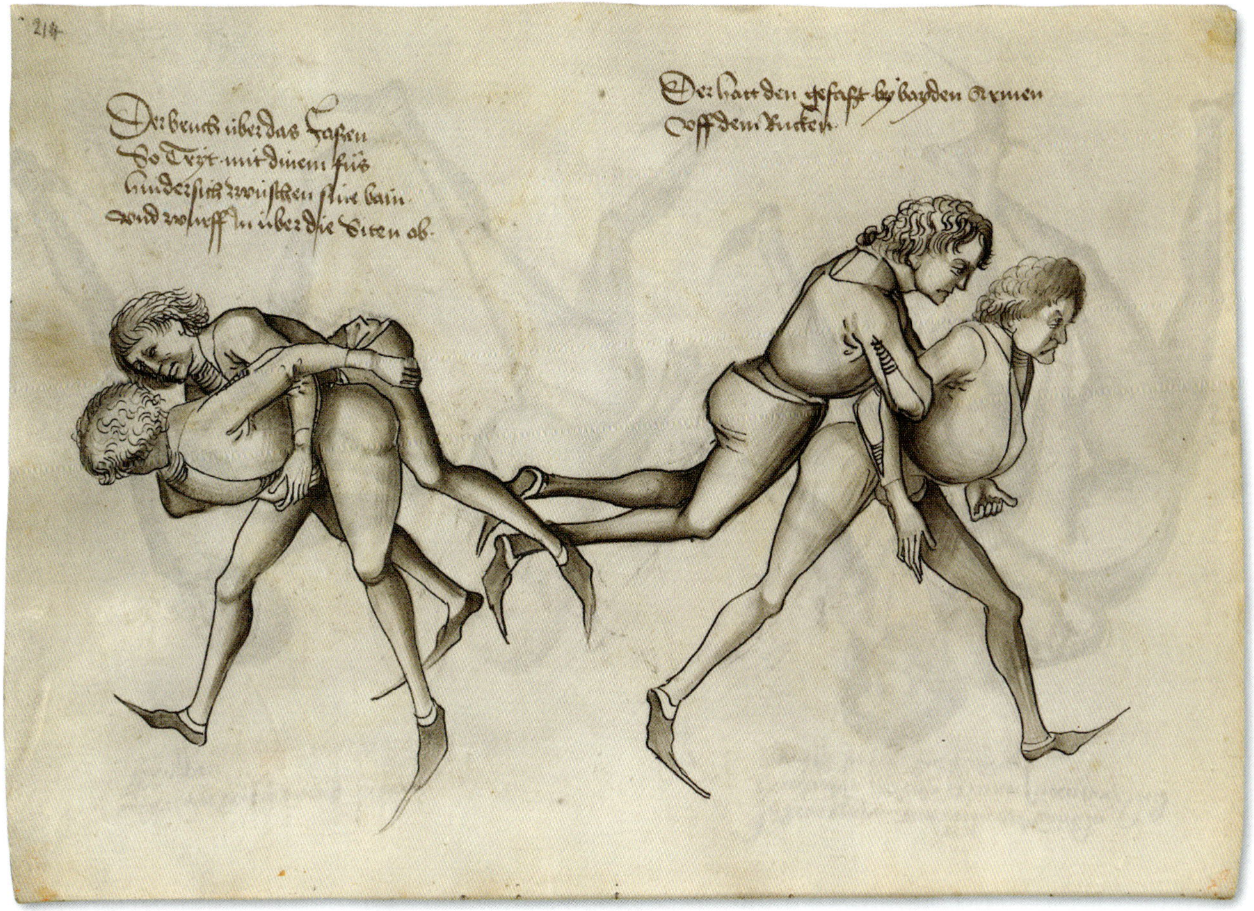

The break against the hold: step backwards with your foot between his legs and throw him to the side.

One has grabbed both of the other's arms behind his back.

Another hold by the neck: one grabs the other at the elbow, steps forward and pushes him over his leg.

A step behind. [Go] with your left arm to his neck and push him over your leg.

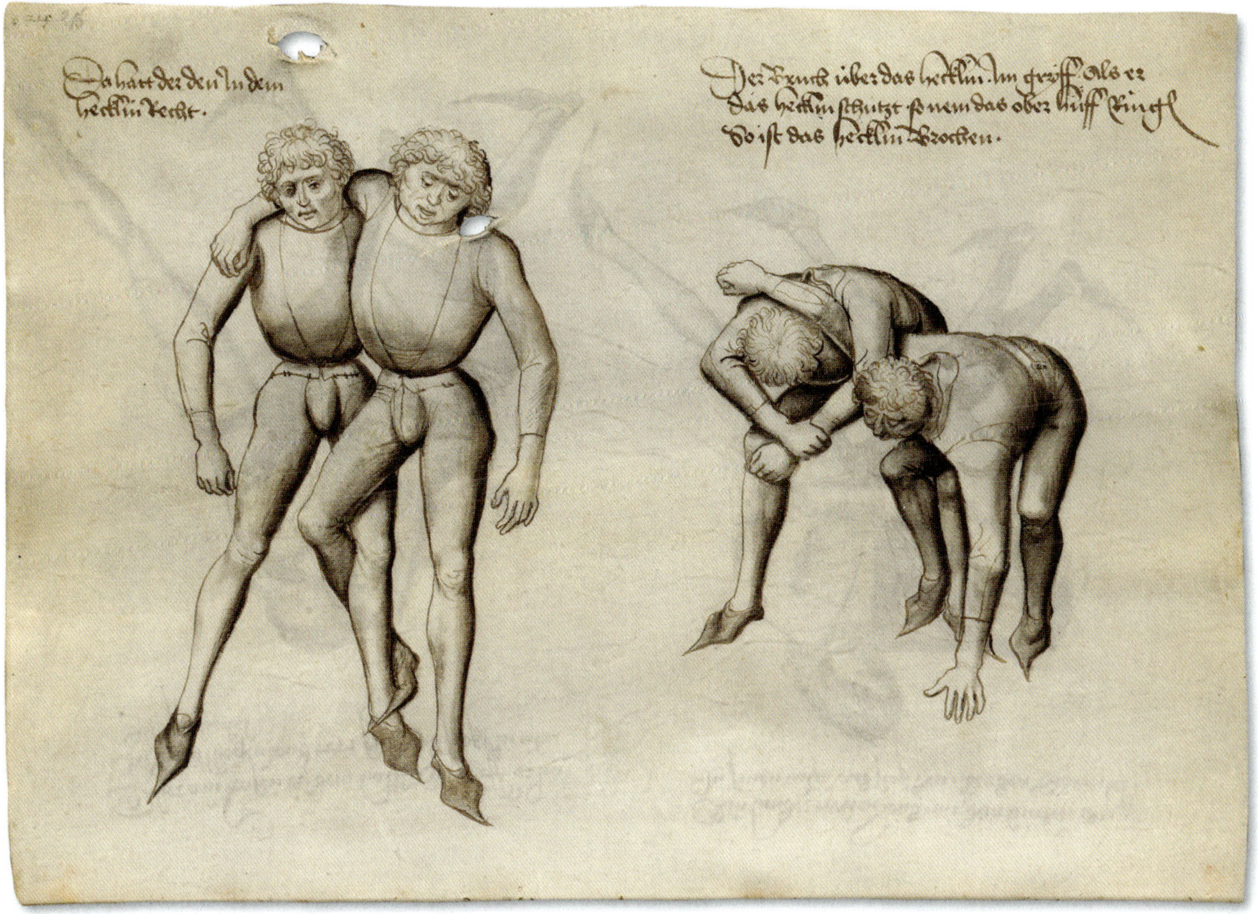

Here, one has the other one well in the hook.

The break against the hook in the hold: as soon as he throws the hook, take the upper hip wrestling posture; thus his hook is broken.

This is a break on the outside against the hook and the upper hip wrestling move.

This is also a break against the hook or hip [wrestling move].

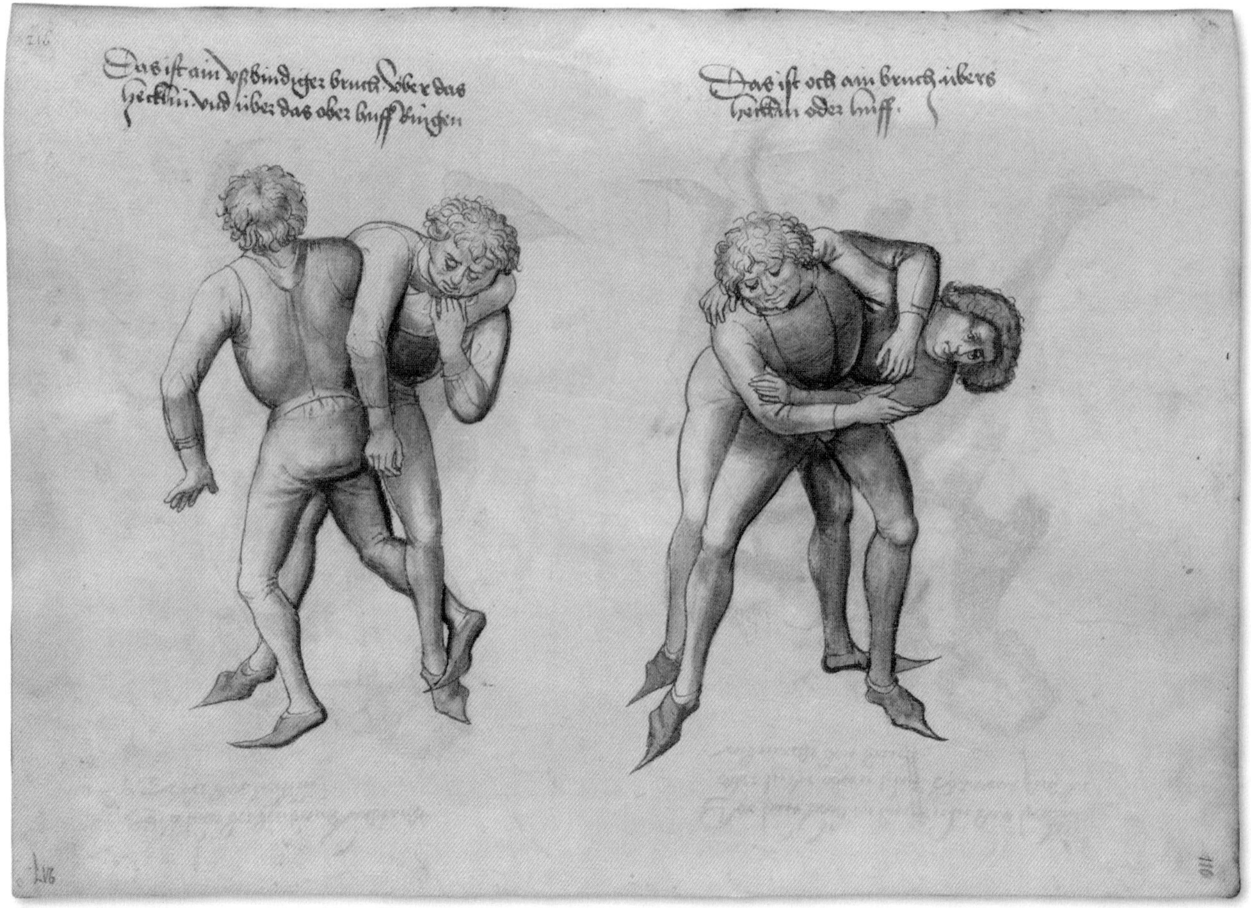

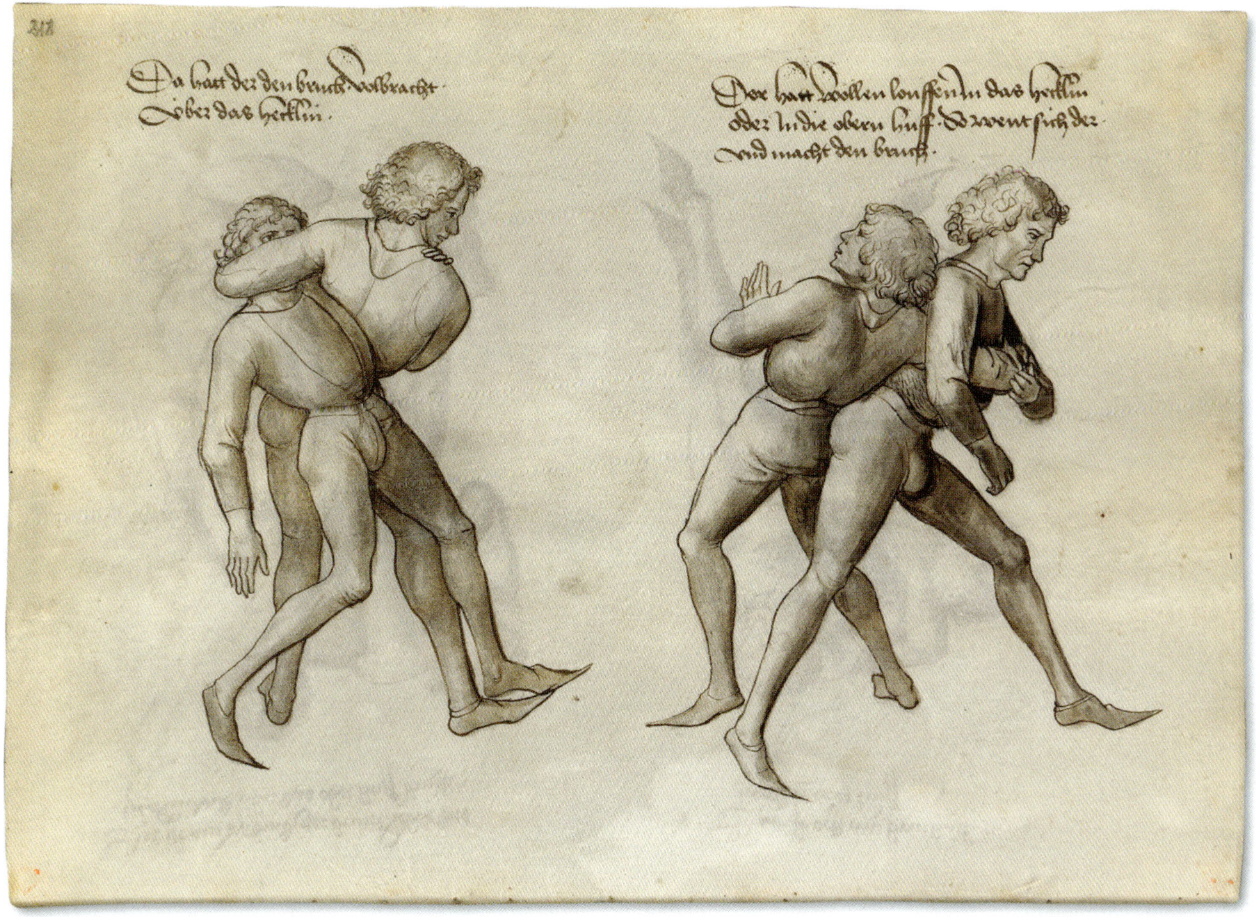

Here, he has accomplished the break against the hook.

~

This one tried to run into the hook or the upper hip [wrestling move]. But the other turns around and makes the break.

From holding the neck, he makes an enclosure.

Here, one has enclosed the other and throws him over his leg.

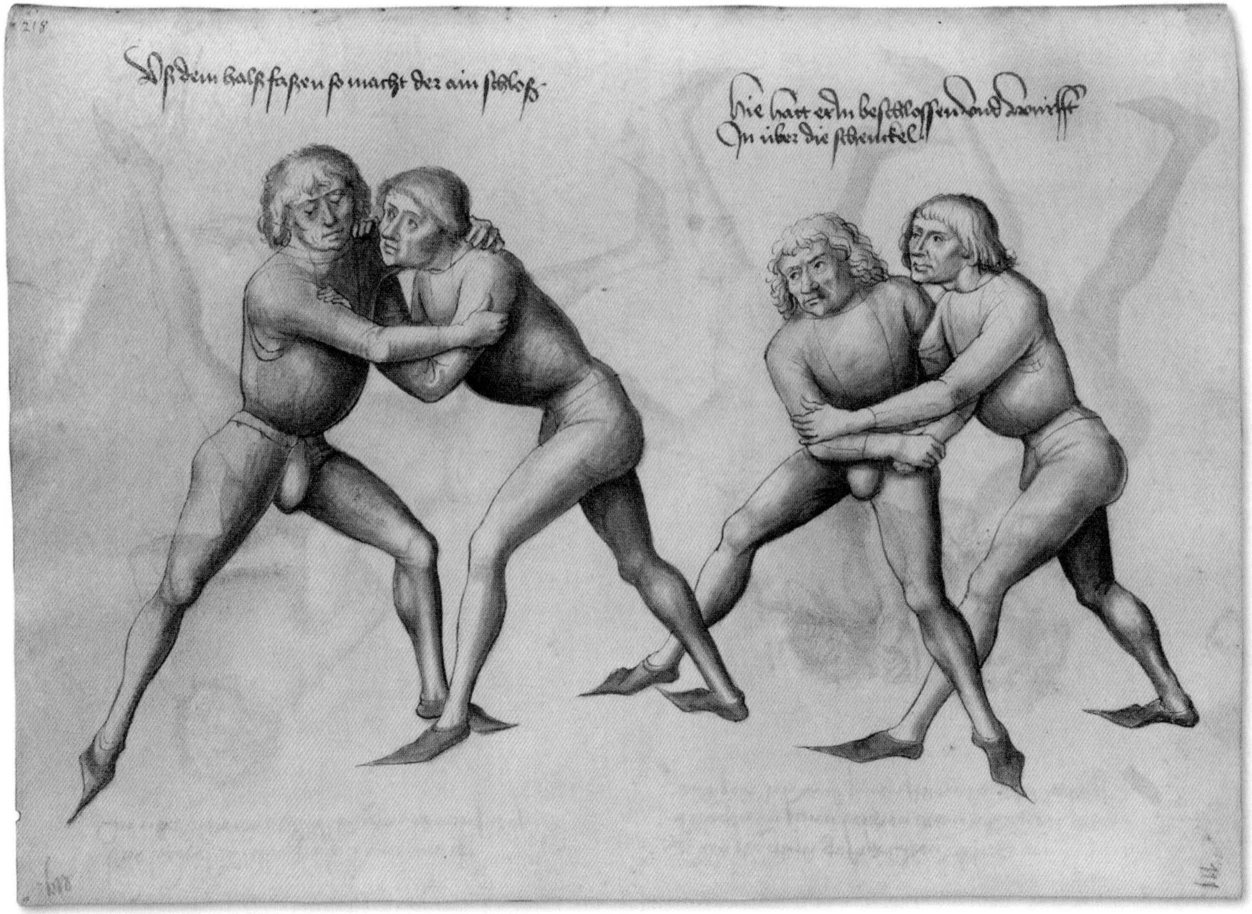

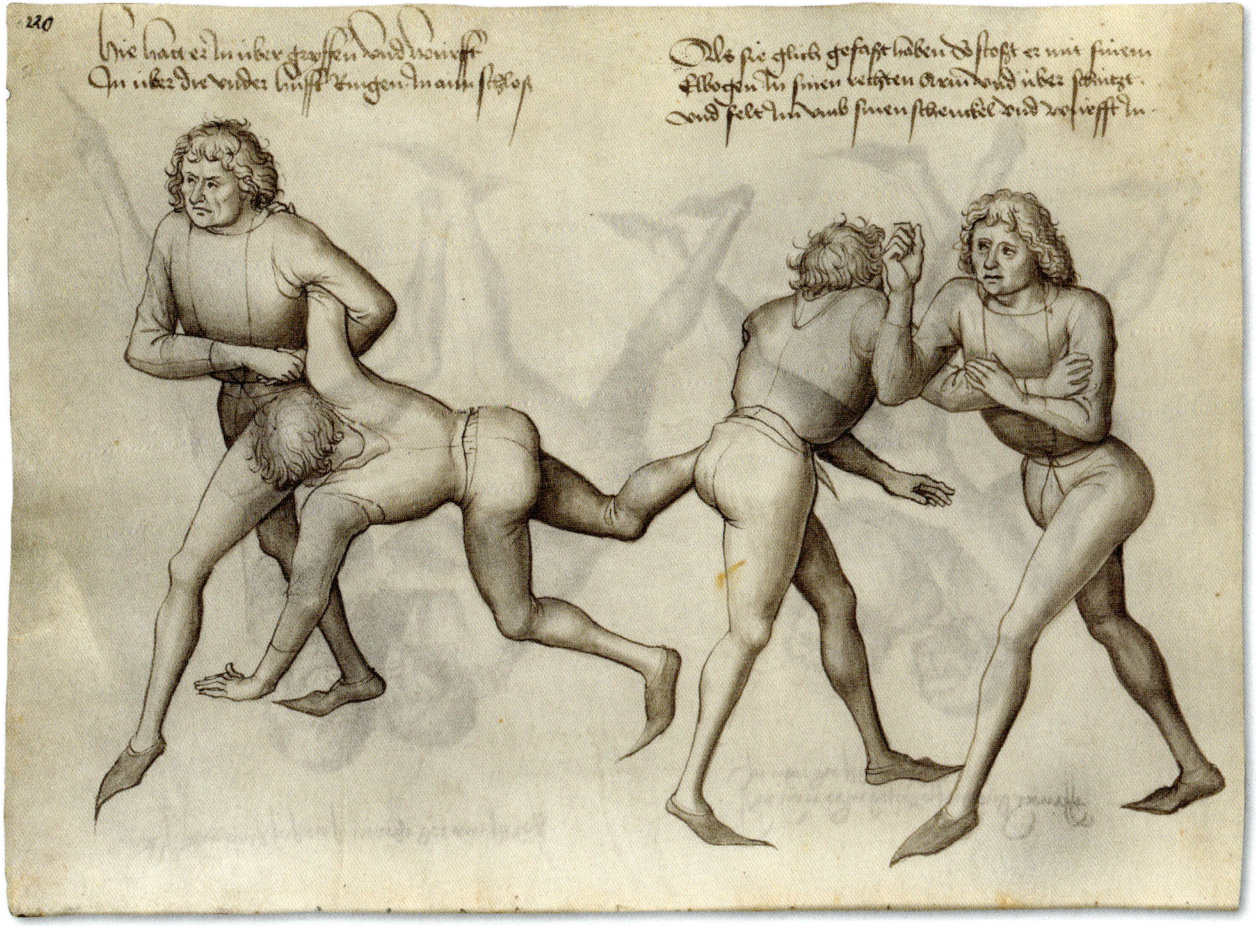

Here, one has grabbed over the other and throws him cnclosed with the lower hip wrestling move.

After they have taken hold of each other in the same way, one pushes his elbow into the right arm of the other and hurls [him] over it. He falls around his leg and throws him.

Here, the throw is completed, as decribed below.

—

Another break against the hook: if he hits you with the hook, fall with your knee into the hollow of his knee.

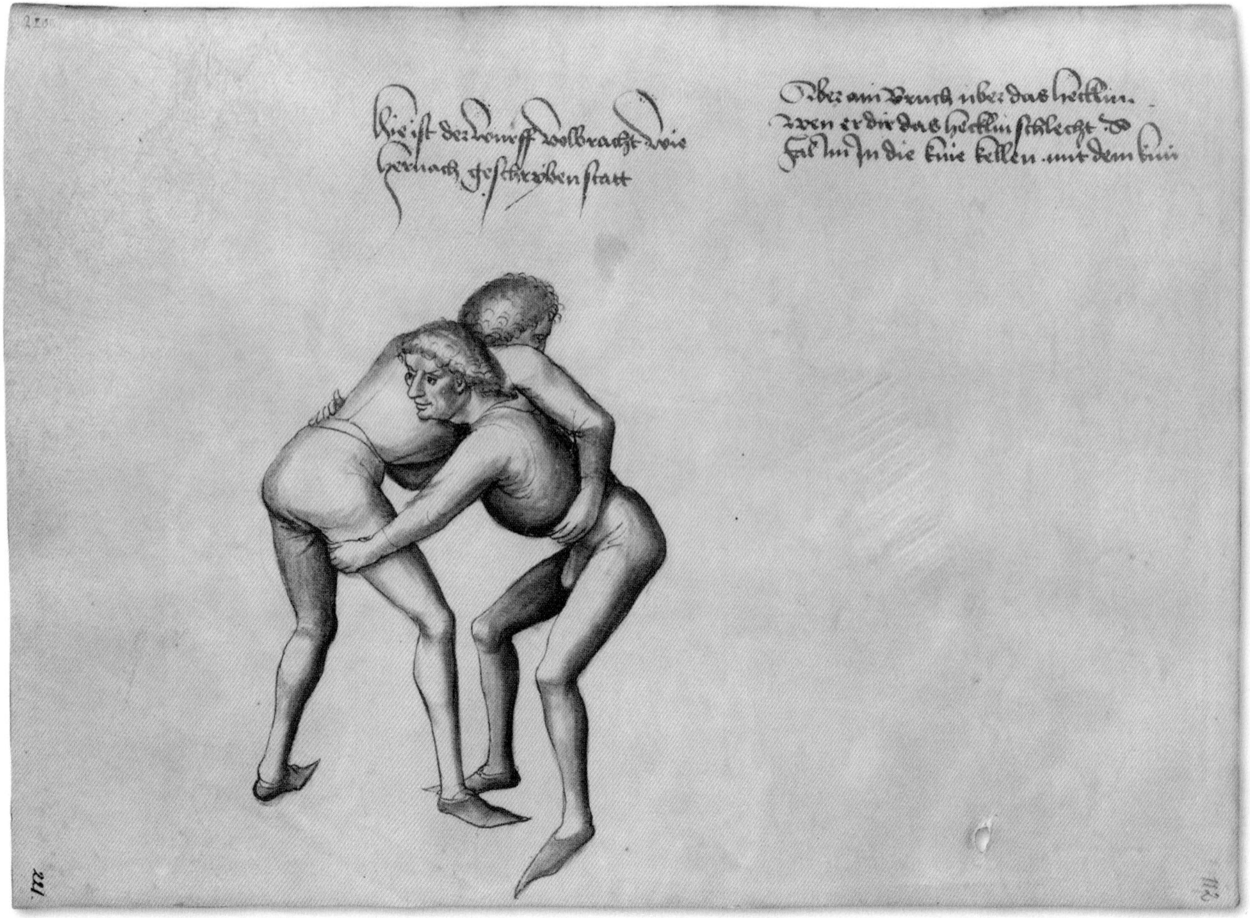

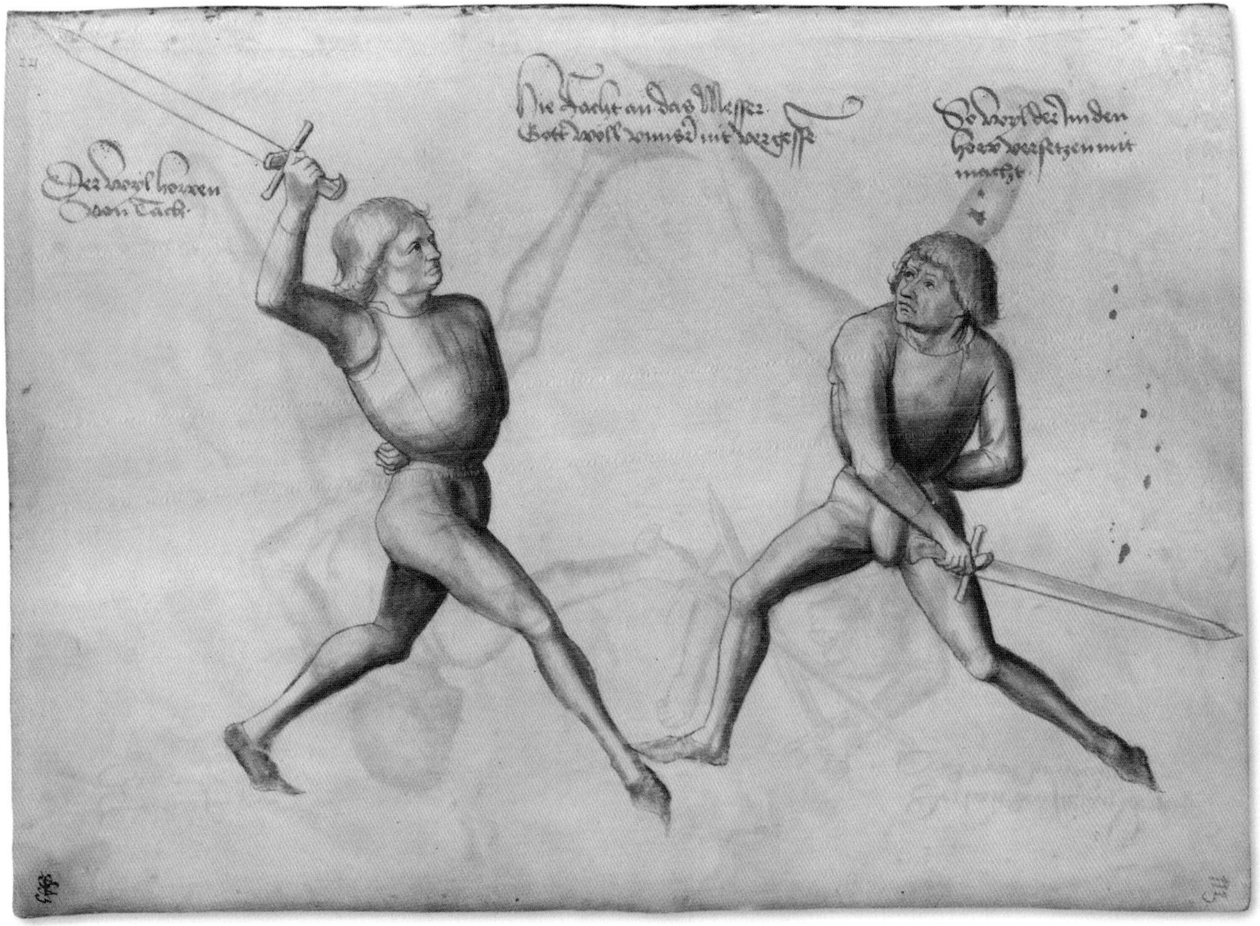

Here, begins the *Messer*. May God remember us.

He strives to strike *vom Tag*.

So the other strives to displace the strike powerfully.

He has accomplished his strike.

One has displaced the strike and strives to grab above the other.

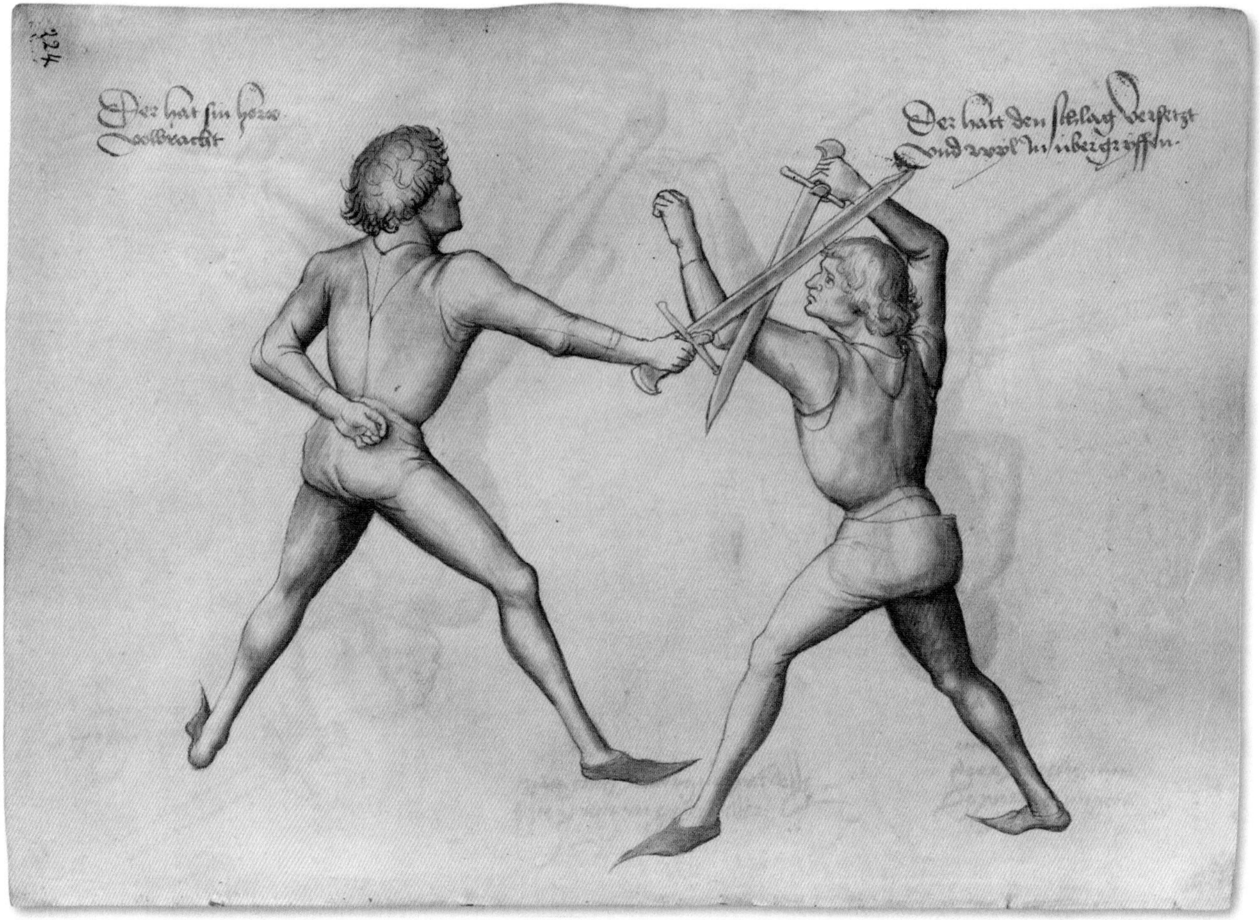

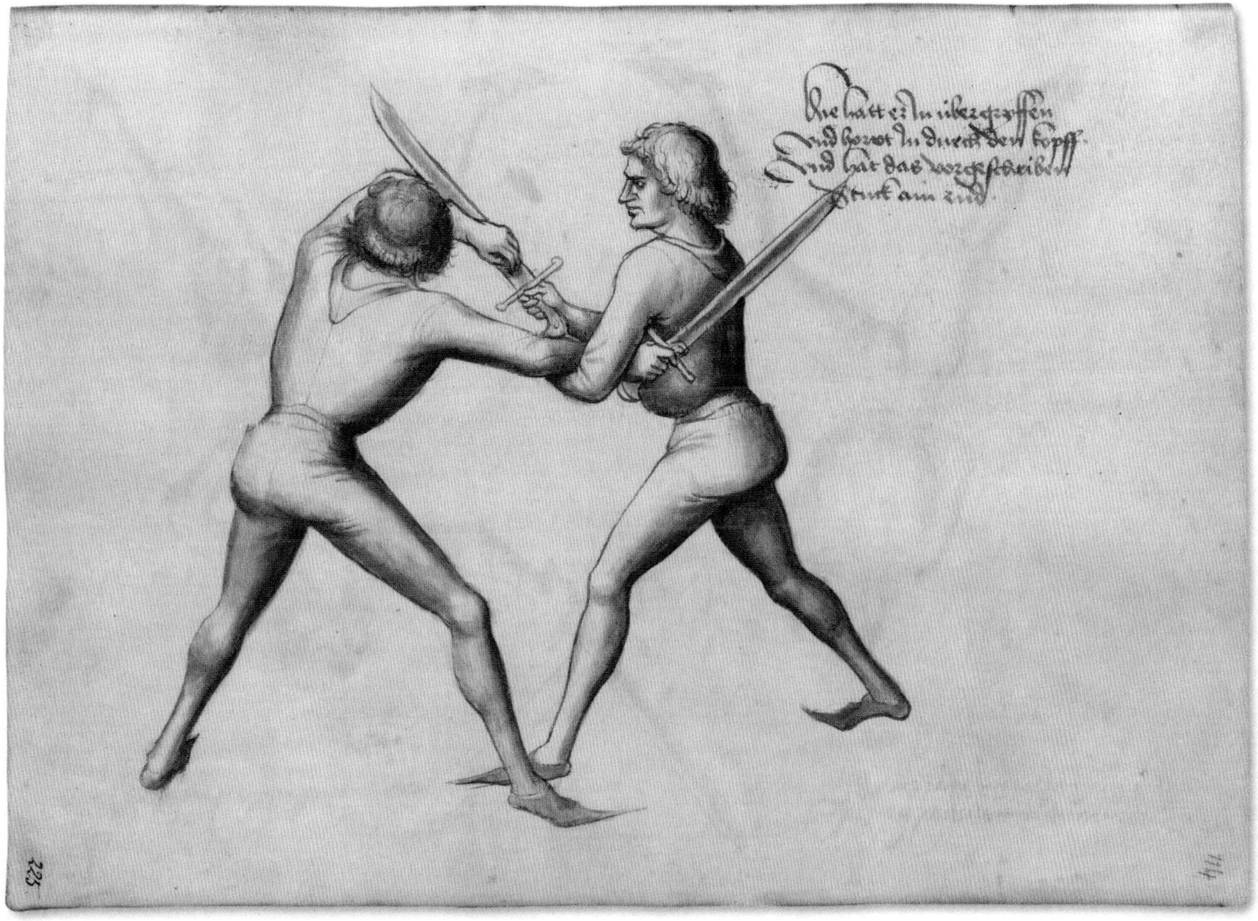

Here, one has grabbed above the other and hits him through the head. This concludes the previous piece.

He strikes freely *vom Tag*.

He has displaced the other with a twisted hand. He strives to step forward and wrench at him.

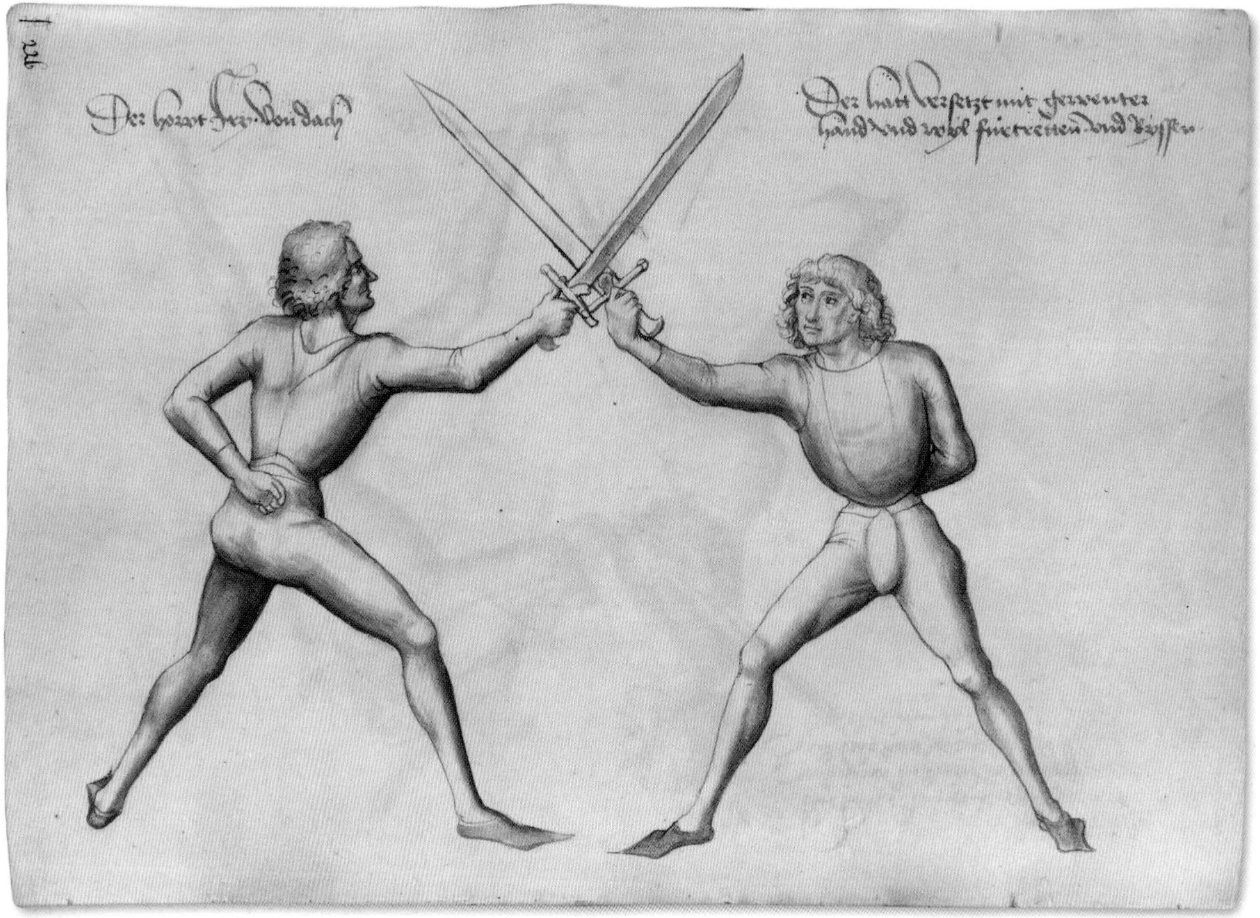

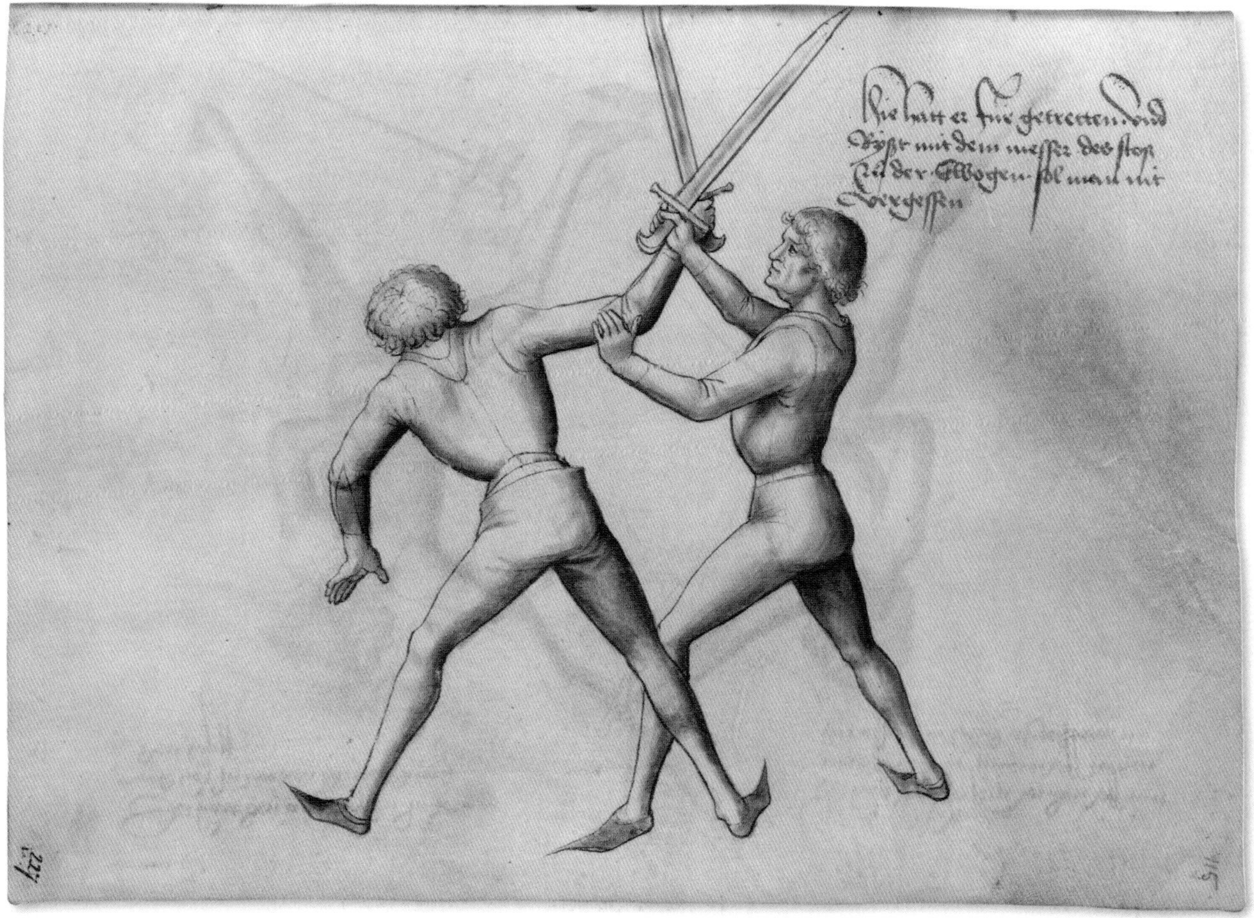

Here, he has stepped forward and is wrenching with the *Messer*. One should not forget the hit to the elbow.

One has run angrily towards the other and tried to strike him in the head.

Thereupon he has displaced his opponent powerfully from below. While going up, he has cut off the other's hand.

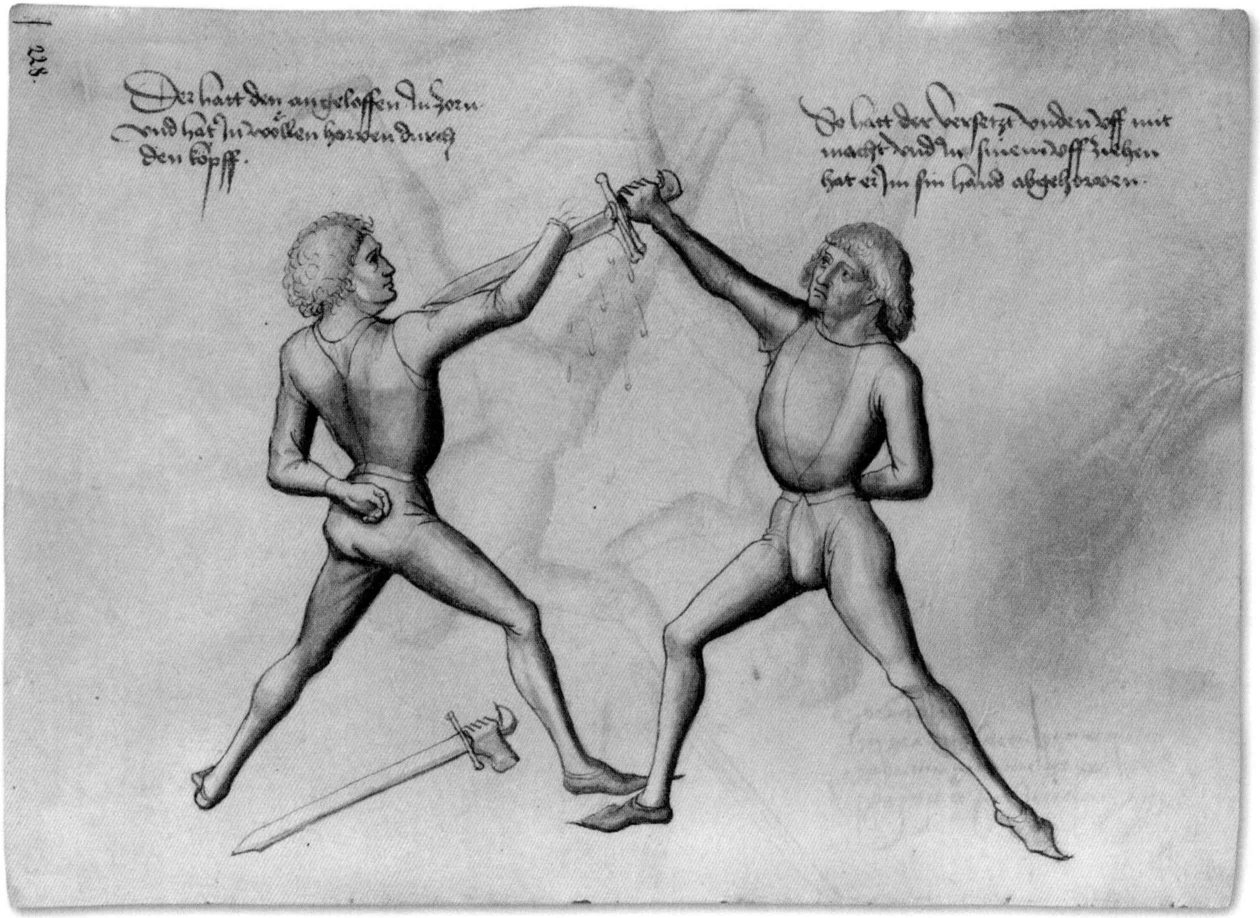

Here, he steps forward and beats the other down to death.

So one has displaced the other and grabbed over him with his arm; he pushes the *Messer* into him.

This one has struck *vom Tag* onto his opponent.

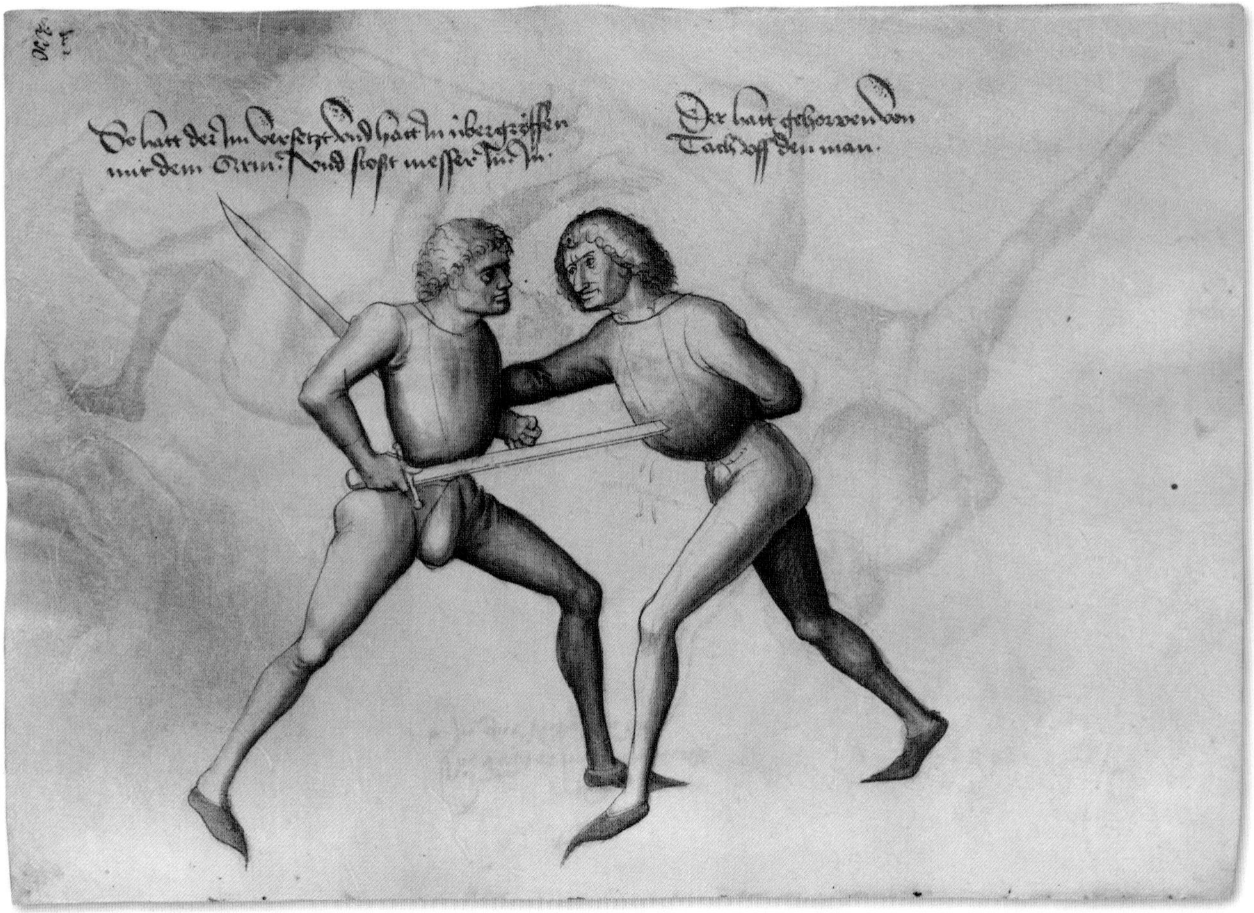

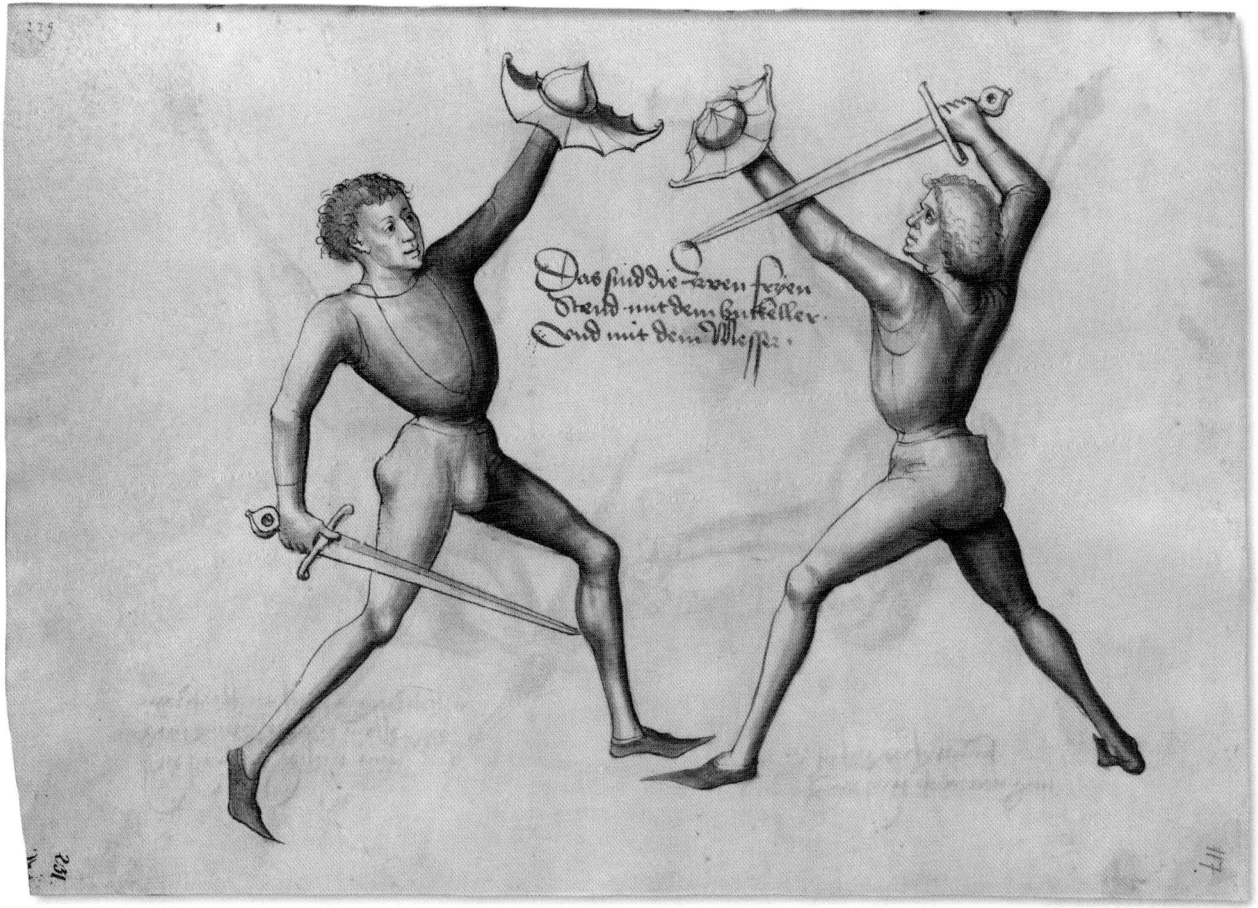

These are the two free stances with the buckler and *Messer*.

Here, he has displaced his opponent with the buckler and *Messer*. He must not forget to grab him from above.

One has struck into the other's displacement.

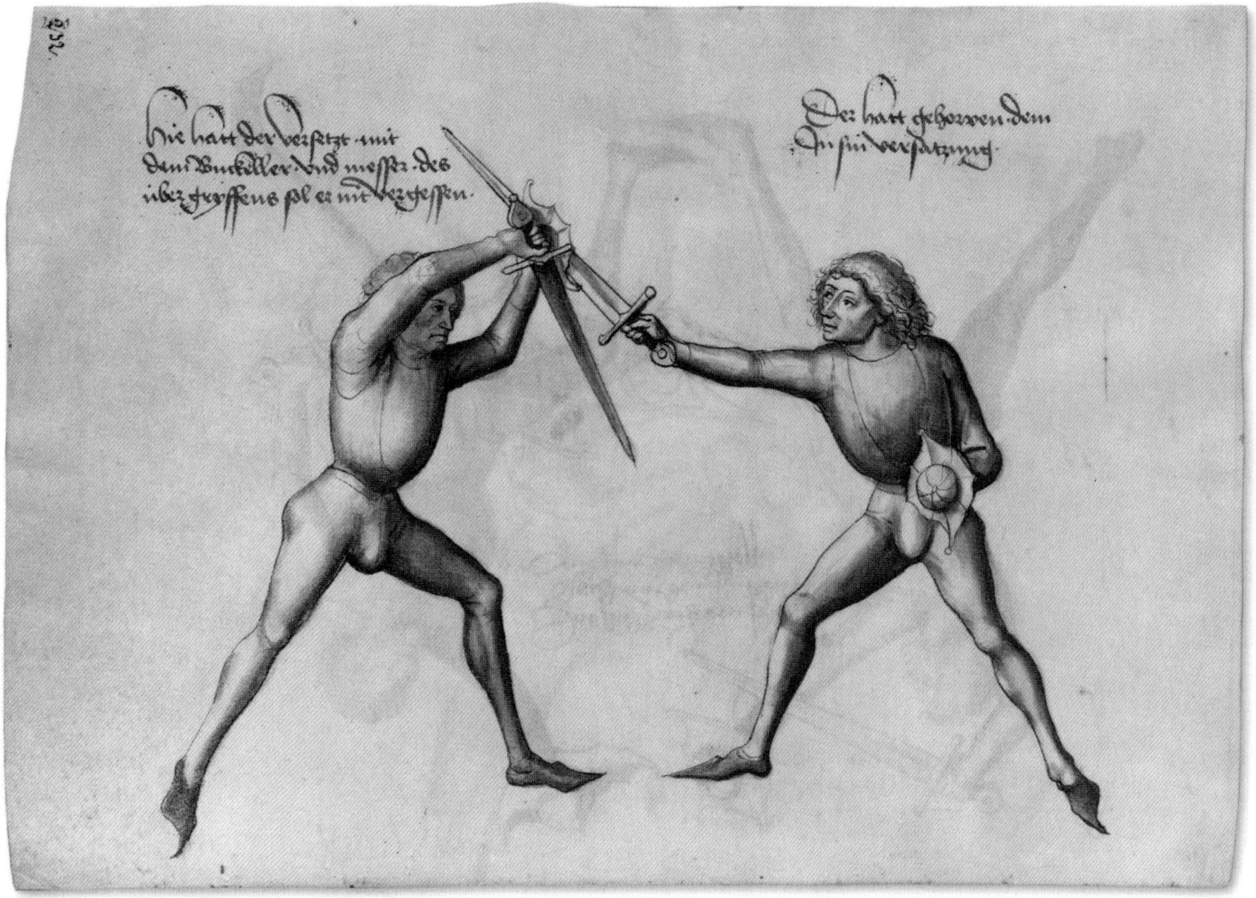

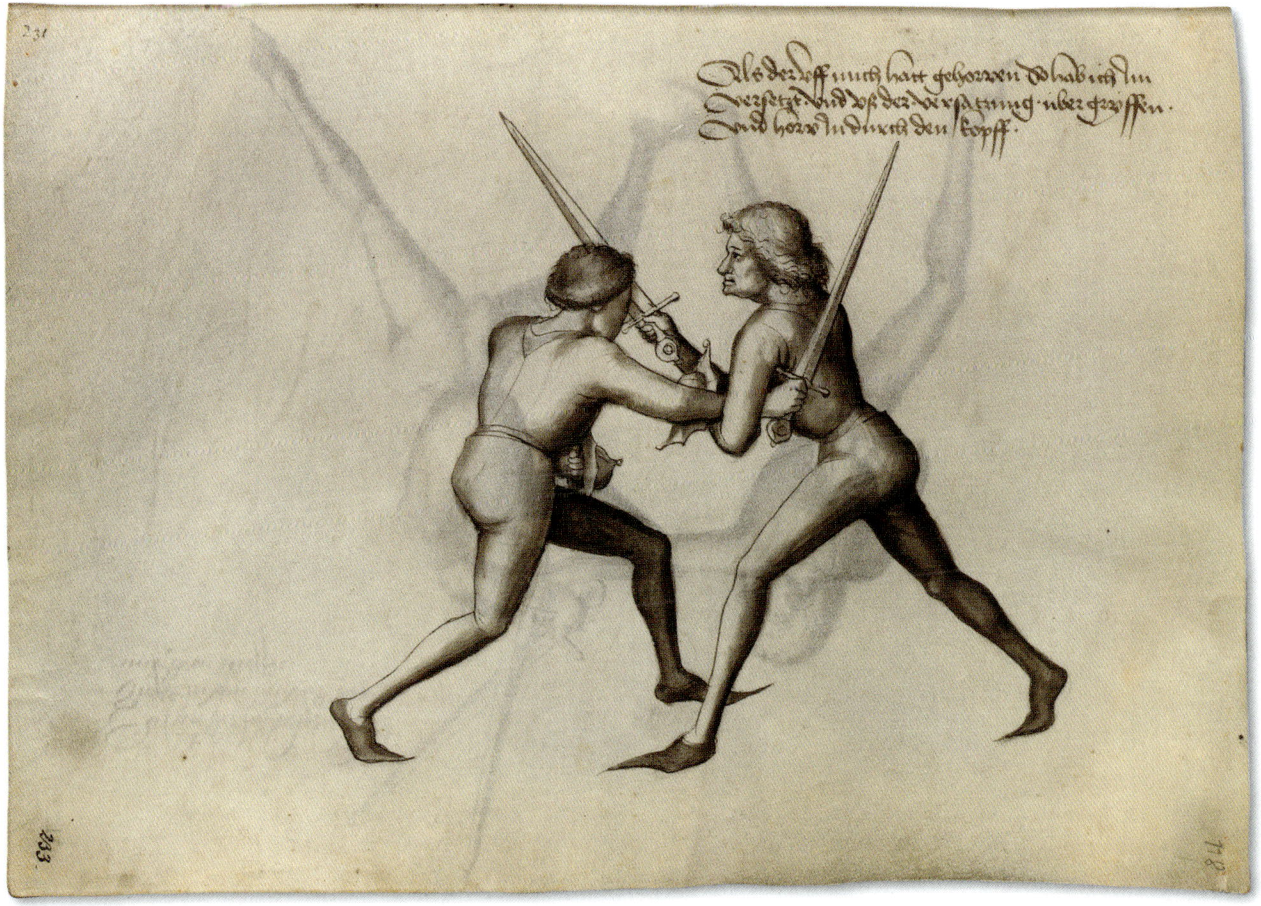

After he has struck at me, I have displaced him, grabbed over him from his displacement, and I strike him through the head.

Here, one has displaced the other, pushes him away from himself and wrenches at him with the *Messer*.

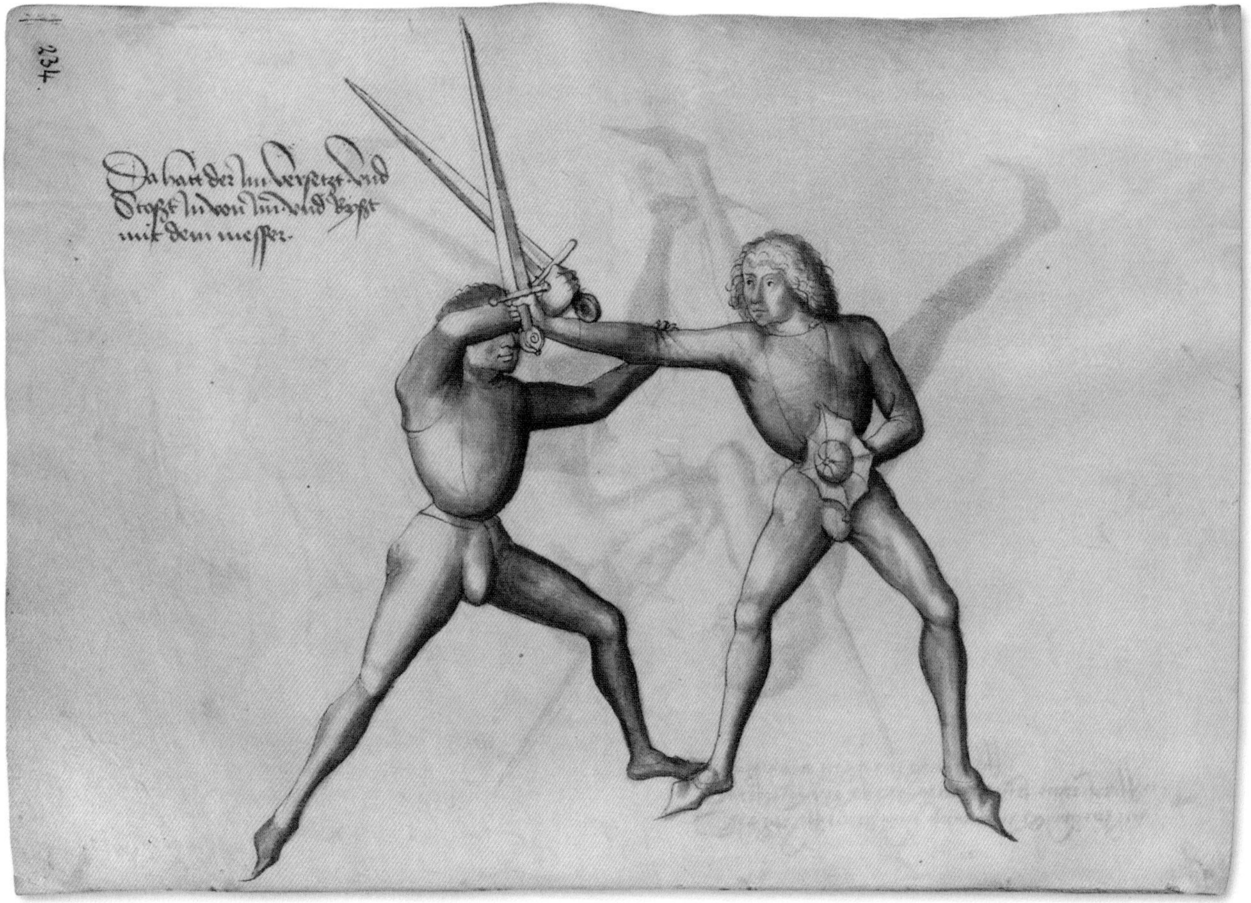

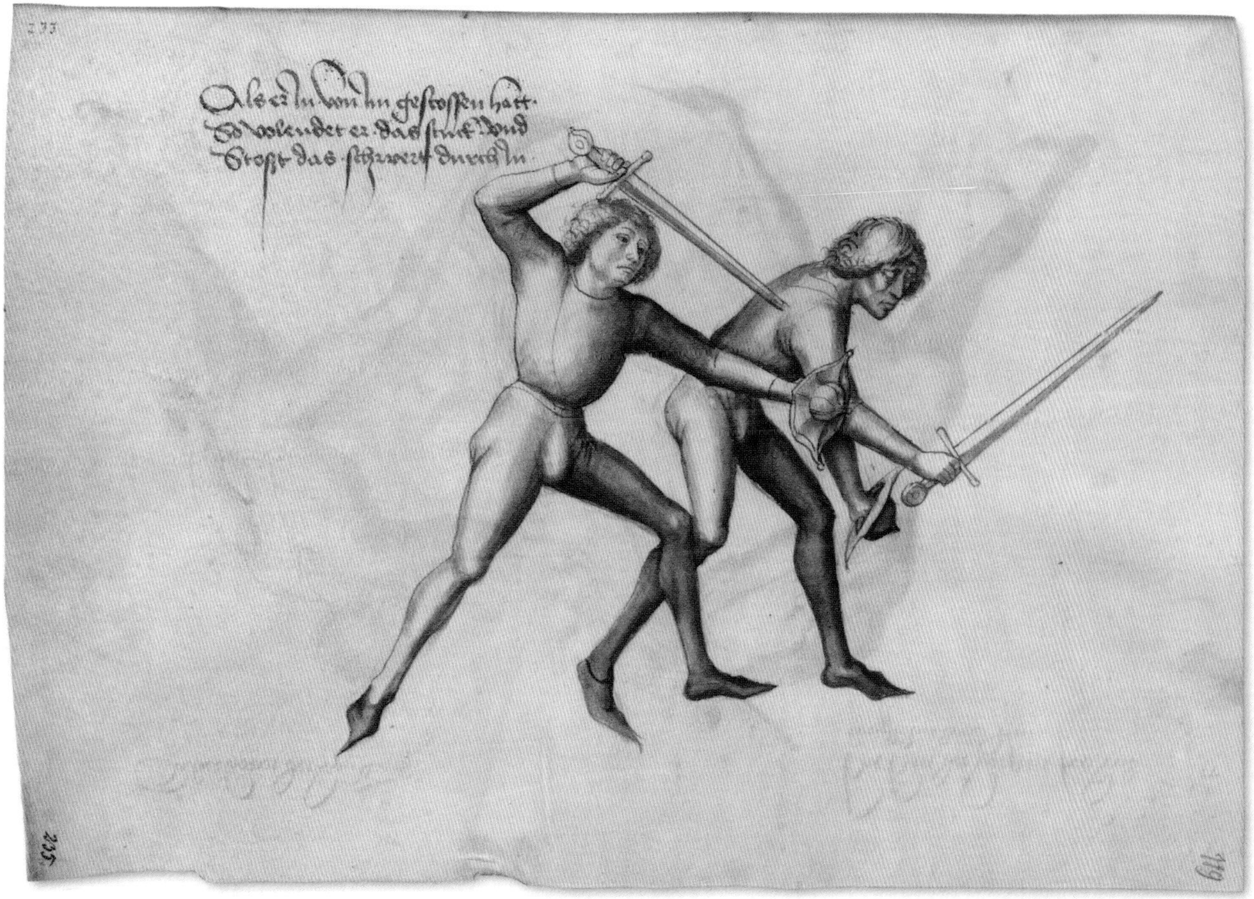

After one has pushed the other from himself, he completes the piece and thrusts the sword through him.

That one also strikes *vom Tag*.

Here, one has displaced the other freely and thus runs in to him.

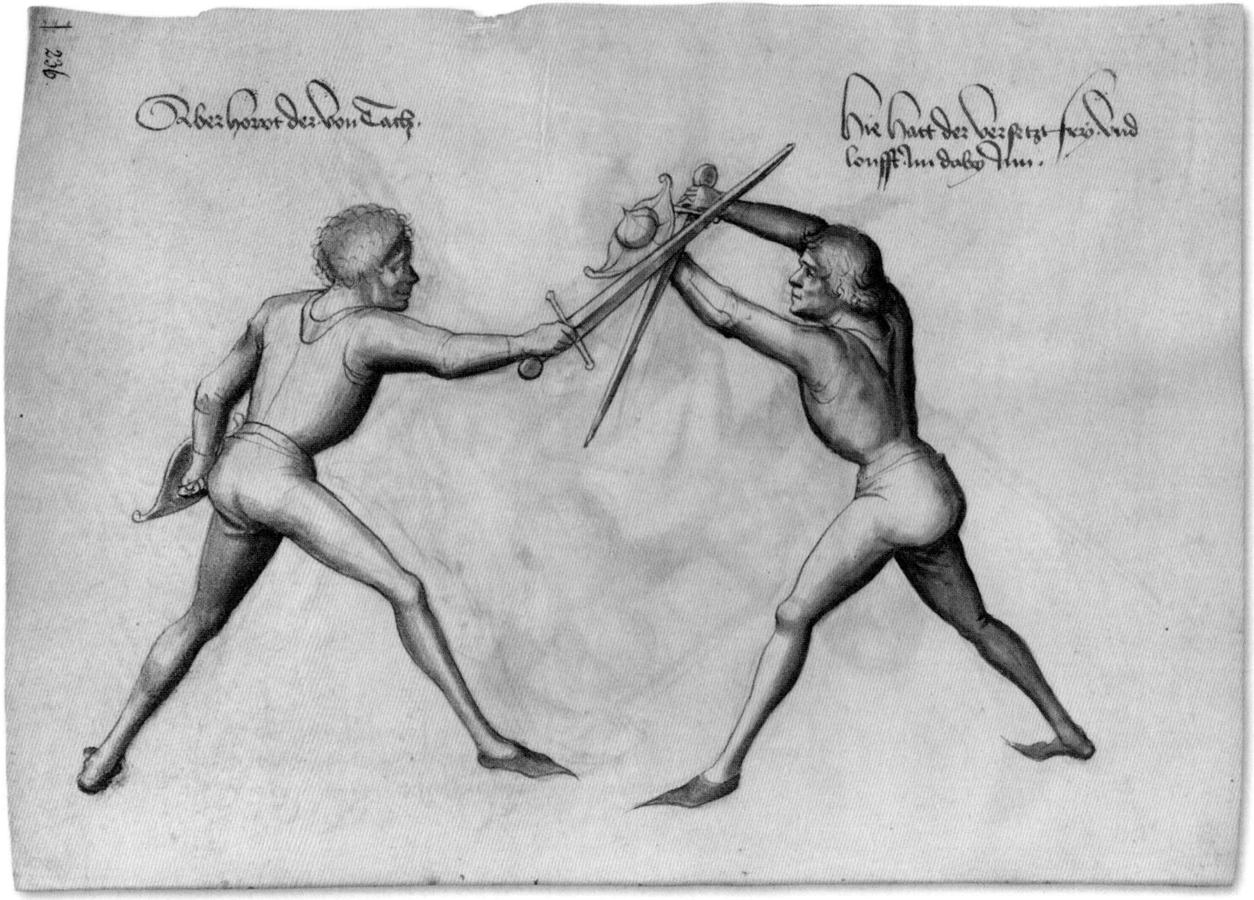

Here, he has run into the other and has pushed the sword through him.

He stands against the thrust.

He intends to thrust.

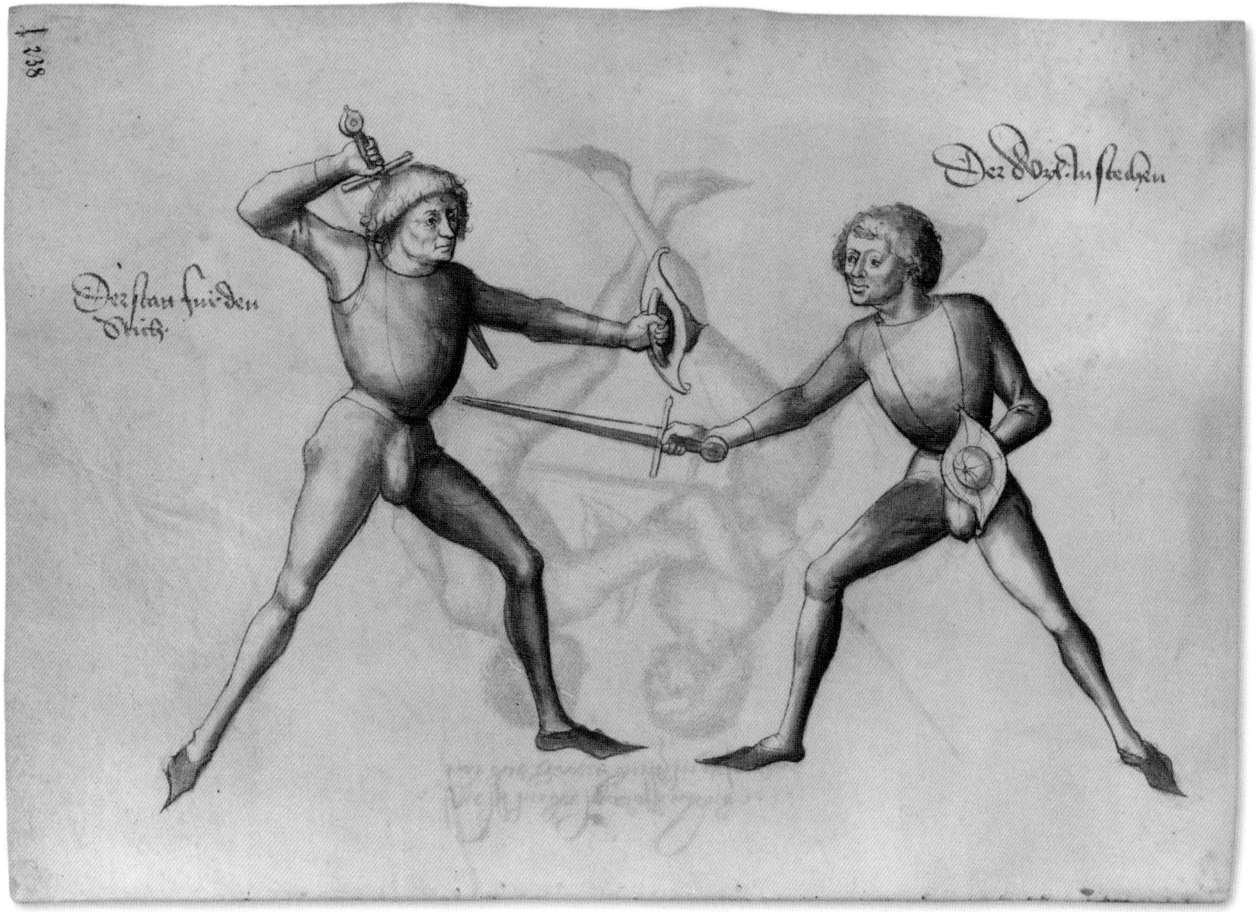

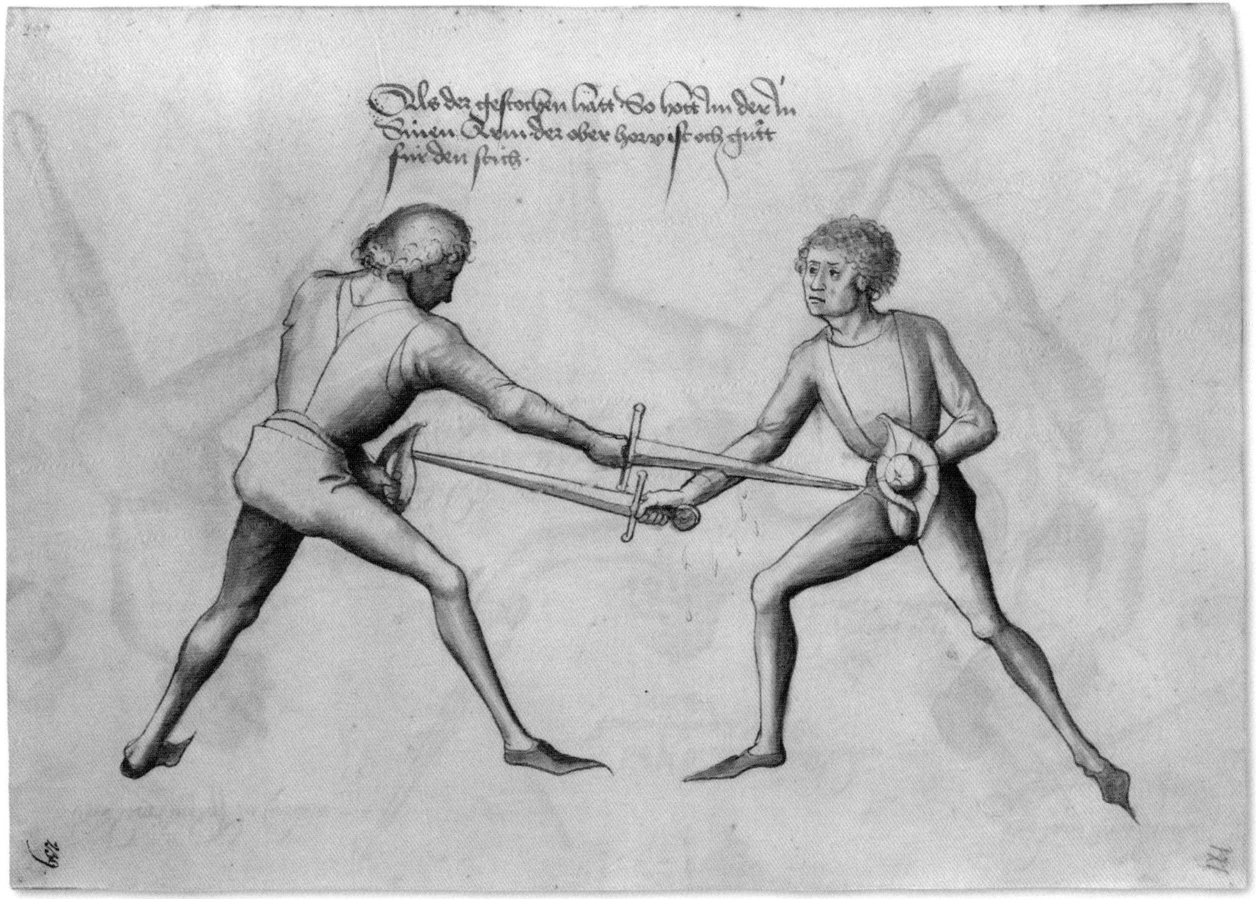

After he has thrust, the other hits him in the arm. The *Oberhau* is also good for the thrust.

This is a crisis stance when you face two adversaries.

Here, I strive to strike at him.

Here, he displaces them with a twisted hand. He will turn around and strike at the one in the back.

Here, he displaces with dagger and buckler.

That one will strike also.

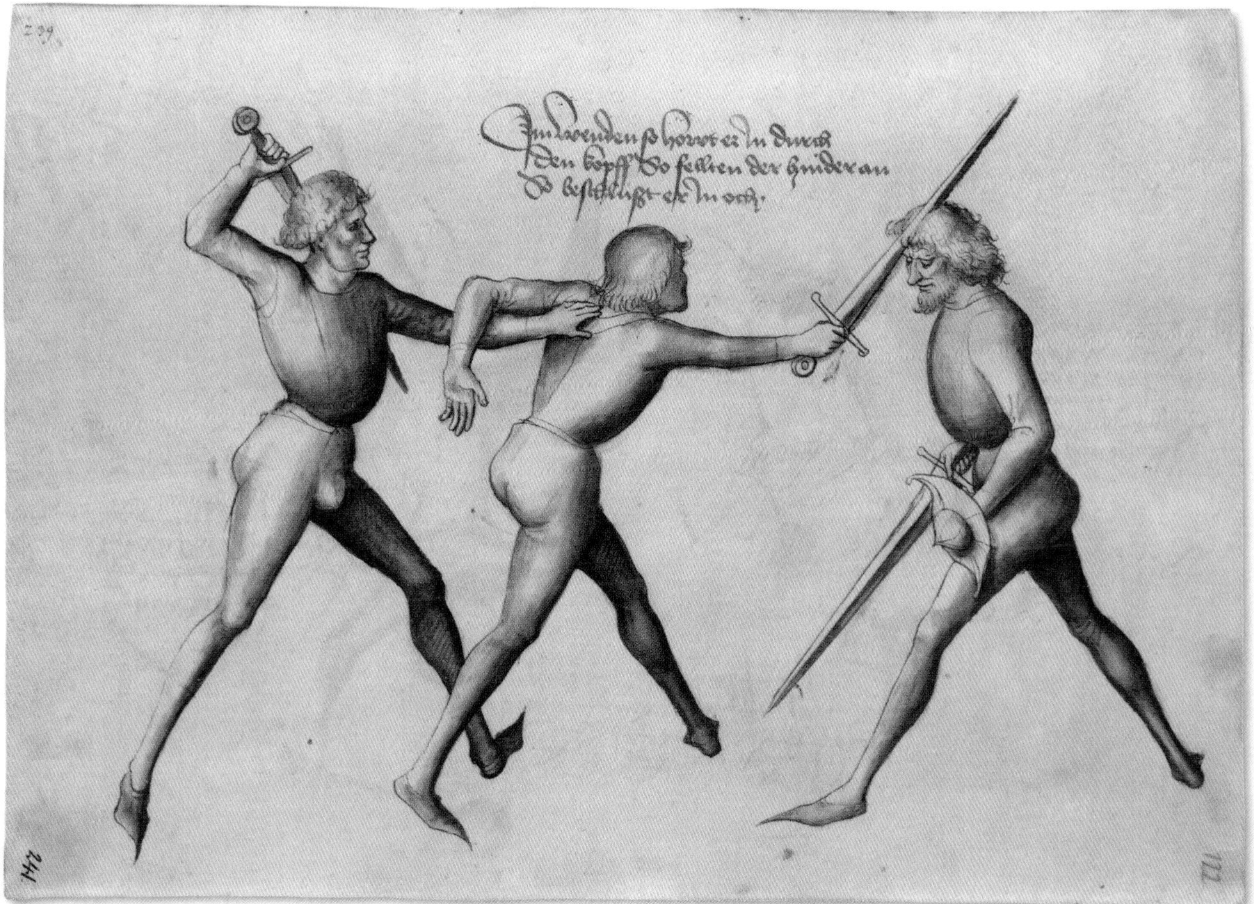

As he turns around, he hits the other one through the head. At the same time, the rear one attacks him, but that one is also enclosed by him.

Here it is stated how man and woman shall fight each other. Here they stand, ready to begin.

Here, the woman stands free and strives to hit. She has a stone in her veil that weighs four or five pounds.

So he stands in the pit up to his groin. The club is just as long as the veil in her hand.

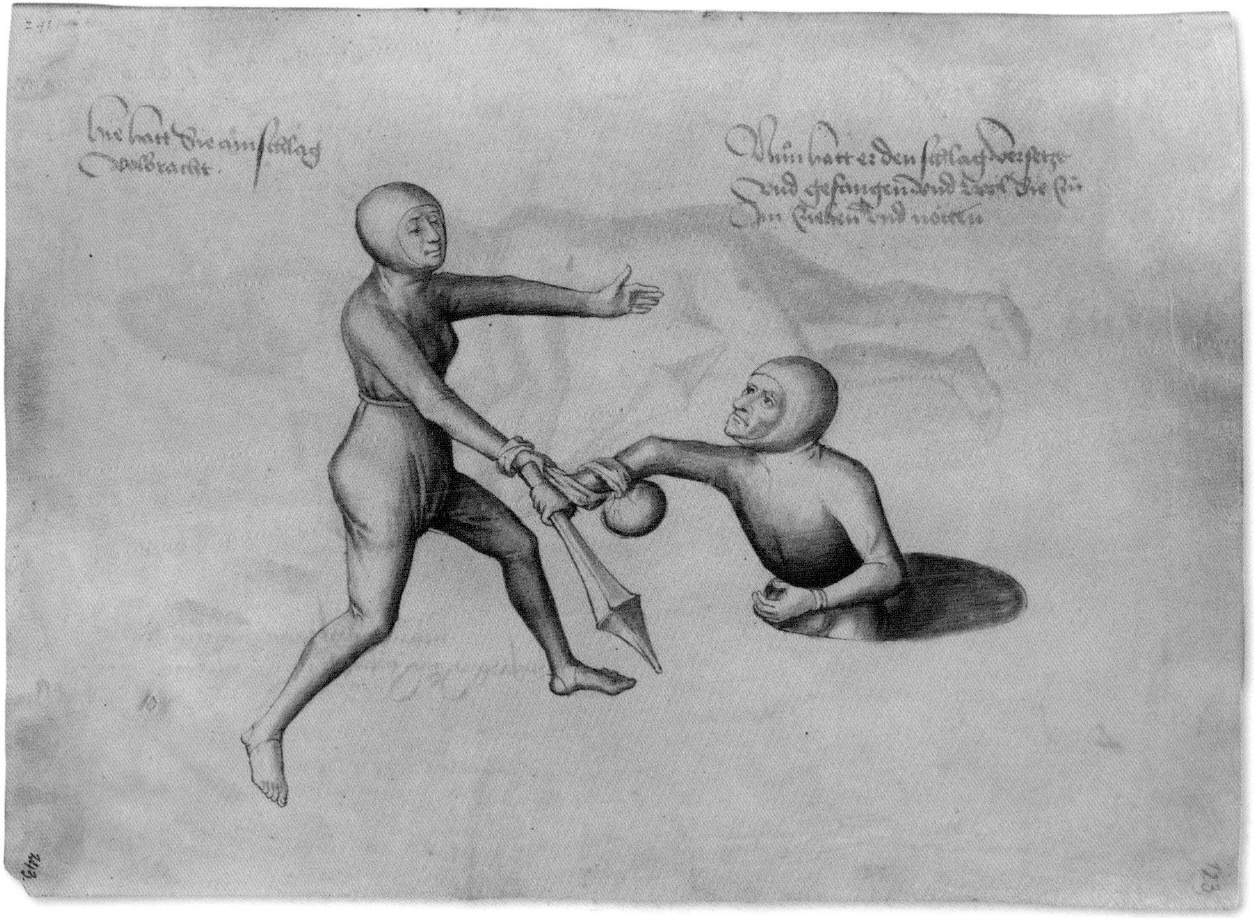

Here, she has struck a blow.

~

Now he has displaced the blow and caught it. He strives to draw her to him and oppress her.

Here, he has drawn her to himself and thrown her down; he strives to strangle her.

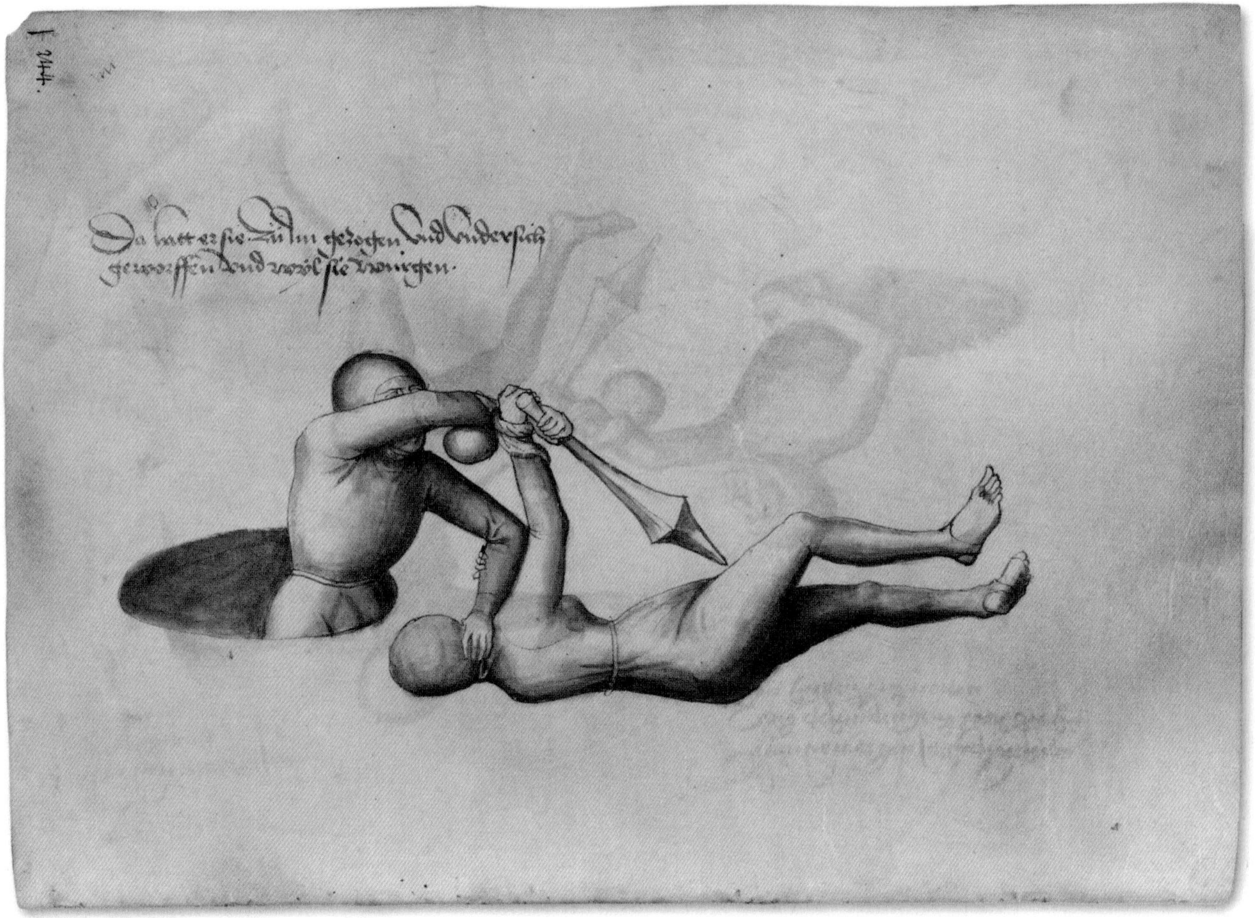

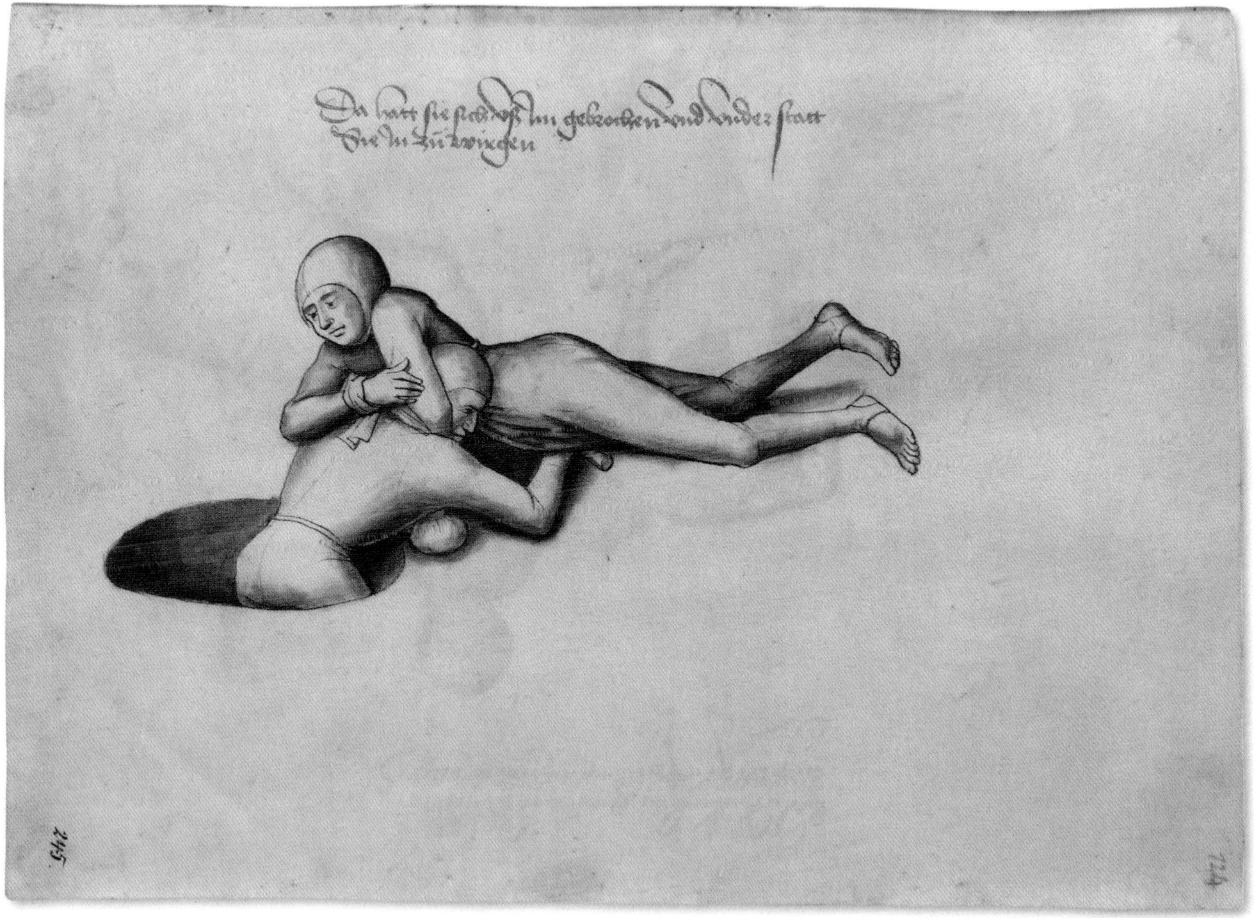

Here, she has broken away from him and is about to strangle him.

Here, she has put him on his back, strives to choke him and pull him out of the pit.

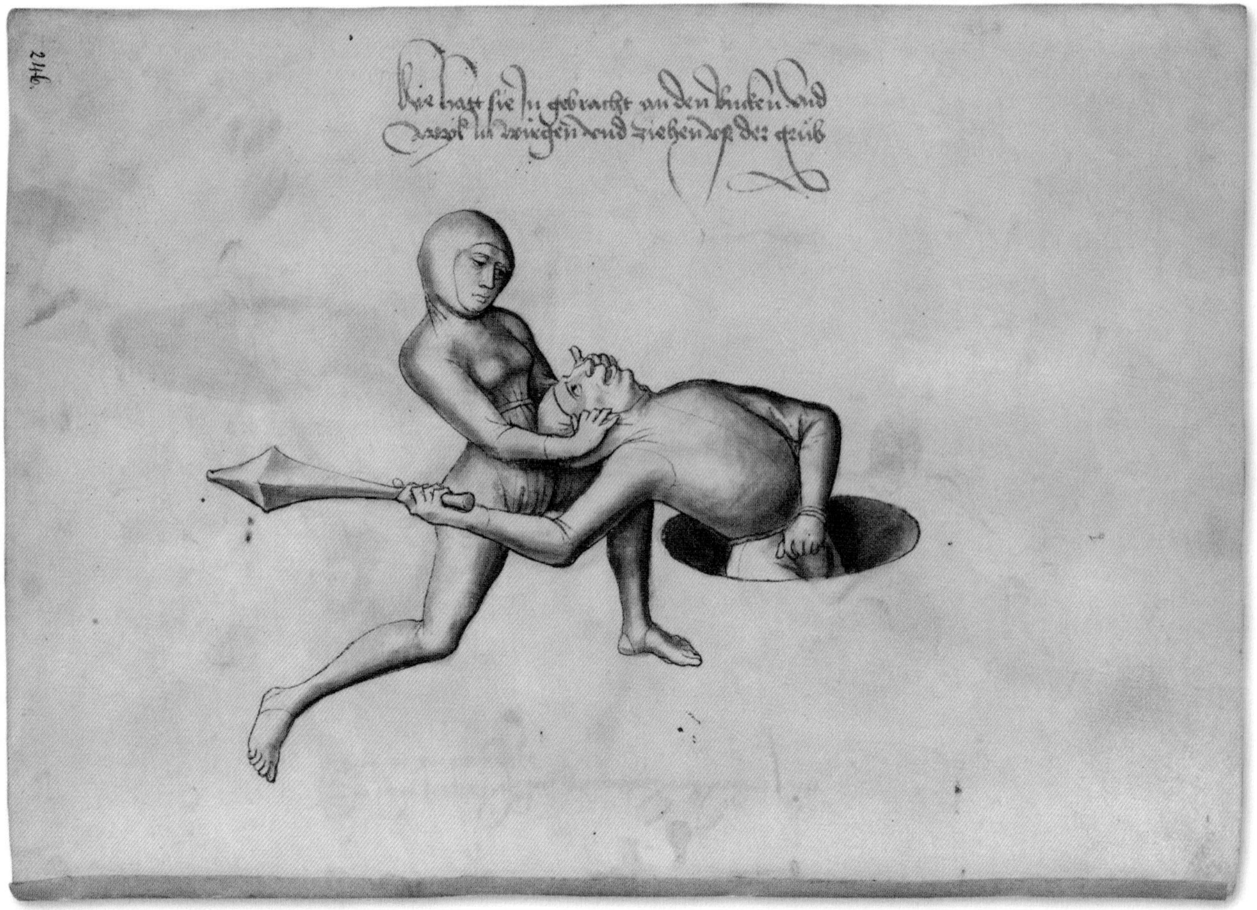

Here, he has pulled her to himself and throws her into the pit.

While she strives to hit him, she gets too close to him, so that he grabs her by the leg and knocks her down.

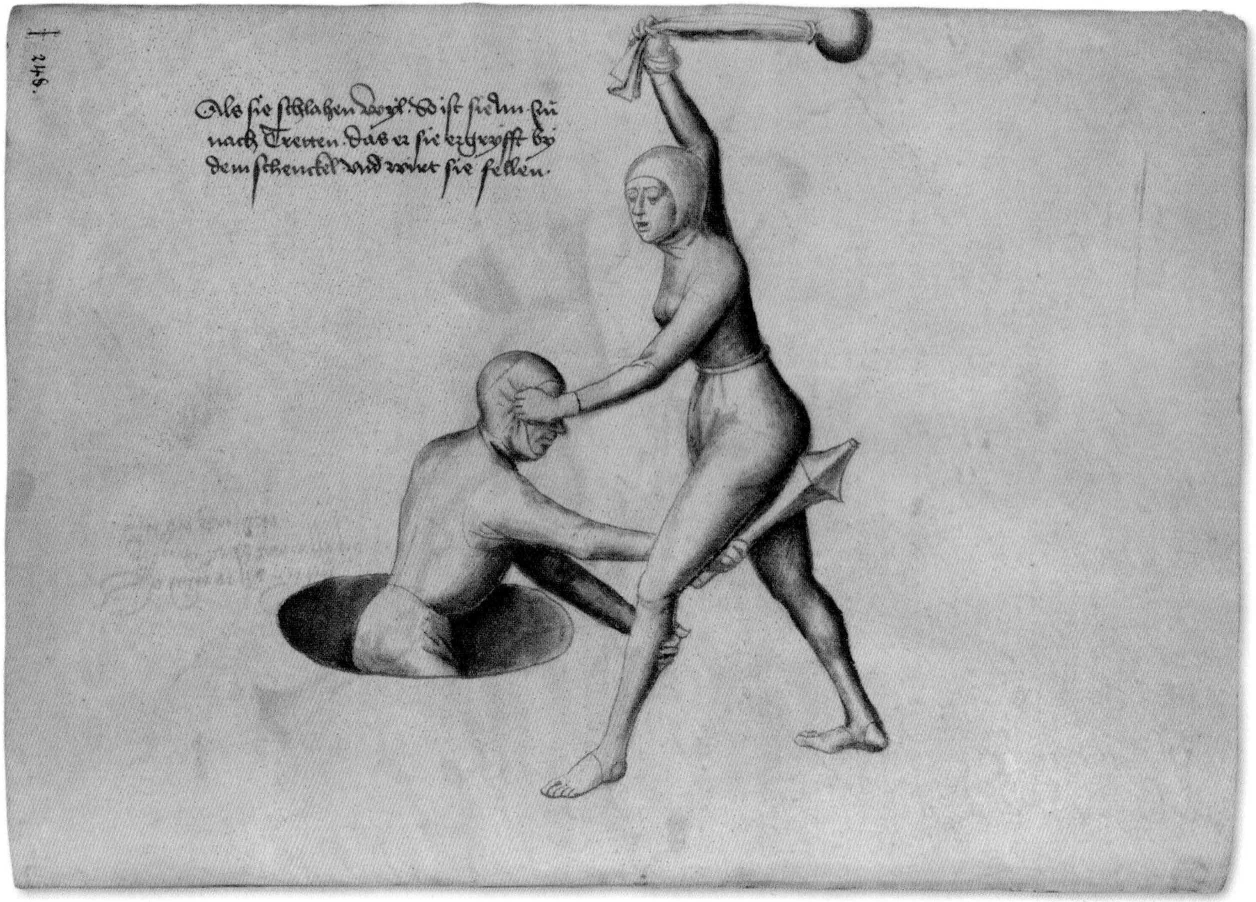

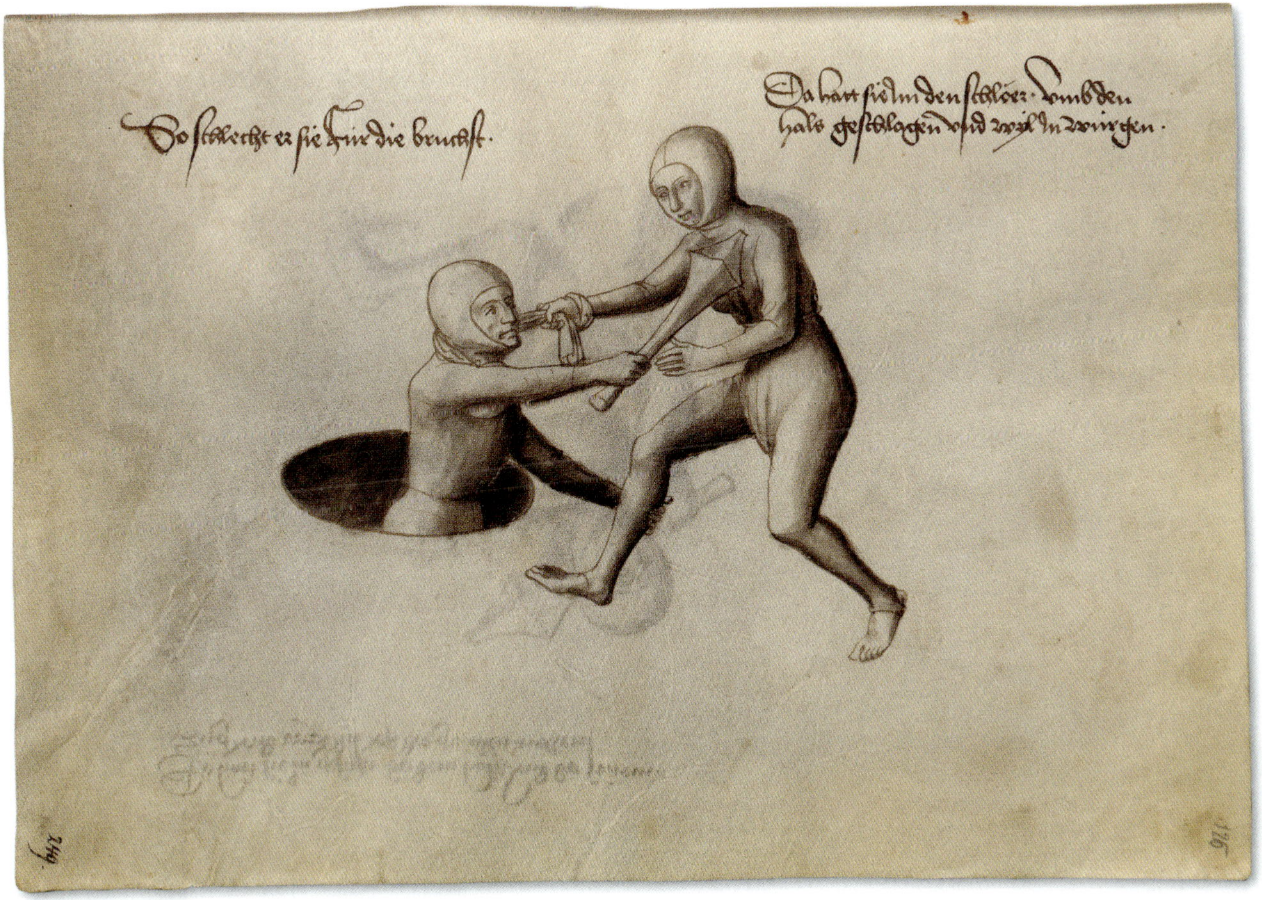

So he strikes her in the chest.

~

Here, she has brought the veil around his neck and strives to strangle him.

Here, she has grabbed him by his neck and his clothes and strives to pull him out of the pit.

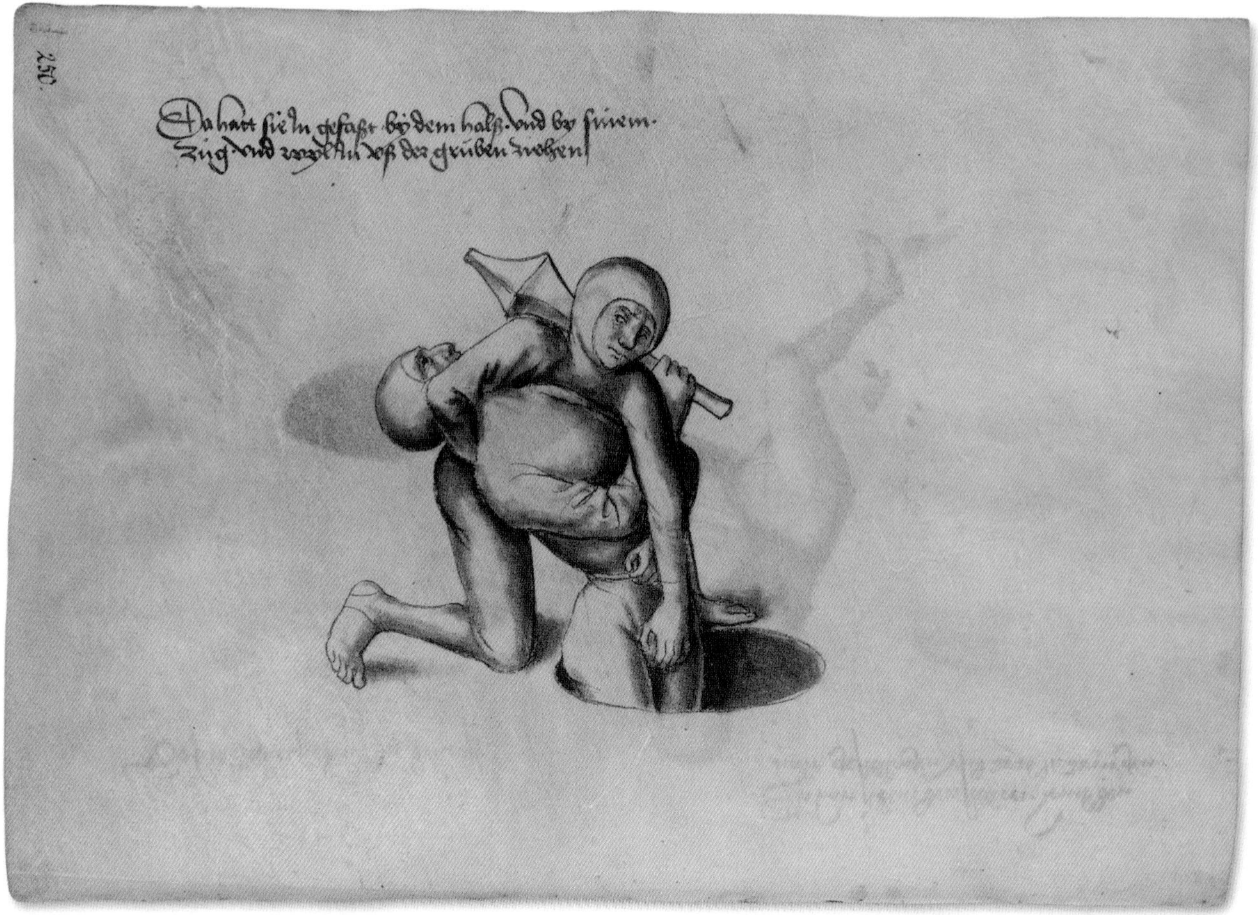

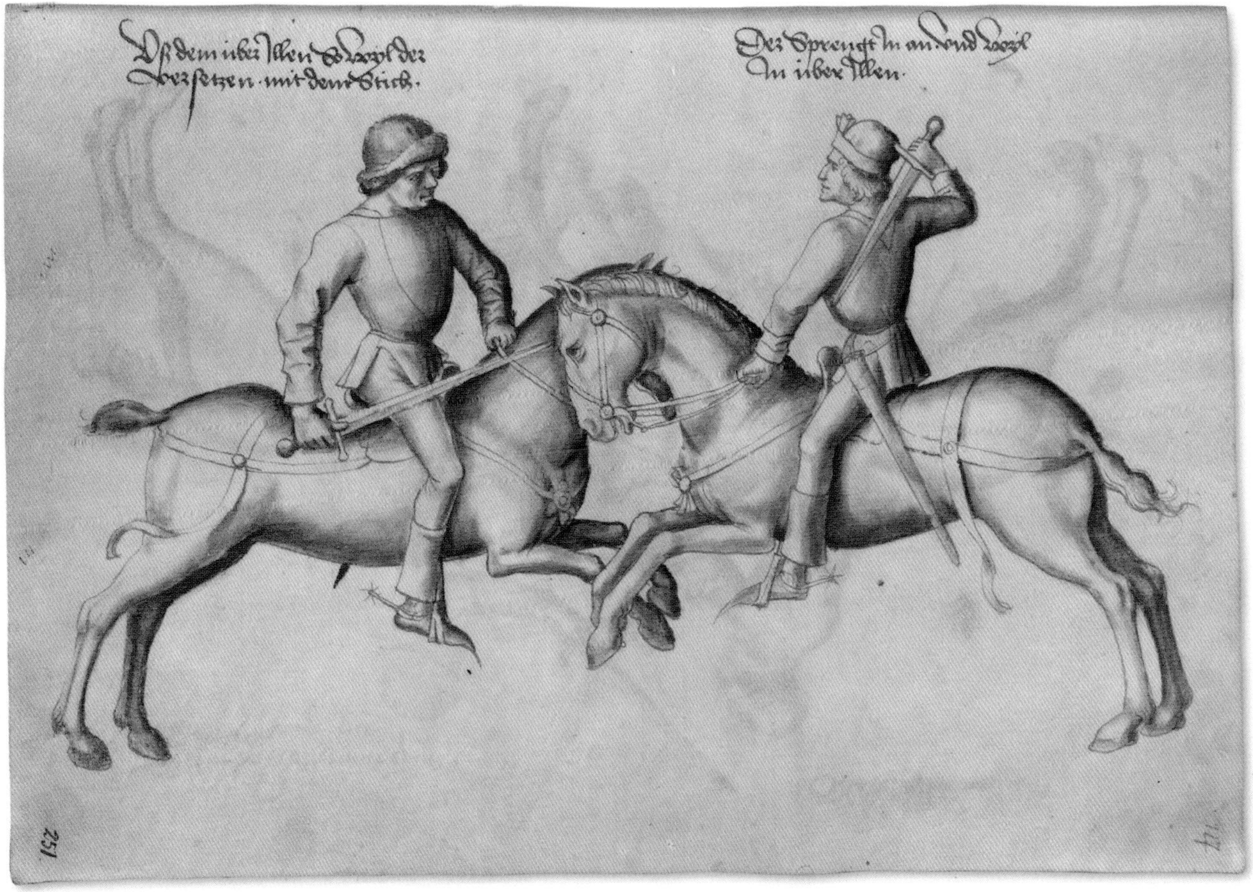

From being hastened, this one strives to displace the other with the thrust.

The other runs towards him and strives to hasten him.

... the other has displaced him with a twisted hand and rushes forwards.

~

After this one has struck ...

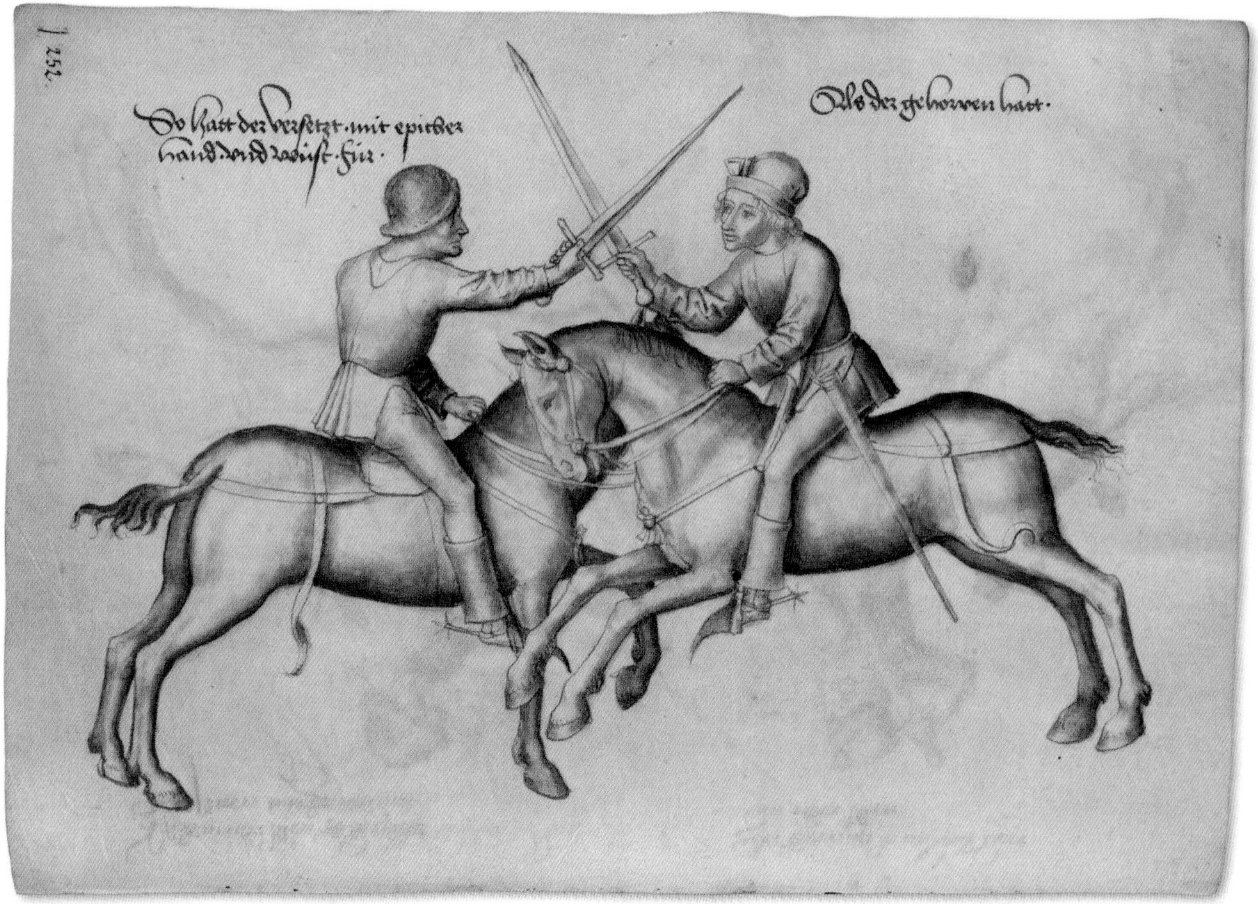

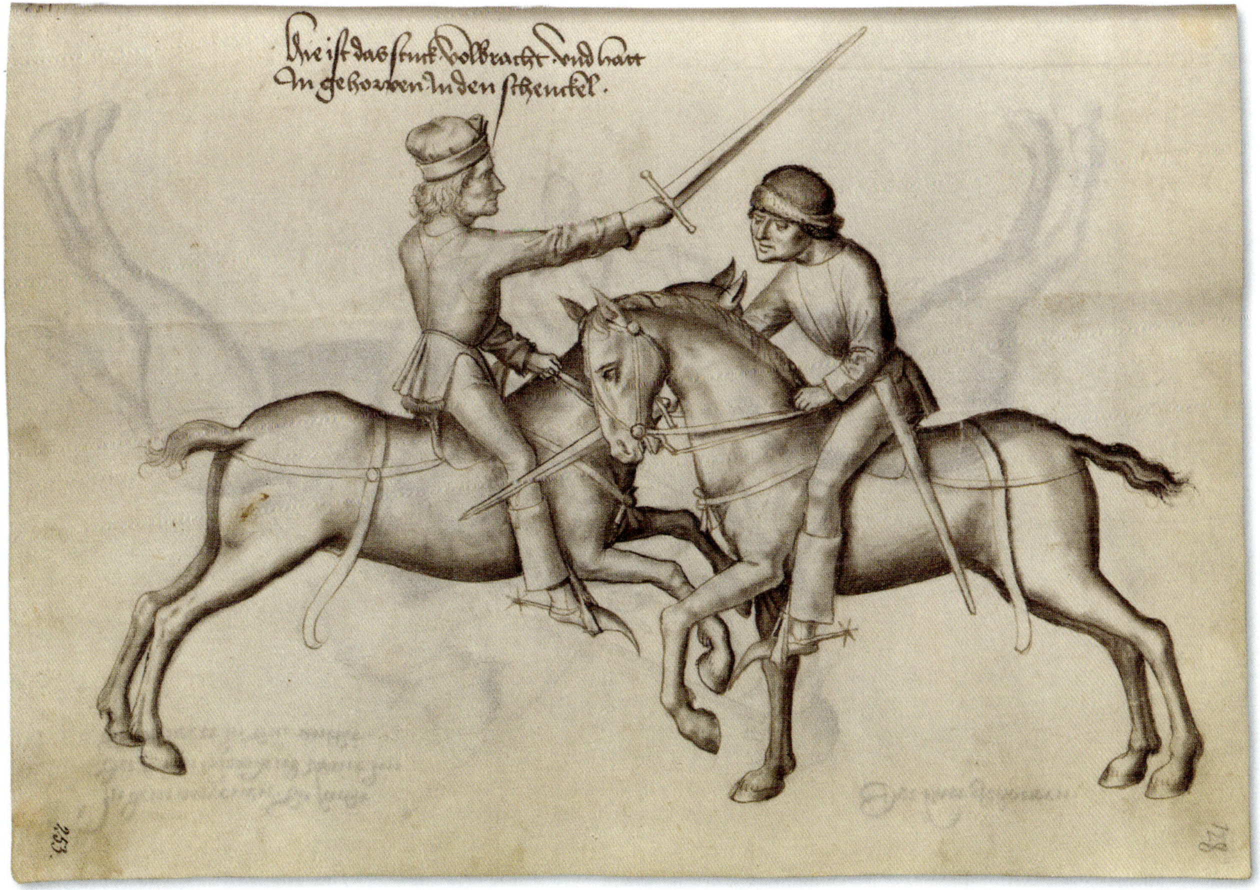

Here, the piece is completed, and the other has struck him in the leg.

While racing towards each other, one catches the strike of the other and turns his sword into his face.

~

That one has struck.

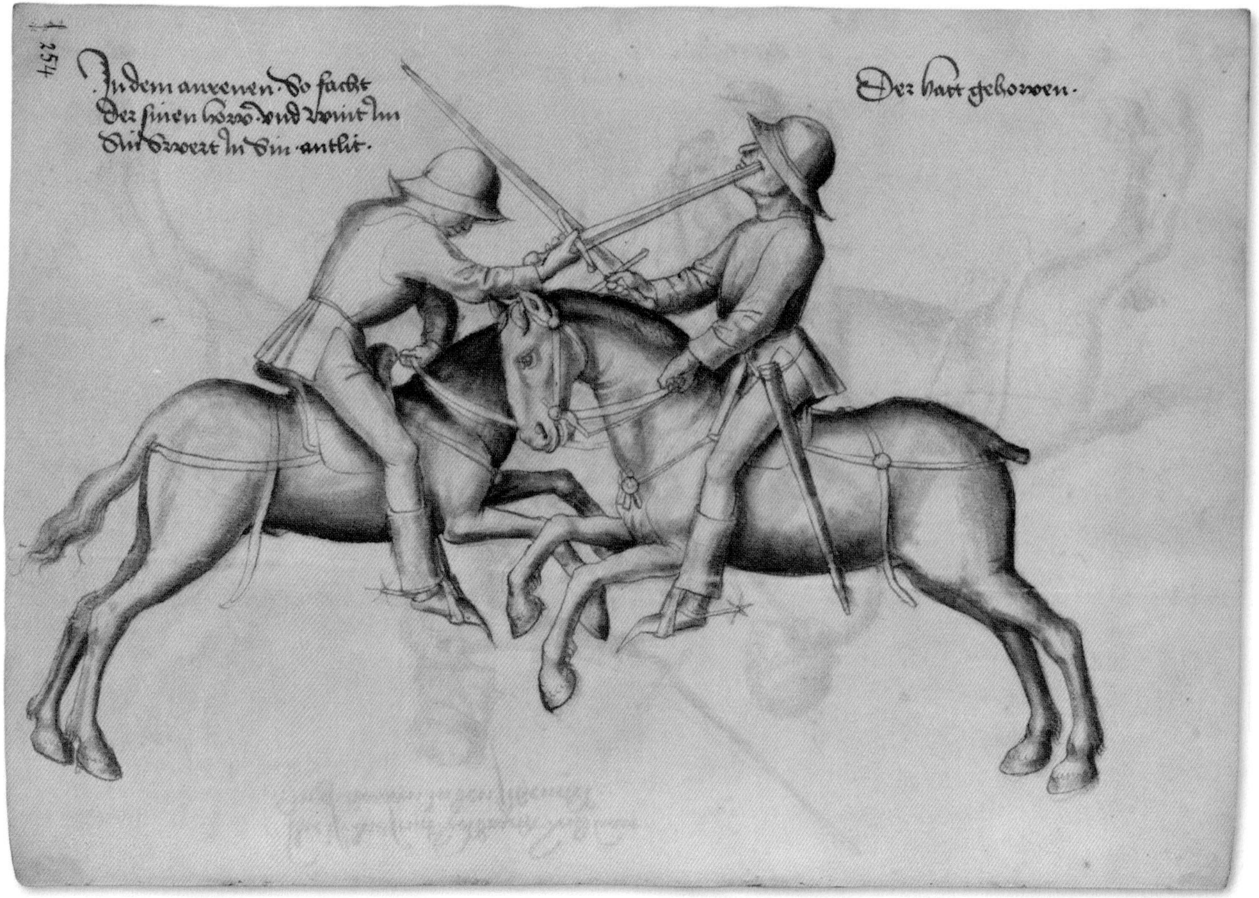

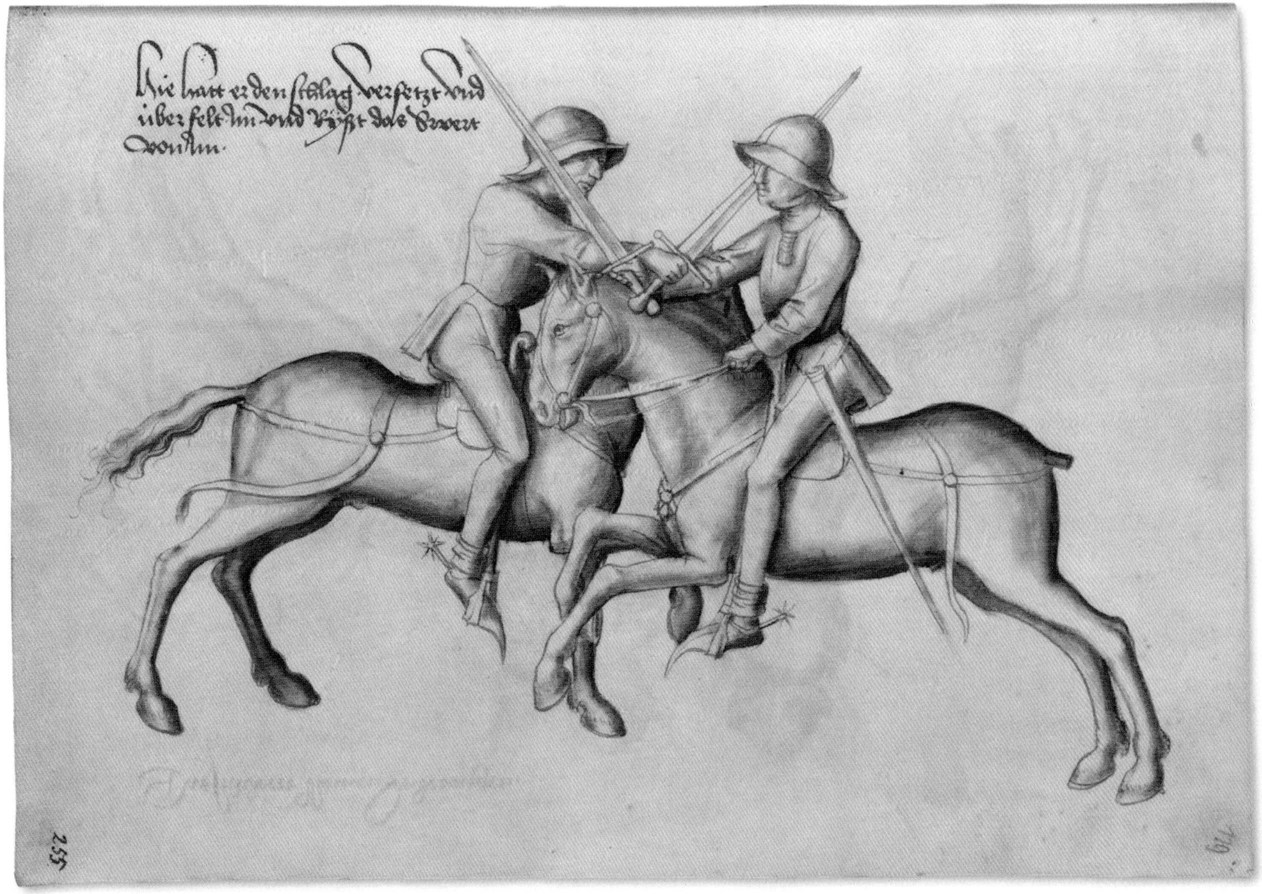

Here, one has displaced the strike, falls over him and snatches the sword from him.

Taking the sword from the front.

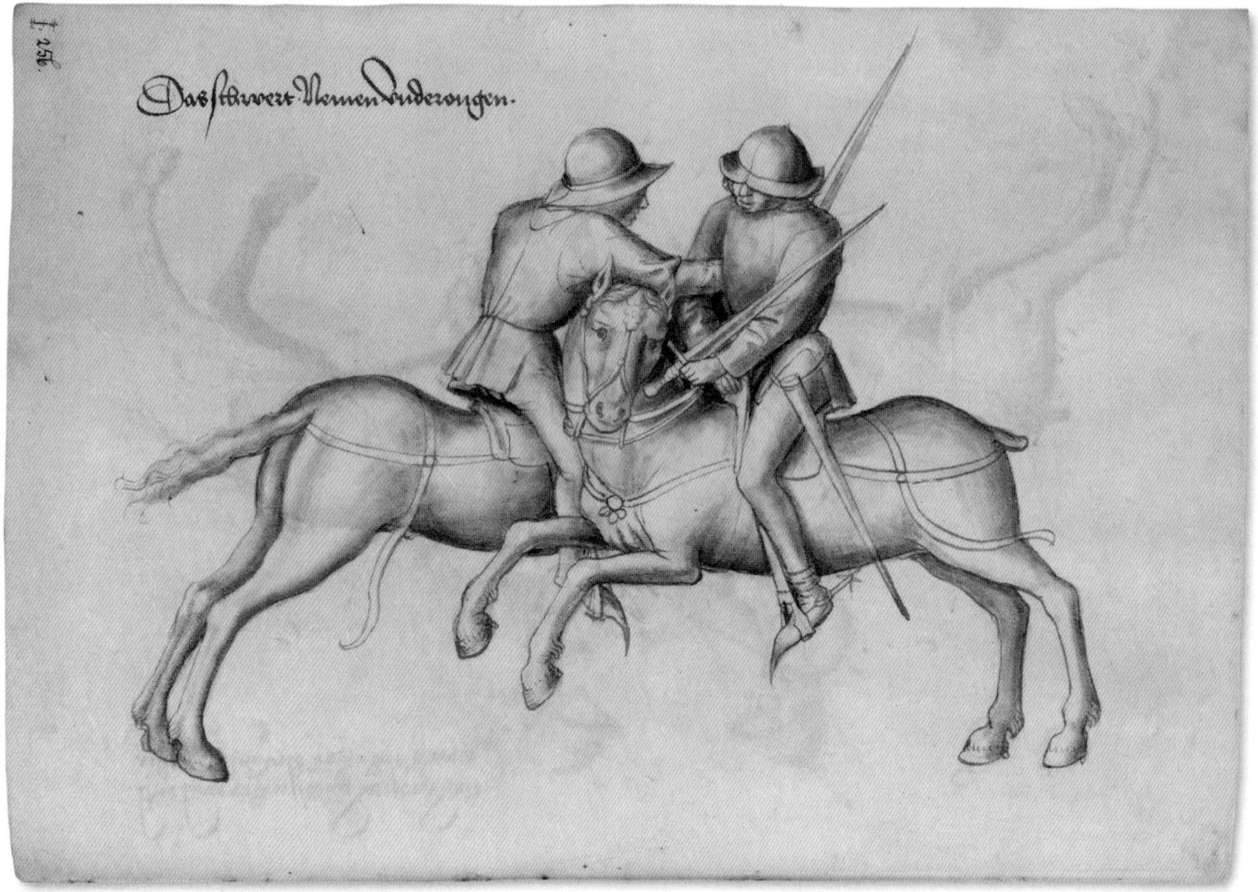

Taking the sword from behind.

After this one has struck …

… the other has displaced with a twisted hand and has grasped him by the neck.

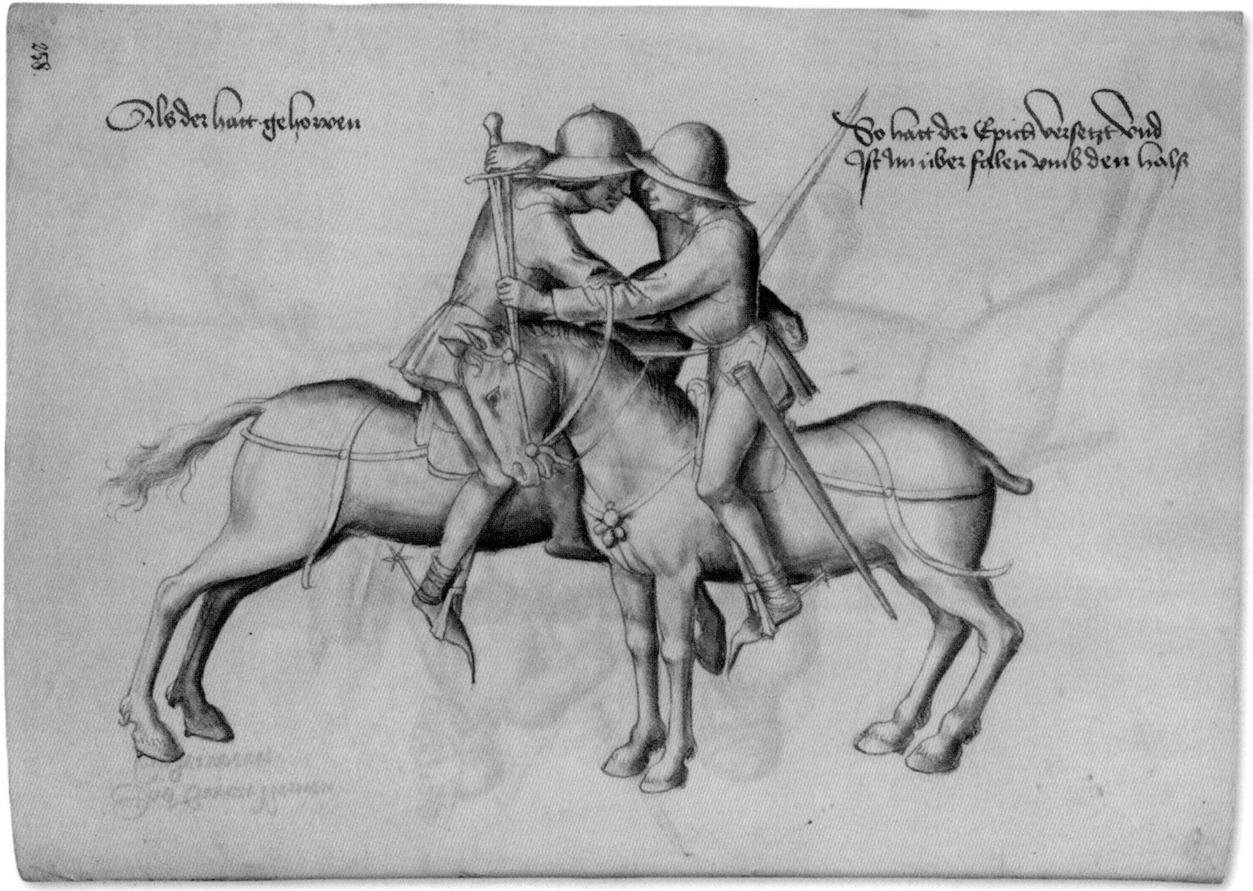

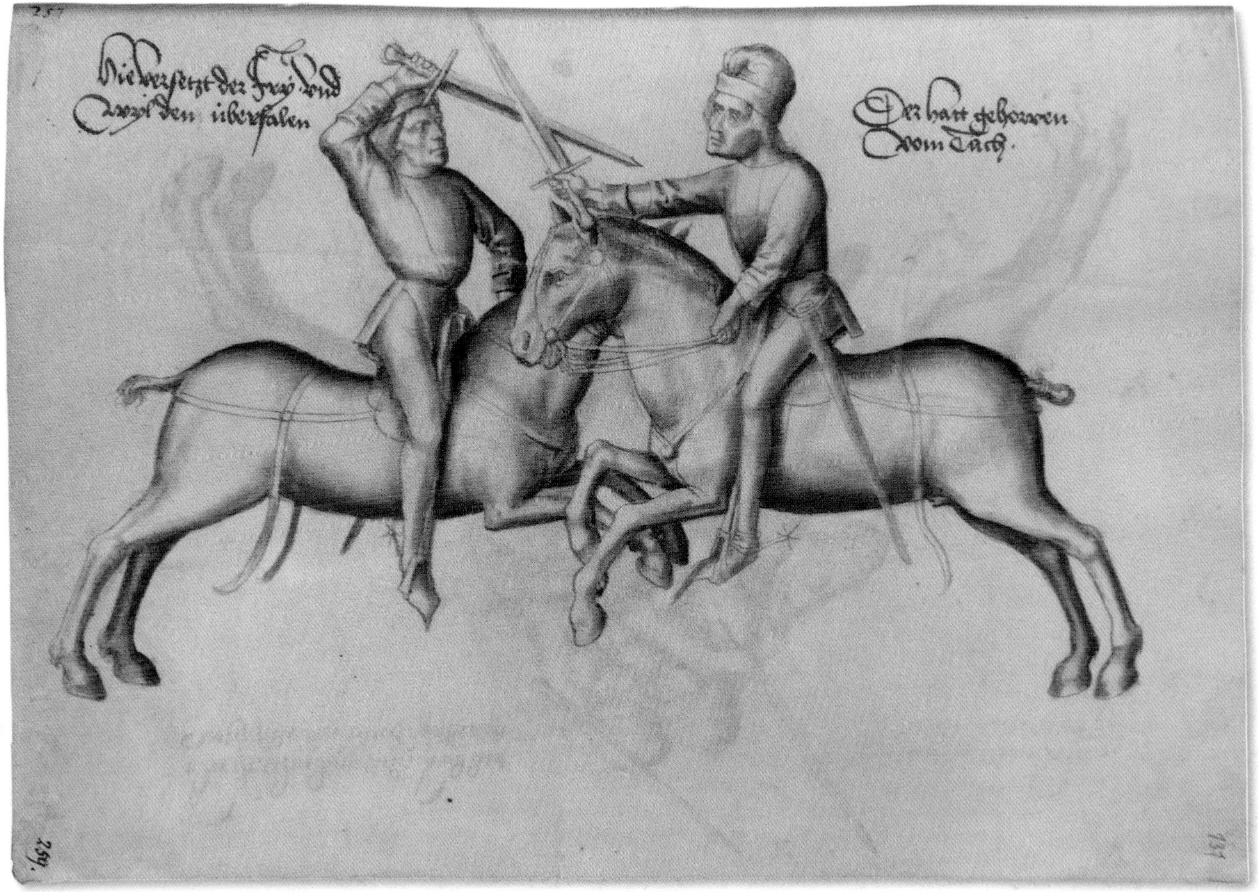

Here, he displaces freely and strives to descend on the other.

That one has struck *vom Tag*.

Here, he has attacked him and completed the piece described above.

Wrestling on horseback.

One has enclosed the other one and grabbed over him at the front.

The break against the described enclosure: this one has torn himself out of the other's hold and [grasps] him at the neck.

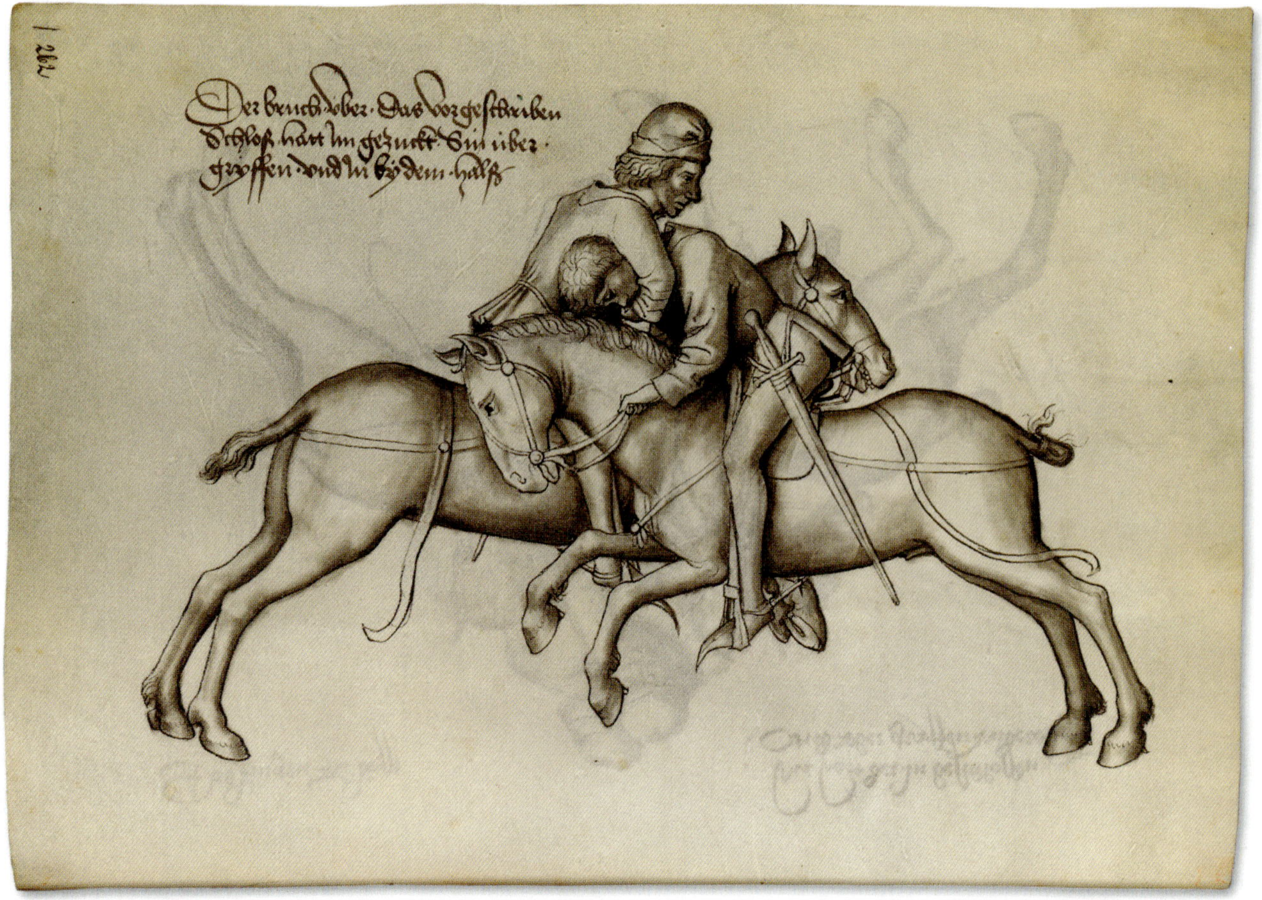

Here, one throws horse and rider.

The grabbing from above and the backward enclosure.

—

One has grabbed the other from behind.

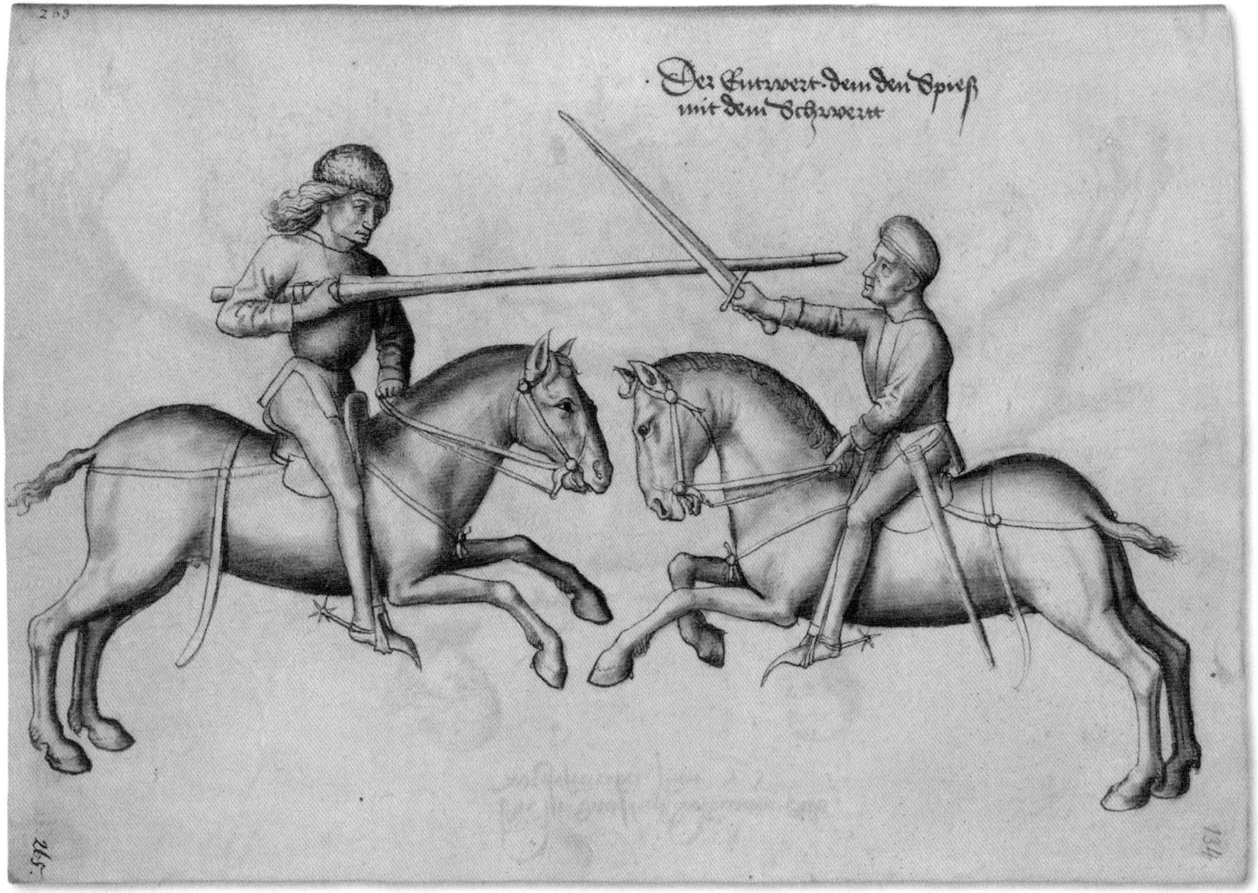

One fends off the other one's lance with his sword.

Here, the previously described piece is finished.

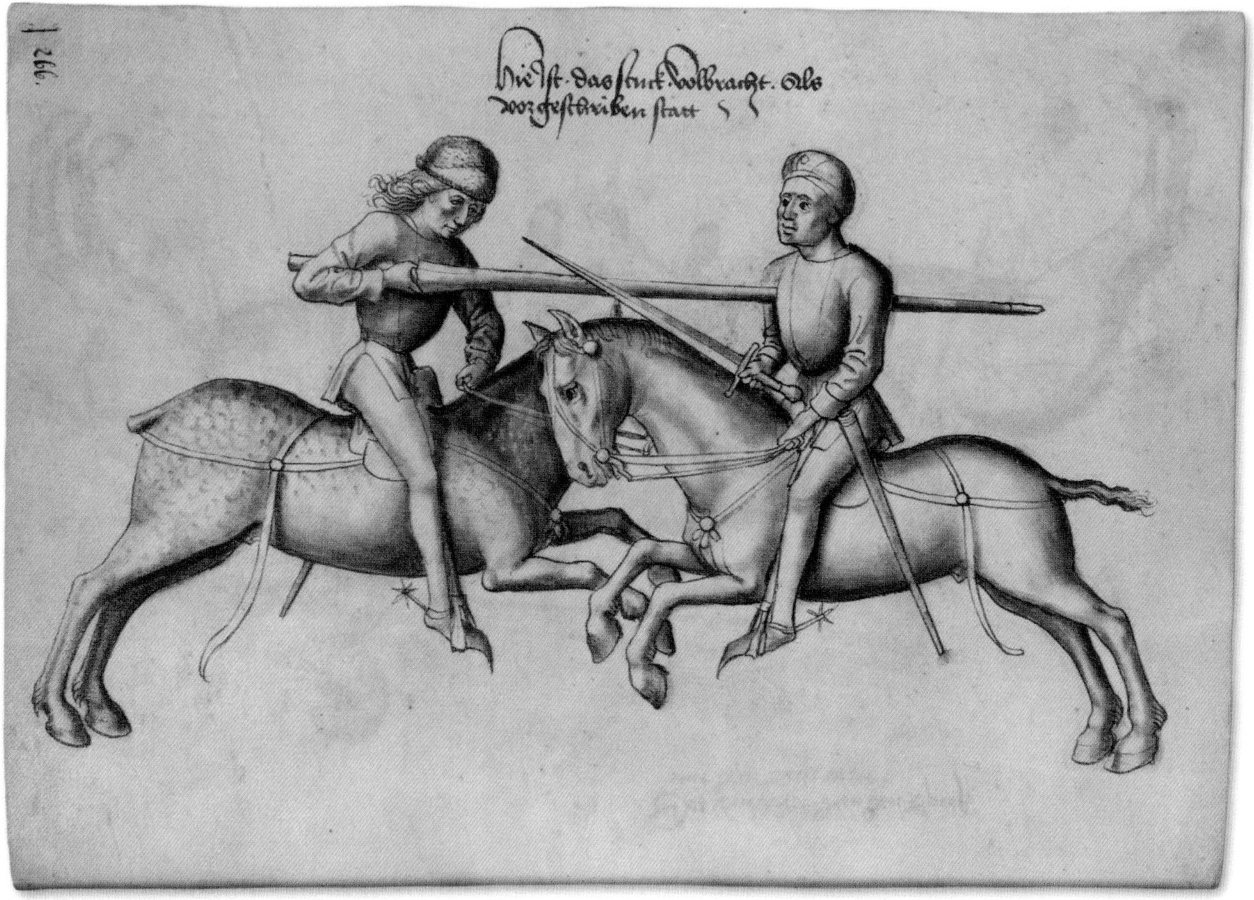

The hasteners.

How to behave while escaping with the crossbow.

One strives to run the other down under the horse.

~

Here, one has hit the other [with the shot] and pushes away the lance with the crossbow. He will grab him by the neck.

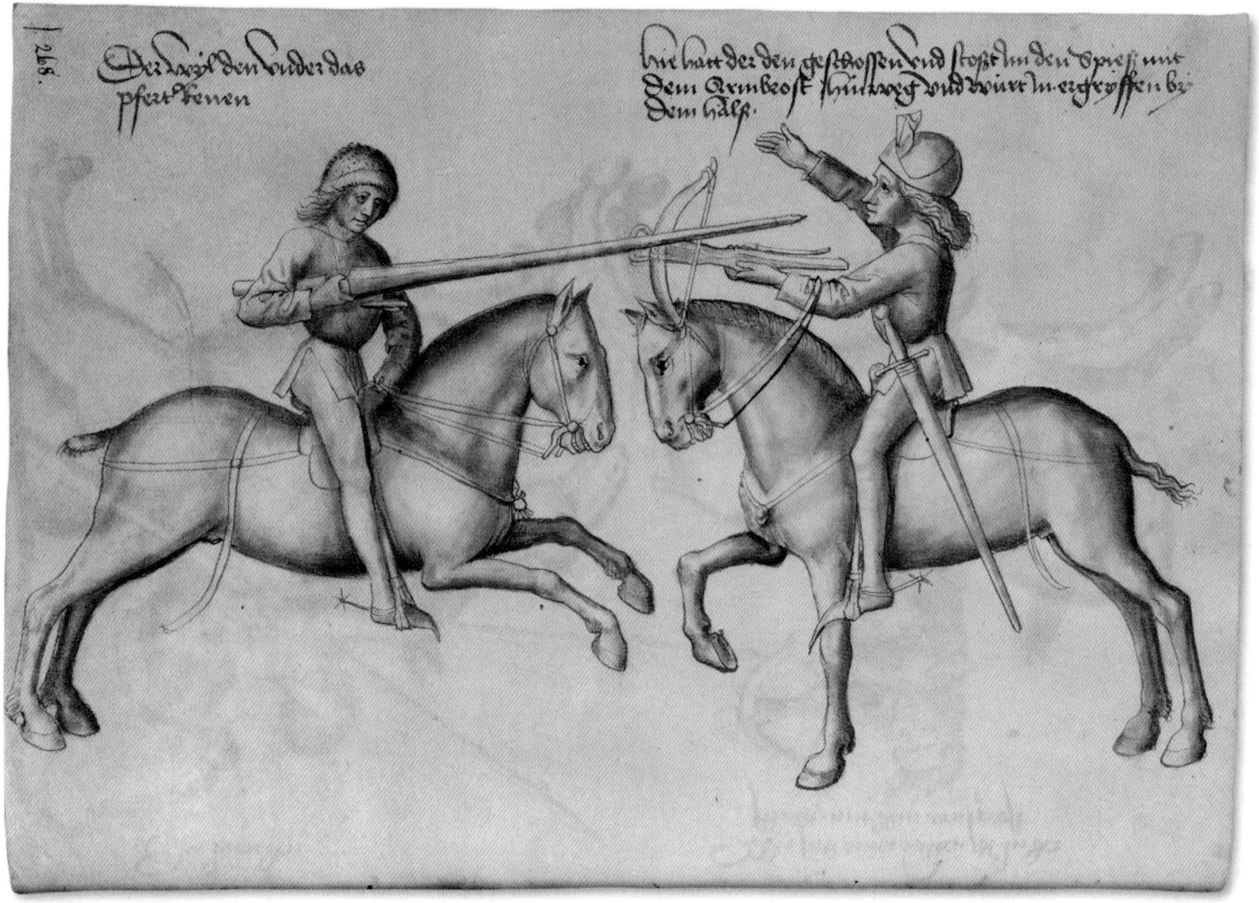

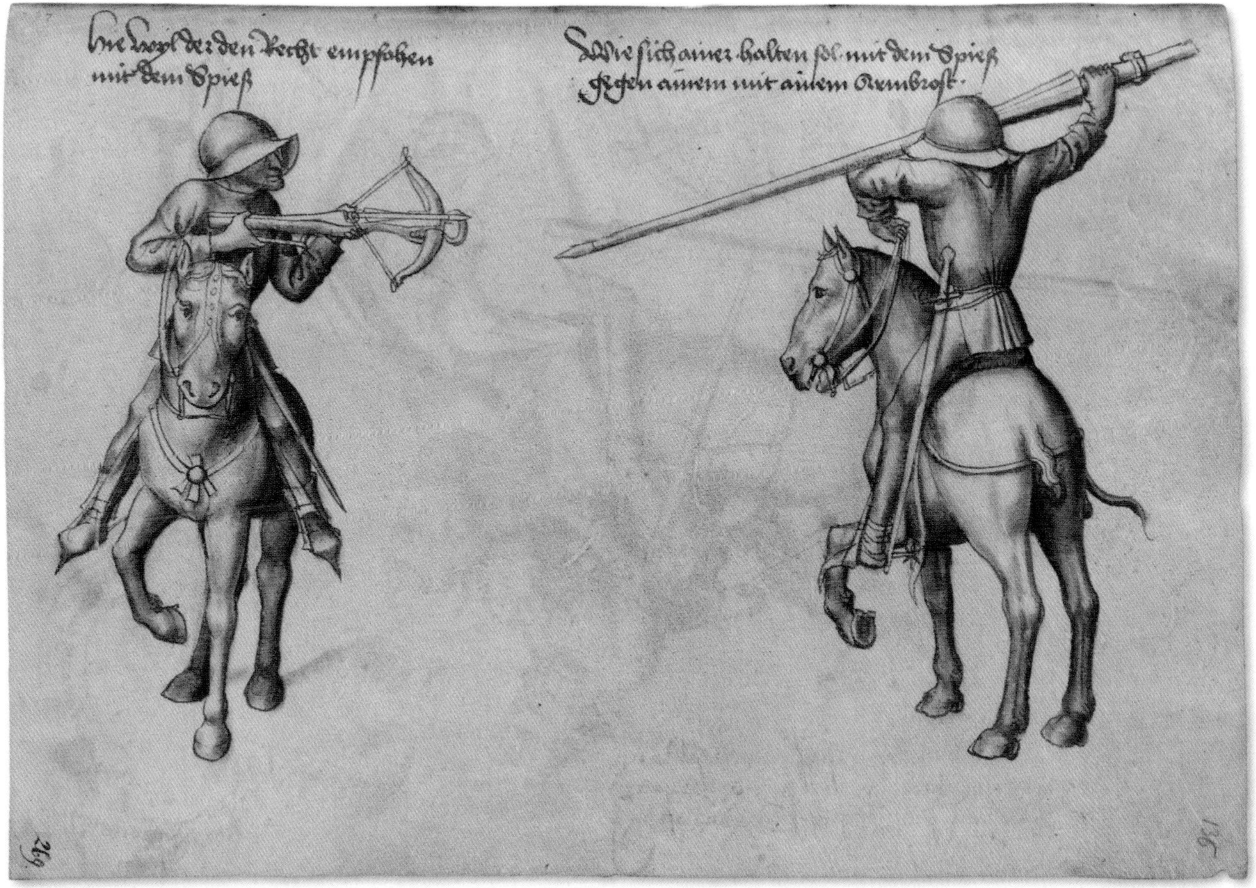

Here, one strives to oppose the other with the lance appropriately.

~

How to use a lance against someone with a crossbow.

Here, the previously described piece with the crossbow and the lance is finished, once one has grasped the other by the neck.

~

Hans Talhoffer commissioned this book and modelled for the pictures.

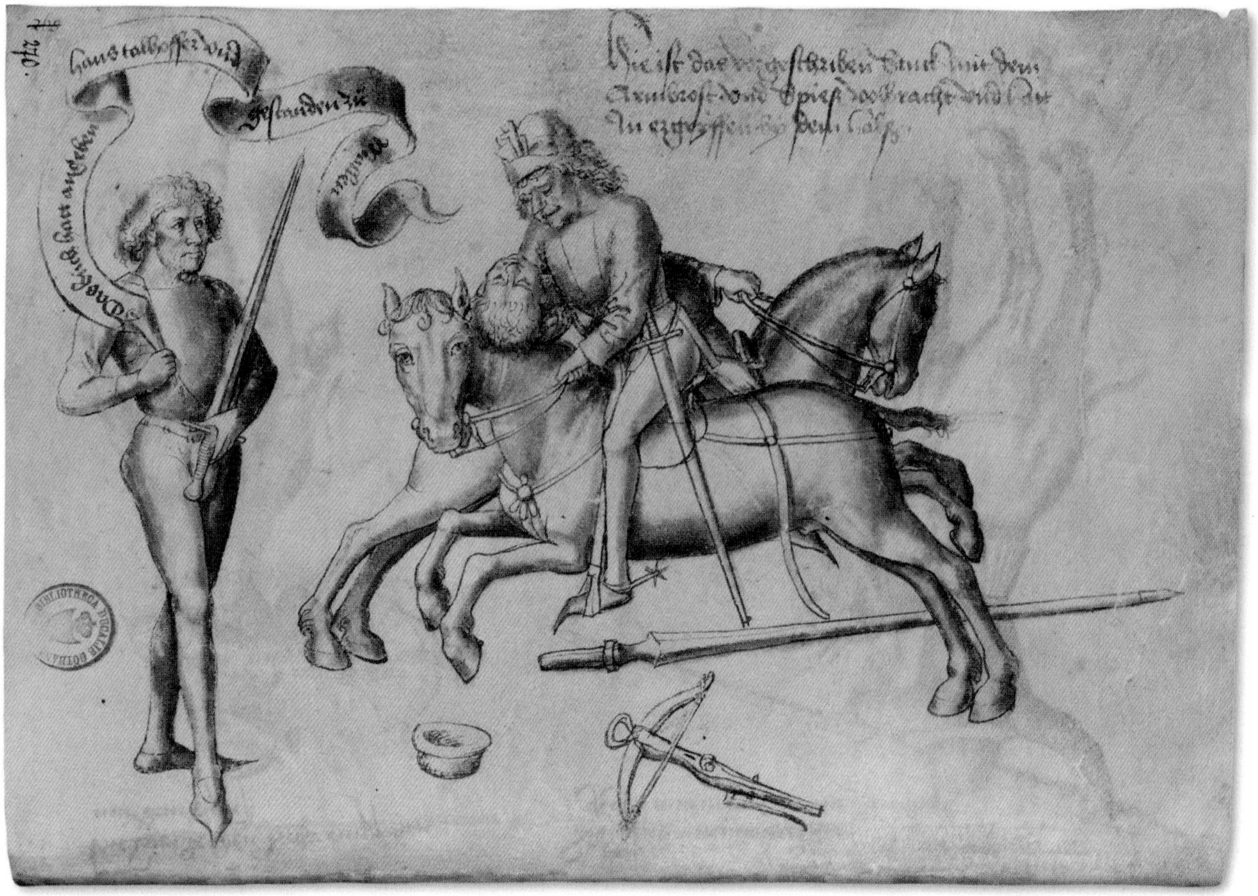

THE TRANSCRIPTION

SWORD – UNARMOURED

2r/1 Oberhow ~ Vnderhow
2v/2 Sturtzhow ~ Wechsselhow
3r/3 Zorn ortt Im drow ~ Aber oberhout
3v/4 Das lang Zorn ortt ~ Darfür ist das geschrenckt ortt
4r/5 Der fry how von Tach ~ Darus das halsfahen mach
4v/6 Ain vnderen blöß ~ ouch ain vnderen blöß
5r/7 Ain schwert niemen
5v/8 Ain zů legen oder eekomen
6r/9 Linck gen rechten das must Starck vechten
6v/10 Ain fryes ortt ~ Das gayßlen
7r/11 Vß dem anbinden hinwegstoßen hinder dem Elbogen vassen
7v/12 Mit dem fryen ortt Inlouffen vnd Tretten In den bůch
8r/13 Zwey ober ortt
8v/14 Die zwo vnderen blöß
9r/15 Das wegbinden oder hinder binden
9v/16 Das gryffen über die Ißny Portt ~ Die yßni Port
10r/17 Den Oberhaw erliegen vnd In die arm slahen
10v/18 Der das geschrenckt ortt macht ~ Der hout von tach
11r/19 Krüm vff behend ~ Das ort widerwend
11v/20 Hie ist die krüm volbracht
12r/21 Der gryfft nach der vnderen bloß ~ Der schnit von oben daryn
12v/22 Die arbeit Im krieg
13r/23 [D]ie geschrenckt schwechin ~ Der stat In der hůt
13v/24 Ain überfallen
14r/25 Stand beid In der hůt
14v/26 Ain gewauppet Infallen vnd swert niemen
15r/27 Der stant Fry Im ort ~ Der hout linck über das ort
15v/28 Vß dem oberhow geuallen In das gewaüppet ort
16r/29 Vß dem anbinden übergryffen vnd werffen
16v/30 Vß dem anbinden aber ain vahen mit gewalt ~ 1467
17r/31 Vß dem anbinden geuallen In die vndere ort
17v/32 Vß dem anbinden stossen hinweg by den Elbogen

HALFSWORD (ARMOURED TECHNIQUES PERFORMED OUT OF ARMOUR)

18r/33 Der mortschlag
18v/34 Der wurff vß dem ortschlag
19r/35 Vß dem fryen how geuallen In das gewauppet ort
19v/36 Vß dem schilher geuallen In das gewauppet ort

20r/37 Vß dem Tunrschlag Ain werffen ~ Vß dem dünrschlag ein Ryßen

20v/38 Nach dem Ryßen der stos In das antlig

21r/39 Das brentschirn ~ Oder das an binden Im kurtzen Schwert

21v/40 Vß dem anbinden Im kurtzen schwert

22r/41 Vß dem brent schüren ein lemen vff den Elbogen

22v/42 Ain Not stand fur den Stich oben vnd vnden

23r/43 Ain gedreng darus Arbaiten

23v/44 Ain hefften In hentschuch

24r/45 Ain hefften In fůs

24v/46 Vom anbinden ain Ryßen by der knüw kelen

25r/47 Den mordtstraich erlogen vnd schlahen In den schenckel

25v/48 Vß dem anbinden ain Halßryßen

26r/49 Hie hand sie die swert begriffen

26v/50 Ain schwert niemen

27r/51 Schwert wechßlen

27v/52 Aber ain sůchen ~ Der wil stoßen vnd Ryßen

28r/53 Hie ist der mordtstraich versetzt vnd ryßt In by dem hals

28v/54 Vß dem brentschiren ain ynschließenn

29r/55 Vß dem swert vassen so wurffs baide von dir vnd vollend

29v/56 Vom dem mordtstraich das vmb kern

30r/57 Der bruch über das vmkern

30v/58 Vß dem Mordtstreich von Im louffen vnd stoßen vnd wider schlahen vnd ryßen

31r/59 Vß dem swert begryffen durch schlupffen

31v/60 Vß dem brentschürn ainschließen vnd werffen

32r/61 Der stant fur stich vnd Straich ~ Der stant verkert fur Stich vnd Streich

32v/62 Aber ain schwert niemen

33r/63 Der will schlahenn ~ Ain Inlouffen

33v/64 Vß dem Inlouffen der wurff

34r/65 Der wurff Ist volbracht

34v/66 Für den stich vnd rysen In Arm

35r/67 Vnd mit für tretten vnd stossen In das antlig

HARNESS FENCING

35v/68 Hie gat er In den schrancken ~ Der tregt Im sin Zug vor

36r/69 Hie sitzent sie beid Im schrancken vnd wartent des anlas vnd hat yeder sin bär hinder Im vnd sin gries warten vor Im

36v/70 Hie sint sie angelaußen vnd stat der In der versatzung für den schutz ~ Der will schiessen mit dem spies

37r/71 Der hat geschossen ~ Der schuß ist versetzt

37v/72 Das versetzen fur den schlag ~ Der anlouff mit dem Mordtschlag

38r/73 Vß der versatzung hinweg stossen ~ Der hant den straich volbraucht

38v/74 Vnd fur treten vnd werffen über den schenckel

39r/75 Ain bruch für ain fryen schlag versetz gewaubet vnd ubergryff den man vnd fall vff den dritten fůs

39v/76 Stuck vnd bruch

40r/77 Der Zwyuelstich ~ Der recht stand In der Nott ~ Der wil arbeiten ~ Der stat In sin vorteil ongewer

40v/78 Da Ringen sie vmb das swert ~ Hie hat der geschlagen vnd der ander versetzt vnd macht ain end

POLLAXE

41r/79 Das erst anbinden mit der axst

41v/80 Vß dem anbinden hat er In werlouß gemacht

42r/81 Der will fry schlahen ~ Der will In hinder binden vnd In werlouß machen vnd sin schlag hinweg helffen

42v/82 Hie hat er dem schlag hinweg geholffen vnd ist das stuck volbraucht

43r/83 Der erlogen straich versetzet vnd darus geryssen

43v/84 Als sie baid geschlagen hand von tach vß dem anbinden so felt er Im vmb den hals vnd ryßt In

44r/85 In dem Ryßen dritt der hinach vnd ergrifft In by dem hals vnd wurfft In über die hufften

44v/86 Vß dem anbinden sint sie Zu Ringen komen vnd hat In vß dem bůben würff geworffen

45r/87 Vß dem fryen schlag vnd anbinden hat er In übergriffen vnd will In werffen

45v/88 Der hat geslagen ~ Hie versetzt der mit dem schafft vnd will arbeiten

46r/89 Hie ist er furtreten vnd das stuck volbracht

46v/90 Der hat geschlagen dem nach den füßen ~ Der hat den schlag versetzt vnd will wyter arbaiten

47r/91 Vß der versatzung schlecht er In an hals vnd will Ryßen

47v/92 Hie hat der den stich versetzt vnd schlecht In vmb den hals vnd will ryßen ~ Der hat gestochen

48r/93 Hie macht er den bruch vnd ryßen sie sich vmb die agst

48v/94 Da ist das stuck volbracht vnd würfft In über den schenckel

49r/95 Vß dem anbinden felt er Im mit der agst hinder den Elbogen vnd schubt In hinweg

49v/96 Da lat er sich gar vmb kern vnd macht den bruch uber das vorgenant stuck

50r/97 Aber ain anlaß ~ Hie will der den noten mit stichen vnd schlahen

50v/98 Hie hat der den stich versetzt vnd will ryßen

51r/99 Ain hinderbinden

51v/100 Der wartet des schlags von Tach ~ Der wyl mit macht Schlahen

52r/101 Hie hat der den schlag versetzt vnd Stotzt In von Im ~ Der hatt mir Tschwech In Sterck geben vnd mich von Im gestossen

52v/102 Hie bin ich von mynem vortail komen ~ Als er In gestossen hatt So Schlecht er Im die agst an halß vnd wurfft In an den Rucken

53r/103 Hie machet er ain endstuck mit Im vnd Sticht In zů tod

DUELLING SHIELD AND CLUB

53v/104 Hie ist der anfang mit dem schilt vnd Ston In mynem vortail Gott geb vns gluk vnd haill ~ Hie Ston Ich nach frenckeschem Rechten

54r/105 Damit lauß ich mich nit erschrecken • Ich wyl mich mit dem schult bedecken ~ Hie ston ich Im wurff

54v/106 Hie In diser stund hast du mich gar bloß funden ~ Hie hatt er den schilt geschrenckt vnd schlecht Im nach Sinem hopt

55r/107 Der wyl Im den schilt Inschlahen ~ So trit er Fur vnd wyl In zů dem hopt schlahen

55v/108 Ich bin Funden blos Ich fircht es werd mir ain stoß ~ Der hatt Im hinder den schilt gebunden

56r/109 Hie macht er mir ain endstoß ~ Hie hon ich den gestossen

56v/110 Mit mynem Tryt hon ich dich entrist vnd schlach gar gewysß ~ Er hatt mir den schilt vmbtretten So wyl Ich werffen

57r/111 Hie wyl der den schilt obnen hinyn stossen ~ Hie wil Ich In hinder binden vnd vmb Sin hertz blos finden

57v/112 Hie ist das stuck volbracht wie vor geschriben Stat

58r/113 Hie Triben Sie ainannder vmb vnd sůcht yeglicher sinen vortail ~ Der hatt den vortail

58v/114 Da hatt er Im den Schilt vß der hend geschlagen vß dem anbinden vnd wyl stossen wa er In mag bloß finden ~ So grifft aber der mit dem Arm vnd wurfft den schilt von Im vnd hilfft Im hinweg

59r/115 Vß sinem anweg helffen kumpt mir myn stosß herwyder ~ Hie bin Ich worden blos des wirt mir ain beser Stosß

59v/116 Vß dem hinweg helffen ist das Stuck gantz volbracht

60r/117 Die Bindent aber Ainannder an

60v/118 Vß dem anbinden So hatt er In hinderbunden Vnd Stoßt Im mit dem schilt oben durch Sinen schenckel

61r/119 Hie statt der In Siner hůtt vnd wartet des Mans ~ So schlecht der den hacken hinder sine*n* Schilt vnd zert hindersich

61v/120 Der Zert hindersich wie vorgeschriben Statt

62r/121 Vß dem hindersich zeren So stoßt er hinwyder Den Schilt In den Vnd ist das Stuck volbracht

62v/122 Hie wurfft er sinen Arm vff vnd übergrifft Im Sin arm vnd latt den Schilt fallen vnd Trytt fur vnd Schlecht Im Sin lincken Arm vmb Sinen halß vnd wurfft In uber die huffen ~ Hie kert er den schilt vmb vnd Schlecht geschrenckt dem zů sinem houpt

63r/123 Hie wirt das Stuck volbracht wie vorgeschriben Statt

63v/124 Der Ist ganntz wörloß vnd wirt geworffen ~ Der hatt Sinen schilt zwuschen Man vnd schilt geschlagen vnd latt den schilt vnd wyl In werffen

64r/125 Hie ist der wurff volbracht vnd Schlecht In dem kolben

64v/126 Da Sint sie komen von den schilten vnd Schlahent ainander mit den kolben ~ Der hatt Im sinen Straich versetzt mit dem kolben vnd uber gryfft In mit dem Arm vnd Schlecht In zu tod ~ Hie hatt das kolben vechten ain end

65r/127 [unfinished sketch of 76v/150]

DUELLING SHIELD AND SWORD

65v/128 Hie Ston Ich Fry Nach Schwebischem Rechten Als man ze hall vicht ~ So ston ich mit schilt vnd schwert vnd hon din lang zů fechten begert

66r/129 Der wyl den Schilt wenden vnd howen ~ So bringt er Sin schwert zwuschen die schilt vnd versůcht wie starck er Stand

66v/130 Der sticht mit geschrencktem Schilt ~ Hie statt der blos vnuersetzt

67r/131 Wir zwen standen wie die obern wenn das wir vns verkert haben vnd Statt yeglicher an des anndern Statt

67v/132 Im verborgen Stand wend Ich min Schilt vnd schryt fur mit dem Stich ~ Hie hab ich myn schilt gewendt vnd bin doch geschent

68r/133 So zuckt ich mynen Schilt vnd Tryt vß Sinem treffen damit So tun ich Im sin Stuck Brechen ~ Ich hon wöllen geschrenckt Stechen hinder Sinen Schilt

68v/134 Vß dem anlouffen So Tryt Ich den schilt von dir vnd hab myn Stich volbracht an dir

69r/135 Da tůn Ich versetzten mit Sterckin wyl ich dir din höw vnd Stich letzen ~ Der hatt den Schilt verkert Das merck

69v/136 So lauß ich mynen schilt fallen vnd how zu der blessin ~ Der hatt wollen zwuschen Schilt Stossen

70r/137 Hie Ist aber ain end ~ So latt aber der sinen schilt och Fallen vnd ergryfft In by den Elbogen vnd Stoßt Schwert durch In

70v/138 So wyll der Schilt vnd Schwert ynschlahen vnd hinderbinden ob er In möcht blos finden ~ Der statt In siner versatzung

71r/139 Er hatt mir gezuckt Das treffen damit er thůt mich Effen ~ Hie hon ich Im gezuckt sin Treffen das er hatt gefelt vnd hon Im nachgerayßt mit dem Stick

71v/140 So hatt er den Stoß vernomen vnd latt In nit dartzů komen ~ Der hatt vß dem bochen schwert vnd schilt zemen genomen ob er Im möcht oben hinyn komen

72r/141 Da bin Ich zwuschen Schilt vnd Swert gedrungen das ich dir zum houpt bin komen ~ Als Ich hon gehowen geschrenckt Da Ist mir mysse lungen

72v/142 Mit schrencken Vnd Schryten wyl er howen ~ Hie Merck wie ich das main das ich hon gezuckt vnd Stich In durch sin bain

73r/143 Der Statt In der versatzung Fry ~ In dem stand verborgen wächselt er das Swert In die lincken hand vnd ersticht In übern schilt

73v/144 In ainem Fryen Stand So bin Ich komen vmb myn hand ~ Das mag Ich wol gedencken das ich hon gehowen mit Schrencken

74r/145 Der stant In dem stand dar Inn der kam vmb Sin hand ~ Aber wyl der vber den schilt winden Ob er In möchte bloß Finden

74v/146 Hie stand wir bayd blos ~ Aber der mag den schilt wol hinyn werffen vnd In von Im stossen

75r/147 Da hatt er In Von Im gestossen vom wurff vnd vom stoß muß er Sich bucken des how Ich In In kopff vnd In Rucken

75v/148 Der hatt wöllen mit schilt Vnd Swert Inbinden vnd Rysen ~ So hatt der hindersich geruckt vnd Im vß sinem Inschlahen gezuckt vnd latt den schilt fallen vnd Tryt herfur vnd howt Im In sinen halß

76r/149 Der hatt wöllen den schilt oben hinyn wunden ~ Den schilt solt Man ynen Sehen vnd der stich geschrenck solt Sin gescheen

76v/150 Hie hon ich versetzt mit macht vnd bin beliben ston vnd hon mynen starcken stich getan ~ Als der sinen schilt Ingeschlagen hatt vnd wolt den wörlouß machen vnd den schilt hon von Im geschlagen das ist Im gebrochen

DUELLING SHIELD

77r/151 So ist der Fůr dretten vnd nempt In by dem halß vnd wurfft In mit dem Schilt ~ Als der den schilt Ingeschlagen hatt

77v/152 Da Ist das stuck volbracht wie vorgeschriben statt

78r/153 Der hatt Ingebunden zwuschen den Schilt vnd den man vnd hatt Im den hacken geslagen vmb sinen fuß vnd wyl Ryssen ~ Darumb er mir ist komen zwuschen mich vnd den schilt Des ich billich engult

78v/154 Von dem anbinden So mag man vnden vnd oben Ryssen zů baiden syten

79r/155 Es ist ain als dem and*er*n vnd oder oben wie er des komen mag ~ Aber vß dem obern anbund oder bochen So mag ain yeder dem anndern griffen In sinen schilt vnd von Im Rysen ob er mag

79v/156 Vß dem Infallen vnd griffen wie vorgeschriben statt welicher denn nit wöll lon so mag es wol an ain Ringen gon

80r/157 Nůn Merck das stuck recht Als Nůn der hatt wöllen schlahen vnd bochen So ist er vast gestanden vnd latt die Rechten hand vom Schilt vnd gryfft dem In den Elbogen vnd stoßt In von Im das er sich verkertt ~ Hie hatt er dem den schilt Ingeslagen vnd hinderbunden

80v/158 Als der den gestossen by dem Elboge*n* hatt Als vorgeschriben statt ~ So kert sich der gar vmb vnd stoßt den Schilt In den vnd ist stuck vnd bruch volbracht züm end

81r/159 In dem anlouffen So hat er In hinder Tretten vnd bracht In In die höin Vnd furcht er werd geworffen ~ Das ist aber ain starcker anlouff Vnd vß dem anlouff So hatt er In hinderbunden vnd wirt In werffen oder er mus wörlouß werden

81v/160 Hie hatt er In bracht züm fall wie vorgeschriben statt vnd macht aber ain end mit dem fryen stoß

82r/161 Hie stond sie bayd zu dem verkerten anbund In zü slahen vnd ain annder Nöten

82v/162 Doch so wyl der geben die schweche In die sterckin och sin arbait daby merckt ~ In dem bochen so bend yeglicher den andern gern vß vnd das er Im kem hinder den Schilt vnd machte In wörlouß

83r/163 Hie hatt er die schwechin volbracht In die sterckin vnd mag stossen oder Rysen ~ Da hatt der die sterckin verlorn vnd wirt nyder geryssen

83v/164 Hie ist das stuck volbracht wie vorgeschriben statt züm end

84r/165 Nun laut sich der Nöten vnd gatt vnden In die schwech vnd Nempt Im den fůß vnd Ryßt ~ Aber vß dem anbinden vnd ain annder Nöten So hat er den oben übertrungen

84v/166 Das sind die zwen fryen Notstend dar vß man suchen mag vnd finden allen vortail vnd Inn louffen vnd werlouß machen vnd vil hüpscher arbait die man daruß machen mag So man sust dehain wer hatt denn den schilt ~ Ich wartt des ends

85r/167 Wurd mir der schlag An sin halß ~ Ker Vmb so vindestu das Stuck volbracht - Vnd hatt Im der das treffen gezuckt vnd Im oben In den Schilt gestochen ~ Hie hatt wöllen der anbinden vnd bochen

85v/168 Hie wurt das stuck volbracht Als vorgeschriben Statt

86r/169 Da hatt das schiltfechten ain end das vns gott allen kumer wend

DAGGER

86v/170 Hie Vacht an der Tegen gott der wöll vnser aller pflegenn ~ Der hatt gestochen von dach ~ Der hatt versetzt mit ainer hand vnd hatt den Tegen von Im gewandt ~ Der wyl obnen Nider Stechen ~ So gatt der vnden vff mit versatzung vnd wyls Im Brechen

87r/171 Der ober schilt Für den Stich ~ Der Sticht fry von Tach ~ So macht der den vndern Schilt mit aim stosß - Der Sticht vnden zů dem

87v/172 Fur den obern stich so wurff din lincken arm vff vnd fach sin Stich vff den lincken arm vnd macht mit dinem Tegen stechen oder slahen ~ Als Ich hon gestochen So ist es mir gebrochen ~ Aber Fur den obern Stich wurff den lincken Arm fur vnd mit dem Tegen vmb sin arm vnd wurff In von Ir ~ Der hatt sin stich volbracht vnd des bruchs nit gedacht

88r/173 Als der oben Inn hatt gestochen So hatt der Es mit dem lincken arm gebrochen vnd grifft Im mit dem Tegen zwuschen die bain vnd zuckt In vff vnd wurfft In von Im ~ Als der oben hinyn hatt gestochen So hatt er mit dem rechten Arm versetzt vnd wint den Tegen vmb sinen arm vnd tryt damit fur vnd wurfft In von Im

88v/174 Da ist der vßgangen vnd der bruch uber das fahenn ~ Als der In beschlossen hatt ~ So hatt der In beschlossen mit der schär ~ Der hat gestochen Oben Inn

89r/175 Fur den obern stich Ain grosser wurff gang mit dem rechten Arm vff vnd fal mit dem lincken Arm vber sinen rechten Arm vnd wurff In von dir ~ Da Ist das stuck vnd wurff volbracht

89v/176 Als der obnen Nyder hatt gestochen so ist er vff gangen mit dem rechten arm vnd grifft Im mit dem lincken hand In sin Elbogen vnd Truckt In von Im so kert sich der gar vmb vnd wurfft In vber Rucken ~ Hie ist volbracht stuck vnd bruch zwiuach

90r/177 Fur den obern stich hatt er sinen rechten Arm vff geworffen vnd sinem Stich hinweg geholffen vnd ersticht In ~ Der hatt gestochen Vnd Ist Im der stich gebrochen ~ Das sol sin ain vahen vß der schär den arm oder den hals

90v/178 Aber Fur den ob*er*n stich ain arm bruch vnd fah sin stich vff den lincken arm vnd griff mit der Rechten hand vnder sinen Arm vnd zuck an dich ~ Aber fur den obern stich fach sin stich vff din Rechten Arm Vnd Tryt fur vnd schlah Im din lincken Arm an sin halß vnd wurff In vber die huffen

91r/179 Aber Fur den obern stich wurff den Rechten Arm vff vnd fach sin stich daruff vnd gryff mit der lincken hand In sin Elbogen vnd truck In von dir fürsich Nyder ~ Der bruch ker dich gar vmb vnd wurff In uber die huffen

91v/180 Fur den vndern stich so schlah mit der rechten fuust In sin arm vnd stoß din Tegen In In ~ Hie hatt einer als gütt Als der annder

92r/181 Aber Fur den vndern stich schlah mit diner rechten hand Sin stich hinweg vnd Slah In mit der lincken hand an sinen hals vnd Tryt mit fur vnd wurff In von dir ~ Fur den vndern stich fah sin stich In din bayd hend vnd wirg Im Sin arm vff die lincken Achsel vnd brich In ab

92v/182 Hie ist der arm bruch wie vorgeschriben statt ~ Aber ain fryer stand mit dem schilt fur den obern stich oder fur den vndern

93r/183 Als der den schilt hat gemacht ~ So hatt der den obern stich verwent vnd Stotzt den tegen mit dem hefft zwuschen sinen arm vber den schilt vnd wurft In von Im ~ Da Verbringt er dasselb Stuck

93v/184 Der hatt den gefangen Vß der Schär ~ Der bricht dem die schär ~ So hatt och der den beschlossen mit der schär vmb den hals ~ So macht der den Bruch vnd wirt In werffen

94r/185 Hie ist er vffgangen mit dem Schilt fur den obern stich ~ So hatt der den stich erlogen vnd slecht vnden vff vnd sticht ~ Der hatt aber den obern Schilt gemacht ~ So hat der aber den Stich erlogen vnd vber gryfft Im bayd Arm vnd tůt In werffen

94v/186 Der hatt dem Äbich angebunden vnd wyl In uber gryffen ~ Der hatt gestochen Oben hinyn ~ Hie hatt er In uber gryffen vnd den beschluß volbracht ~ Als der gestochen hatt

95r/187 Hie hatt er versetzt mit der schär vnd mag In hinweg Stossen mit welicher hand er wyl ~ Der hatt sin stich volbracht ~ Der statt In der Fryen hutt ~ So halt sich der Im zwiffel stich vnd hat den Tegen vff dem Ruck*en* vnd mag stechen mit welicher hand er wyl

95v/188 So hatt der versetzt mit sinem lincken arm vnd hatt In oben uber gryffen vnd hatt Im vnder Sinem schenckel vnder gryffen vnd wirt In werffen ~ Der hatt gestochen oben Inn mit macht vnd ain fryen Stich volfürt ~ Der Maister hatt sich uber seuhen vnd ist bayde ain stuck

96r/189 Als der mich uberloffen hatt mit aim fryen stich oben Nyder So hab ich den lincken arm vff geworffen vnd versetzt vnd mit mym rechten Arm hab ich In vmb sin hals gefaßt vnd wirg In Nider ~ Hie hatt er In Nyder gewirckt vnd macht ain end

WRESTLING

96v/190 Der an Fall Vß den Armen zu Ringen ~ So hatt der versetzt mit dem lincken Arm vnd Sticht och Von Tach ~ Als der gestochen hatt von Tach

97r/191 Die habent birlichs gefaßt ain Arm vnden den anndern oben vnd ligend In dem streb ~ In dem Streb So schlefft er sin hopt durch sin Arm vnd zuckt In vff

97v/192 Als der gefaset hatt Vß den Armen ~ So stoßt der mit sinem Rechten Elbogen vff sinen Rechten arm vnd mit dem stoß So gryfft er Im vnder sinen rechten fuß vnd zuckt In vff vnd wurfft In uber vsß ~ Der hatt es nit gůtt ~ Der wyl den werffen vber den schenckel

98r/193 Hie wyl der den bringen In Trapen ob er Im möcht Brechen den Arm ~ Hie hat der Im den arm Bracht vff die achsel vnd wöllen Im den abbrechen

98v/194 Vß dem Fasen So wurfft er In vber den Schenkel ~ Hie hatt Er In hinder Tretten vnd Stoßt In über den schenckel

99r/195 Aber ain glichs Byrlichs Fasen ~ Vß dem Byrlichs fasen So Zuckt Er den vndern Arm vnd ergryfft In by dem hopt vnd Swenckt In von Im

99v/196 In dem Fasen So wyl der durch gon ~ Hie Ist er Im durch gangen vnd wurfft In uber den Rucken

100r/197 In dem durch gon So ist das der Bruch vnd wirckt In by dem halß ~ Aber ain Bruch vber das durch gon In dem durch gon So zuck den füß hindersich vnd ergryff In by dem halß

100v/198 Das hecklin Fursich ~ Das hecklin der Bruch mit dem Schlag

101r/199 Wen ainer ain ergryfft hinderwertligen So wartt ob dir möcht Sin schenckel werden vnd zuch In her durch vnd Richt dich vff ~ Aber so ainr ain faßt hinderwertz so Tryt mit dinem füß hindersich zwuschen Sine bain vnd gryff Im nach dem schenckel

101v/200 Das ist ain heben hinderwertz der Bruch dar uber Nem In by dem har vnd zuch In uber die achsel ~ Das haißt ain beschulß wyl er denn ledig werden So brich Im die finger oder Buck sich vnd werff In vber Rucken

102r/201 Wen ainr ain ergryfft vnd bayd arm vnden hatt vnd dich vff zucken wyl So gryff Im mit bayden henden In sin Antlit vnd schüb In von dir so latt er dich ~ Der latt sich mit wylen Faßen vnd er mag In dennocht nit von Im werffen er wurfft In

102v/202 Der wirfft den mit dem halben hecklin vnd vß dem halben hufft Ringen ~ Hie hatt er In geworffen vß der selben Arbaitt

103r/203 Ain ober huff Ringen ~ Vber das ober huff Ringen so Sucht er ain Bruch

103v/204 In dem Ringen So hatt der In hinder Tretten vnd In uber den Fus ~ In dem streben so wyl der dem den fus vnden vß Slahen

104r/205 In dem Ringen hatt er In vffzuckt vnd lofft mit Im vmb das haißt ain Schwindel Ringen ~ Der hatt den gefaßt mit bayden henden by dem halß vnd erschitt In so stoßt der in von Im

104v/206 Ain verkert Ringen Ist Aber Swer zů heben ~ Ain Ringen vß dem Tegen mit dem arm wirgen von Im kern

105r/207 Das haißt ain Bůben Ringen ~ Der bruch daruber Fall Im mit dem knu In sin macht

105v/208 Der hatt den gefanngen vnd fůrt In by dem arm ~ Hie latt er sich fürn vnd gatt gern vnd In dem gon so zuckt er In vff vnd wurfft In hindersich an Rucke*n*

106r/209 Der hatt wöllen durch gon So hatt er Im zuckt vnd ergryfft In vnd truckt In Nyder ~ Das haißt ain Achsel Ringen vnd kumpt von dem halß fassen

106v/210 Vß dem faßen So Tryt der dem den Schenckel ab ~ Wen dich ainr Nötten wyl In dem Ringen vnd fast hept So heb mit Im vnd In dem heben So gib dich In die Schwechin vnd heng Im nach so wirt er dir In ain schlosß

107r/211 Der hat den by dem goller gefaßt Vnd Schlecht Im vnden den fůß vß ~ Vß der vorigen arbait So wurfft Er In uber die vndere huffen

107v/212 Ain arm winden ob er In möcht bringen zwuschen die Bain ~ Hie hatt er das vorgeschrybe*n* Stuck volbracht

108r/213 Der hatt den gebracht In den Trappen ~ In dem Trapen zuckt er In an Sich In dem zucken schlecht er In an sinen halß vnd wurfft In uber die huffen

108v/214 Der bruch uber das Faßen So Tryt mit dinem füs hindersich zwuschen sine bain vnd wurff In uber die Siten ab ~ Der hatt den gefaßt by bayden Armen vff dem Rucken

109r/215 Aber ain Fasen by dem halß So gryfft er Im In die Elbogen vnd tryt fur vnd stoßt In uber den schenckel ~ Ain hinder tretten vnd mit dem lincken Arm In sinen halß vnd stoß den uber den Schenckel

109v/216 Da hatt der den In dem hecklin Recht ~ Der Bruch uber das hecklin Im gryff Als er das hecklin schutzt so nem das ober huff Ring*en* So ist das hecklin Brochen

110r/217 Das ist ain Vßbindiger bruch vber das hecklin vnd uber das ober huff Ringen ~ Das ist och ain bruch ubers hecklin oder huff

110v/218 Da hatt der den bruch volbracht Vber das hecklin ~ Der hatt wollen louffen In das hecklin oder In die obern huff So went sich der vnd macht den bruch

111r/219 Vß dem halß faßen so macht der ain schloß ~ Hie hatt er In beschlossen vnd wurfft In uber die schenckel

111v/220 Hie hatt er In uber gryffen vnd wurfft In uber die vnder hufft Ringen In ain schloß ~ Als sie glich gefaßt haben So stoßt er mit sinem Elbogen In sinen rechten Arm vnd über schutzt vnd felt Im vmb sinen schenckel vnd wurfft In

112r/221 Hie ist der wurff volbracht wie hernach geschryben statt ~ Aber ain Bruch uber das hecklin wen er dir das hecklin schlecht So Fal Im In die knie kellen mit dem knu

112v/222 [blank]

LANGES MESSER

113r/223 Hie Facht an das Messer Gott wöll vnnser nit vergessen ~ Der wyl howen von Tach ~ So wyl der Im den how versetzen mit macht

113v/224 Der hat sin how volbracht ~ Der hatt den schlag versetzt vnd wyl In ubergryffen

114r/225 Hie hatt er In ubergryffen vnd howt In durch den kopff vnd hat das vorgeschriben Stuck ain end

114v/226 Der howt Fry Von dach ~ Der hatt versetzt mit gewenter hand vnd wyl furtretten vnd Ryssen

115r/227 Hie hatt er Für getretten vnd Rysßt mit dem messer des stoß zů der Elbogen sol man nit vergessen

115v/228 Der hatt den angeloffen In zorn vnd hat In wöllen howen durch den kopff ~ So hatt der versetzt vnden vff mit macht vnd In sinem vff ziehen hat er Im sin hand abgehowen

116r/229 Hie Tryt er fur vnd Slecht In gar Nider zü tod

116v/230 So hatt der Im versetzt vnd hatt In ubergryffen mit dem Arm vnd stoßt messer In In ~ Der hatt gehowen von Tach vff den man

SWORD AND BUCKLER

117r/231 Das sind die Zwen fryen Stend mit dem buckeller vnd mit dem Messer

117v/232 Hie hatt der versetzt mit dem Buckeller vnd messer des ubergryffens sol er nit vergessen - Der hatt gehowen dem In sin versatzung

118r/233 Als der vff mich hatt gehowen So hab ich Im versetzt vnd vß der versatzung ubergryffen vnd how In durch den kopff

118v/234 Da hatt der Im versetzt vnd Stoßt In von Im vnd Ryßt mit dem messer

119r/235 Als er In von Im gestossen hatt So volendet er das stuck vnd Stoßt das schwert durch In

119v/236 Aber howt der von Tach ~ Hie hatt der versetzt fry vnd loufft Im daby Inn

120r/237 Hie Ist Im der Ingeloffen vnd hat das Swert durch In gestochen

120v/238 Der statt Fur den Stich ~ Der Wyl In stechen

121r/239 Als der gestochen hatt So hott Im der In Sinen Arm • der ober how ist och gůtt fur den stich

ONE AGAINST TWO

121v/240 Das Ist der Not stand wen zwen uber ain Sind ~ Hie wyl ich vff In howen ~ Da versetzt der mit Eppicher hand vnd wirt sich wenden vnd howen zů dem hindern ~ Da versetzt er mit dem tegen vnd buckeller ~ Hie wil der och howen

122r/241 Im wenden so howt er In durch den kopff So fellten der hinder an So beschlußt er In och

MAN AGAINST WOMAN

122v/242 Da Statt Wie Man Vnd Frowen mit ainander kempffen söllen vnd stond hie In dem anfanng ~ Da statt die Frow Fry vnd wyl schlahen vnd hat ain stain In dem Sleer wigt vier oder find pfund ~ So statt er In der gruben bis an die waichin vnd ist der kolb als lang Als Ir der Schleer von der hand

123r/243 Hie hatt Sie ain schlag volbracht ~ Nůn hatt er den schlag versetzt vnd gefangen vnd wyl Sie zu Im ziehen vnd nötten

123v/244 Da hatt er sie Zu Im gezogen vnd vndersich geworffen vnd wyl sie wurgen

124r/245 Da hatt sie sich vß Im gebrochen vnd vnder statt Sie In zü wirgen

124v/246 Hie hatt sie In gebracht an den Rucken vnd wyl In wirgen vnd ziehen vß der grüb

125r/247 Da hatt er si Zü Im gezuckt vnd wurfft sie In die grüben

125v/248 Als sie schlahen wyl So ist sie Im zu nach Tretten das er sie ergryfft by dem schenckel vnd wirt sie fellen

126r/249 So schlecht er sie Fur die bruchst ~ Da hatt sie Im den schlöer vmb den hals geschlagen vnd wyl In wurgen

126v/250 Da hatt sie In gefaßt by dem halß vnd by seinem zug vnd wyl In vß der grüben ziehen

FIGHTING ON HORSEBACK ~ SWORD

127r/251 Vß dem uber Illen So wyl der versetzen mit dem Stich ~ Der Sprengt In an vnd wyl In uber Illen

127v/252 So Hatt der versetzt mit epicher hand vnd wust fur ~ Als der gehowen hatt

128r/253 Hie ist das stuck volbracht vnd hatt In gehowen In den schenckel

128v/254 In dem anrenen So facht der sinen how vnd wint Im Sin Swert In Sin antlit ~ Der hatt gehowen

129r/255 Hie hatt er den schlag versetzt vnd uberfelt Im vnd Ryßt das Swert von Im

129v/256 Das schwert Nemen vnderoungen

130r/257 Das Swert Nemen hinderwertz

130v/258 Als der hatt gehowen ~ So hatt der Epich versetzt vnd Ist Im uber falen vmb den halß

131r/259 Hie versetzt der Fry vnd wyl den uberfalen ~ Der hatt gehowen vom Tach

131v/260 Hie ist er Im Ingefallen vnd das vorgeschriben Stuck volbracht

FIGHTING ON HORSEBACK ~ WRESTLING

132r/261 Das Ringen zů Rosß ~ Hie hatt der In beschlossen vnd vber gryffen vnderougen

132v/262 Der bruch vber Das vorgeschriben Schloß hatt Im gezuckt Sin uber gryffen vnd In by dem halß

133r/263 Da wurfft der Rosß vnd Man

133v/264 Das ubergryffen vnd der beschluß hindersich ~ Der hatt den hinderwertz ergryffen

FIGHTING ON HORSEBACK ~ LANCE

134r/265 Der Entwert dem den Spieß mit dem Schwertt

134v/266 Hie Ist das stuck volbracht Als vorgeschriben statt

FIGHTING ON HORSEBACK ~ CROSSBOW AGAINST LANCE

135r/267 Die Illenden ~ Wie sich ainer halten sol In der flucht mit dem armprost

135v/268 Der wyl den vnder das pfert Renen ~ Hie hatt der den geschossen vnd stoßt Im den Spieß mit dem Armbrost hinweg vnd wurt In ergryffen by dem halß

136r/269 Hie wyl der den Recht empfahen mit dem Spieß ~ Wie sich ainer halten sol mit dem Spieß gegen ainem mit ainem Armbrost

136v/270 Hie ist das vorgeschriben Stuck mit dem Armbrost vnd Spieß volbracht vnd hatt In ergryffen by dem halß ~ Das buch hatt angeben hans talhoffer vnd gestanden zu Mallen

GLOSSARY

Armoured (hand): A way of holding the sword in armoured combat, in which the range of the sword is shortened by grabbing the middle of the blade with the left hand, which is armoured with a gauntlet, in order to more powerfully and accurately make a thrust into the opponent's → opening.

Beating: The shields striking against each other.

Binding: The clash of the blades.

Binding from behind: A → binding from the back side that is turned away from the stroke, i.e. usually the left side.

Break: Counter technique.

Brentschirn, brent schüren: Brant is an old term for 'sword', and 'schüren' is similar to the modern German *scheuern* (to chafe, to grind): The blades are in contact in a → binding and they move ever so slightly against each other.

Bubenringen, Bubenwurf (literally: boy's wrestling, boy's throw): A series of wrestling techniques.

Clinging: The adhering of the tip of the sword in an → opening of the opponent.

Displacement: Defence, parry.

Donnerschlag (literally: thunder strike): See → *Mordschlag.*

Falling in: A fast, powerful advancement to oppress the opponent.

Falling over: One fencer falls over the opponent's arm and weapon.

Final piece: A → piece in which one opponent loses his life – a circumstance, however, that was not absolutely necessary in a judicial duel.

Free: Unhindered, without a → binding.

Grieswart: Referee in a judicial duel.

Guard: Position with the weapon, in which one stands protected and from which one can perform attacks; see also → iron gate, → *vom Tag.*

Hasten: A fast approach to the opponent in horse fencing.

Hau: Strike.

Iron gate: A → guard.

Krummhau: A particular, so-called 'hidden' → *Hau,* which originates from the teachings of Johannes Liechtenauer.

Lists: The combat area, sometimes enclosed by a wooden parapet.

Manner of farmers: A wrestling grip in which both opponents hold each other in the same way.

Mordschlag (literally: murder strike): A strike with the pommel of the sword, with the blade held in both hands, also called → *Donnerschlag.*

Oberhau: A strike from above in which the blade moves from top to bottom.

Oberstich: A thrust from above.

Opening: An unprotected spot in the opponent's cover.

Piece: Technique, sequence of movements.

Point: Generally, the tip of the sword; for Talhoffer, usually synonymous with the → thrust.

Pressing: A situation in armoured combat in which both opponents have put their blades in the other's → opening.

Running in: A rapid approach with the aim of defeating or bringing down the opponent.

Running to: Running towards each other at the beginning of the fight.

Schielhau/Schieler: A special, so-called 'hidden' strike, which originates from the teachings of Johannes Liechtenauer.

Scissors: Enclosing the opponent's hand with scissor-like crossed hands, holding the dagger at both ends.

Shield: A protective defensive position in dagger fencing.

Short sword: Taking the sword with the → armoured hand.

Strength: The part of the weapon with a favourable balance of power for the fighter; the part of the sword from the handle to the middle. See also → weakness.

Struggle: a wrestling grip; striving, moving hard, fighting.

Sturzhau (literally: plunging strike): Combination of an → *Oberhau* from the right, followed by a second one from the left, the latter being executed with the back edge of the sword.

Thrust: Attack with the point of the sword; here often called → point.

Trappe: Wrestling grip of unclear meaning. A *Trappe* can be a wooden frame for the grapes in wine making, a type of bird (bustard in English) or a staircase; *trappen* (v.) also means stomping. The idea that *Trappe* might be related to the English 'trap' is an intriguing one, but little evidence has been found to support it.

Travelling after: A follow-up attack following the direction of the opponent's failed attack.

Twitching: Pulling or tearing away abruptly, evading something.

Unterhau: A strike in which the blade moves upwards from below.

Unterstich: A thrust performed from below.

Vom Tag: A → guard with the → point showing upwards.

War: The → work in the → binding; an attempt to hit the enemy while the weapons are still in contact.

Weakness: The part of the weapon with an unfavourable balance of power for the fighter; the part of the sword from the middle to the tip. See also → strength.

Wechselhau (literally: changing strike): An → *Unterhau* following an → *Oberhau*, striking back in the same direction.

Work: Attack; an attempt to oppress and hit the opponent.

Zornort: Follow-up technique of the *Zornhau* (one of fencing master Johannes Liechtenauer's 'hidden strikes'), in which the tip of the sword is brought vigorously and rapidly against the opponent.

BIBLIOGRAPHY

MANUSCRIPTS

Augsburg, Universitätsbibliothek, *Cod.I.6.2°.1*

Berlin, Kuperstichkabinett der Stiftung Preußischer Kulturbesitz, *78 A 15*

Bologna, Biblioteca Universitaria, *Ms. 1825*

Coburg, Kunstsammlungen der Veste Coburg, *Hz.014*

Copenhagen, Det Kongelige Bibliotek, *Thott 290 2°*

Cracow, Biblioteka Jagiellonska, *Ms. Germ. Quart. 16*

Dresden, Sächsische Landesbibliothek, *Mscr. Dresd. C 487*

Glasgow, R. L. Scott Collection, *E.1939.65.341*

Gotha, Forschungsbibliothek Schloss Friedenstein, *Ms. Chart. A558*

Gotha, Forschungsbibliothek Schloss Friedenstein, *Ms. Chart. B1021*

Göttingen, Niedersächische Staats- und Universitätsbibliothek, *2° Cod. Ms. philos. 61*

Kassel, Landesbibliothek, *2° Ms. iurid. 29*

Königseggwald, Gräfliches Schloss, *Hs. XIX, 17-3*

Leeds, Royal Armouries Museum, *I.33*

Munich, Bayerische Staatsbibliothek, *Cgm 1507*

Munich, Bayerische Staatsbibliothek, *Cod. icon 394a*

Munich, Bayerische Staatsbibliothek, *Cod. icon 394*

Munich, Bayerische Staatsbibliothek, *Cod. icon 395*

Nuremberg, Germanisches Nationalmuseum, *Cod. Hs. 3227a*

New York, Metropolitan Museum of Art, *26.236*

Rome, Biblioteca dell'Accademia Nazionale dei Lincei e Corsiniana, *44A8*

Vienna, Albertina, *Hs. 26-232*

Vienna, Kunsthistorisches Museum, *KK 5342*

Wolfenbüttel, Herzog August-Bibliothek, *Cod. Guelf. 125.16 Extrav*

Private collection, olim: Vienna, Österreichische Nationalbibliothek, *Cod. Vindob. Ser. Nov. 2978*

ADDITIONAL SOURCES

BURKART (2014)

ERIC BURKART: 'Die Aufzeichnung des Nicht-Sagbaren. Annäherung an die kommunikative Funktion der Bilder in den Fechtbüchern des Hans Talhofer', in JASER, CHRISTIAN/ISRAEL, UWE (EDS): *Zweikämpfer. Fechtmeister – Kämpen – Samurai* (Das Mittelalter. Perspektiven mediävistischer Forschung, Band 19, Heft 2), De Gruyter, Berlin, 2014

Du Boulay (1983)
F. R. H. Du Boulay: *Germany in the Later Middle Ages,* Athlone Press , London, 1983

Hagedorn (2016)
Dierk Hagedorn: 'German Fechtbücher from the Middle Ages to the Renaissance', in Daniel Jaquet/Karin Verelst/Timothy Dawson (eds): *Late Medieval and Early Modern Fight Books,* Brill, Leiden/Boston, 2016

Hergsell (1887)
Gustav Hergsell: *Talhoffers Fechtbuch aus dem Jahre 1467,* Prague, 1887

Hils (1983)
Hans-Peter Hils: 'Die Handschriften des oberdeutschen Fechtmeisters Hans Talhoffer', in *Codices manuscripti,* Hollinek, Vienna 1983

Hils (1985)
Hans-Peter Hils: *Meister Johann Liechtenauers Kunst des langen Schwertes,* Peter Lang, Frankfurt am Main/Bern/New York, 1985

Israel (2017)
Uwe Israel: 'Die Fechtbücher Hans Talhofers und die Praxis des gerichtlichen Zweikampfs', in Elisabth Vavra/Matthias Johannes Bauer (eds): *Die Kunst des Fechtens,* Universitätsverlag Winter, Heidelberg, 2017

Leiske (2018)
Patrick Leiske: *Höfisches Spiel und tödlicher Ernst – Das Bloßfechten mit dem langen Schwert in den deutschsprachigen Fechtbüchern des späten Mittelalters und der frühen Neuzeit,* Thorbecke, Ostfildern, 2018.

Leng (2008)
Rainer Leng, Hella Frühmorgen-Voss, Norbert H. Ott, Ulrike Bodemann, Peter Schmidt, Christine Stöllinger-Löser: *Katalog der deutschsprachigen illustrierten Handschriften des Mittelalters* Band 4/2, Lfg. 1/2: 38. Fecht- und Ringbücher, C. H. Beck, Munich 2008

Müller (1992)
Jan-Dirk Müller: 'Bild – Vers – Prosakommentar am Beispiel von Fechtbüchern. Probleme der Verschriftlichung einer schriftlosen Praxis', in Hagen Keller, Klaus Grubmüller, Nikolaus Staubach (Eds): *Pragmatische Schriftlichkeit im Mittelalter.* Wilhelm Fink, Munich, 1992

Oman (1924)
C. W. C. Oman, *A History of the Art of War in the Middle Ages,* 1924, Methuen & Co., London, 1991

Paurenfeindt (1516)
Andre Paurenfeindt: *Ergrundung Ritterlicher kunst der Fechterey,* Hieronymus Vietor, Vienna, 1516

Stangier (2009)
Thomas Stangier: 'Ich hab hertz als ein leb … – Zweikampfrealität und Tugendideal in den Fechtbüchern Hans Talhoffers und

Paulus Kals', in *Ritterwelten im Spätmittelalter – Höfisch-ritterliche Kultur der Reichen Herzöge von Bayern-Landshut*. Museen der Stadt Landshut, 2009

WIERSCHIN (1965)
MARTIN WIERSCHIN *Meister Johann Liechtenauers Kunst Des Fechtens*. C. H. Beck, Munich, 1965

ONLINE SOURCES

(Accessed 23 March 2018)

wiktenauer.com/wiki/Hans_Talhoffer
talhoffer.wordpress.com/category/a-life-like-that-of-talhoffer/biography/

Also available from Greenhill Books

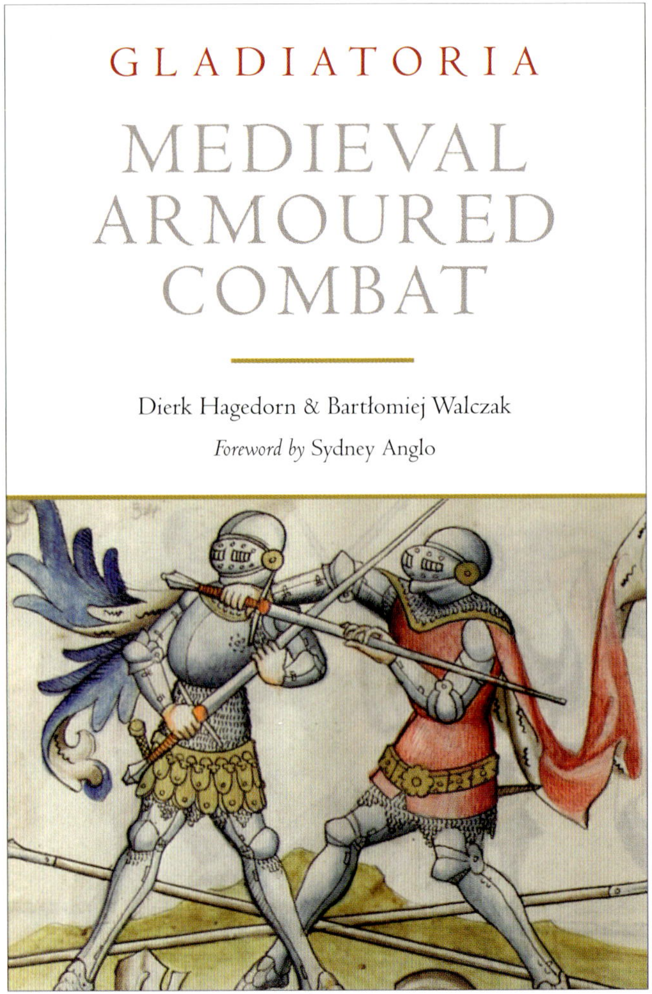